# Uncertainty Underground

# Uncertainty Underground

Yucca Mountain and the Nation's High-Level Nuclear Waste

edited by Allison M. Macfarlane and Rodney C. Ewing

The MIT Press
Cambridge, Massachusetts
London, England

MIT Press books may be purchased at special quantity discounts for business or sales promotional use. For information, please e-mail special_sales@mitpress.mit.edu or write to Special Sales Department, The MIT Press, 55 Hayward Street, Cambridge, MA 02142.

This book was set in Sabon by SNP Best-Set Typesetter Ltd., Hong Kong and was printed and bound in the United States of America. Printed on recycled paper.

Library of Congress Cataloging-in-Publication Data

Uncertainty underground : Yucca Mountain and the nation's high-level nuclear waste / edited by Allison M. Macfarlane and Rodney C. Ewing.
    p.   cm.
Includes bibliographical references and index.
ISBN 0-262-13462-4 (alk. paper)—ISBN 0-262-63332-9 (pbk. : alk. paper)
1. Radioactive waste disposal in the ground—Nevada—Yucca Mountain. 2. Radioactive waste sites—Nevada—Yucca Mountain—Evaluation. 3. Geology—Nevada—Yucca Mountain Region. 4. Radioactive waste disposal—Government policy—United States.
I. Macfarlane, Allison. II. Ewing, Rodney C.

TD898.12.N3.U5226   2006
363.72'890979334—dc22
                                                        2005056218

10 9 8 7 6 5 4 3 2 1

# Contents

# Preface

Despite substantial effort during the past several decades, there is, at present, no geologic repository receiving spent nuclear fuel or high-level nuclear waste. A few countries, particularly Finland and Sweden, have made considerable progress in siting and developing a geologic repository; others are in the earliest stages of site investigations. Although there is broad consensus that geologic disposal is the safest available solution to the management and disposal of these highly radioactive waste materials (National Research Council 2001b), implementation has proven to be technically and politically difficult. In the United States, although Yucca Mountain in Nevada was identified in the early 1980s as a potential site for nuclear waste disposal, it is still not open, and its fate hangs on a number of regulatory and judicial decisions—each with an important scientific and technical basis—that have yet to be made.

Much has already been written about the political difficulties of siting high-level nuclear waste repositories (see, for example, Carter 1987; Colglazier and Langum 1988; Dunlap et al. 1993; Gerrard 1994; Flynn and Slovic 1995; Metlay 2000). Over the years, a huge number of technical reports, journal papers, proceedings volumes, and monographs have been published as well (see, for example, vol. 38, no. 1–3, and vol. 62–63 of the *Journal of Contaminant Hydrology*; vol. 17, no. 7, of the journal *Applied Geochemistry*; the *Proceedings of the International High-Level Radioactive Waste* conferences; the *Proceedings of the Materials Research Society's Symposium* on the scientific basis for nuclear waste management; Winograd 1981; National Research Council 1957, 1983, 1995, 2001a, 2001b, 2003; Hanks et al. 1999).

Despite this large knowledge base, substantial funding (over seven billion dollars spent on the Yucca Mountain site during the past decades), and a large number of talented scientists and engineers engaged in every aspect of this problem, there are continued delays. We maintain that one important reason for the difficulties and the delay is that the scientific and engineering communities have underestimated the effort required to characterize this site and model its behavior over the long periods judged to be appropriate for nuclear waste disposal—tens to many hundreds of thousands of years. The key to appreciating the scale of the scientific challenge has been the relatively recent recognition of the large uncertainties inherent in such analyses.

In this book, *Uncertainty Underground*, we have asked experts to address these uncertainties in the context of the specific and relevant processes of interest at Yucca Mountain. We have limited the number of pages available to each author, and asked them to describe the relevant issues in a way that is accessible to the broader audience of the interested public and policymakers. We also have attempted to include authors from a wide range of disciplines who hold differing views on the suitability of Yucca Mountain as a repository site. This book is not a judgment of the suitability of Yucca Mountain as a repository for spent nuclear fuel and high-level nuclear waste. We leave that judgment to the reader. Rather, we have tried to present some of the complexities and highlight the sources of uncertainty in the scientific analysis. We also have endeavored to place this discussion into its historical and regulatory context. There is probably no better example of science and engineering being so thoroughly mixed in a cauldron of politics, policy, and regulation than Yucca Mountain.

Even as this book was written, there have been a number of developments that have had an important impact on Yucca Mountain as a potential geologic repository. In July 2004, the U.S. Court of Appeals for the District of Columbia Circuit remanded a part of the existing U.S. Environmental Protection Agency (EPA) standard, ruling that the EPA had not followed the instructions of Congress as given in the 1992 Energy Policy Act, which required the EPA to promulgate a standard that was consistent with the recommendations of the National Academy of Sciences. In the academy report, *Technical Bases for Yucca Mountain Standards* (National Research Council 1995), the committee argued that there was no scientific basis for selecting a ten-thousand-year, or any other period, time of compliance; therefore, they recommended that the period of compliance be extended to the time of peak dose, which for Yucca Mountain is on the order of some hundreds of thousands of years. The academy report clearly recognized that such calculations involved considerable uncertainties, which are the subject of this book. Despite the recommendations of the academy, in 2001, the EPA promulgated a standard that included a time of compliance of ten thousand years, and it is this part of the standard that has been remanded by the court. The EPA is in the process of developing a new standard, which will inevitably be a subject of controversy.

As this book was in the final stages of preparation, the EPA issued draft revised radiation protection standards for the Yucca Mountain repository. The newly proposed standards incorporate multiple compliance criteria applicable at different lengths of time. For periods up to ten thousand years, the dose limit is 150 microsieverts, as it was in the previous standard. For periods after ten thousand years and up to one million years after disposal, the standard is 3.5 millisieverts. The proposed standard states that "The primary means for demonstrating compliance with the standards is the use of computer modeling to project the performance of the disposal system under the range of expected conditions" (EPA 2005, p 33). The *uncertainty* in these projections is, of course, the central issue addressed in the following chapters.

At the same time, plans move ahead for an interim spent fuel storage site near Salt Lake City, Utah. Private Fuel Storage, a consortium of eight nuclear utility companies, joined the Skull Valley band of the Goshute Indians to site a forty-thousand-metric-ton nuclear waste storage facility on land owned by the Goshute Indians. In February 2005, the Atomic Safety Licensing Board of the U.S. Nuclear Regulatory Commission (NRC) ruled that the facility would be able to withstand an airplane crash with acceptable risk (Johnson 2005). The commission has ruled to give a license to the facility, though opening will likely be delayed by pending lawsuits from the state of Utah and some of the members of the Goshute people.

Another recent development was the publication of a National Academy of Sciences report on the security of spent fuel pools at nuclear power reactors (National Research Council 2005). At the prompting of Congress and building on previous work by an independent group (Alvarez et al. 2003), the academy reviewed the security of spent fuel at nuclear power plants in the event of a terrorist attack. It found that fuel in some pool designs may be vulnerable to certain types of attack, at worst causing a zirconium-cladding fire that could release a large amount of radioactivity. The academy suggested that immediate measures should be taken at plants to mitigate loss-of-coolant accidents and that the NRC should conduct further research on pool attacks. It proposed dry cask storage as a safer and more secure alternative to pool storage for older spent nuclear fuel. What the NRC will do with these suggestions and the report's impact on the high-level nuclear waste program remains unclear.

The final development that requires comment occurred within the U.S. Department of Energy's (DOE) Yucca Mountain Project. In March 2005, the U.S. Geological Survey revealed the existence of e-mail messages exchanged among its workers that allegedly suggest scientific fraud may have occurred. The messages were from scientists who developed the models of water transport in the unsaturated zone, above the water table, at Yucca Mountain during the years 1998–2000. In the messages, the scientists complain about the lack of personnel, funding, and time to complete their assigned work, and suggest that to complete the work, they had to make up scientific notebook entries, dates on computer model runs, and the like. The messages focus on the DOE's Quality Assurance program and the burdens the scientists felt in meeting its requirements. The inspector generals from the DOE, the Department of the Interior, the U.S. Geological Survey, and the Federal Bureau of Investigation are currently examining the e-mail messages.

Amid all of these developments, the DOE has announced that the submission of the license application to the NRC has once again been delayed. In 2004, the NRC ruled that the License Support Network did not meet the requirements for a license application submission. The previous schedule had anticipated the license application by the end of 2004, and the first receipt of waste in 2010. Considering that the NRC may take up to four years to review the license application, the date for the first receipt of nuclear waste may slip significantly.

Although many of these recent events have changed the context of some of the book's chapters, which were written several years ago, the scientific and technical issues remain essentially unchanged. Thus, we believe that this book remains a timely contribution to the discussion of the general concept of geologic disposal and the specific issues faced by the Yucca Mountain site.

## References

Alvarez, R., Beyea, J., Janberg, K., Kang, J., Lyman, E., Macfarlane, A., Thompson, G., and von Hippel, F. (2003) Reducing the Hazards from Stored Spent Power Reactor Fuel in the United States. *Science and Global Security* 11, pp. 1–51.

Carter, L.J. (1987) *Nuclear Imperatives and Public Trust: Dealing with Radioactive Waste.* Washington, DC: Resources for the Future, 473 p.

Colglazier, E.W., and Langum, R.B. (1988) Policy Conflicts in the Process for Siting Nuclear Waste Repositories. *Annual Review of Energy* 13, pp. 317–357.

Dunlap, R., Kraft, M., and Rosa, E. (1993) *Public Reactions to Nuclear Waste: Citizens' Views of Repository Siting*. Durham, NC: Duke University Press, 332 p.

Environmental Protection Agency (EPA) (2005) 40 CFR Part 197, *Public Health and Environmental Radiation Protection Standards for Yucca Mountain*, Nevada, Proposed Rule.

Flynn, J., and Slovic, P. (1995) Yucca Mountain: A Crisis for Policy: Prospects for America's High-Level Nuclear Waste Program. *Annual Reviews of Energy and the Environment* 20, pp. 83–118.

Gerrard, M.B. (1994) *Whose Backyard, Whose Risk: Fear and Fairness in Toxic and Nuclear Waste Siting*. Cambridge: MIT Press, 335 p.

Hanks, T., Winograd, I., Anderson, R.E., Reilly, T., and Weeks, E. (1999) *Yucca Mountain as a Radioactive-Waste Repository*. U.S. Geological Survey. Circular 1184.

Johnson, K. (2005) A Tribe, Nimble and Determined, Moves Ahead with Nuclear Waste Plans. *New York Times*, February 28, p. 15.

Metlay, D. (2000) From Tin Roof to Torn Wet Blanket: Predicting and Observing Groundwater Movement at a Proposed Nuclear Waste Site. In *Prediction: Science, Decision Making, and the Future of Nature*, ed. Sarewitz, D., Pielke, R. Jr., and Byerly, JR. Washington, DC: Island Press, pp. 199–228.

National Research Council (1957) *The Disposal of Radioactive Waste on Land*. Washington, DC: National Academy Press, 146 p.

National Research Council (1983) *A Study of the Isolation System for Geologic Disposal of Radioactive Wastes*. Washington, DC: National Academy Press, 345 p.

National Research Council (1995) *Technical Bases for Yucca Mountain Standards*. Washington, DC: National Academy Press, 205 p.

National Research Council (2001a) *Conceptual Models of Flow and Transport in the Vadose Zone*. Washington, DC: National Academy Press, 374 p.

National Research Council (2001b) *Disposition of High-Level Waste and Spent Nuclear Fuel: The Continuing Societal and Technical Challenges*. Washington, DC: National Academy Press, 212 p.

National Research Council (2003) *One Step at a Time: The Staged Development of Geologic Repositories for High-Level Radioactive Waste*. Washington, DC: National Academy Press, 201 p.

National Research Council (2005) *Safety and Security of Commercial Spent Nuclear Fuel Storage*. Washington, DC: National Academy Press, 114 p.

Winograd, I.J. (1981) Radioactive Waste Disposal in Thick Unsaturated Zones. *Science* 212, pp. 1457–1464.

# Acknowledgments

This book has been a long time coming. The basic outline literally began as the proverbial list on a napkin, drawn up over breakfast between the two editors in 1999 (Allison still has it). The earliest efforts resulted in several special sessions organized at the spring 1999 meeting of the American Geophysical Union in Boston. Based on the success of these sessions, we were convinced that there was a need for this book.

We were fortunate to secure funding from two foundations, the John Merck Fund in Boston and Rockefeller Financial Services in New York, as well as an anonymous donor with Rockefeller Financial. Frank Hatch of the John Merck Fund provided invaluable advice along the way, and we owe him a great debt of gratitude. He introduced us to a superb publicist, Henry Miller of Goodman Media. Rodney thanks the John Simon Guggenheim Memorial Foundation and Allison thanks the MacArthur Foundation for support during the period when this book was written.

We thank Clay Morgan, our editor at The MIT Press, for his amazing patience. Herb Brody provided excellent editorial support in molding the chapters into a consistent tone and voice. We appreciate his willingness to stick with the project much longer than originally planned.

We thank all the authors for their patience during this long process (we first contacted them in 2001). They met for the first time in 2001 at a workshop held at the Lawrence Berkeley National Laboratory in Berkeley, California. Bo Bodvarsson graciously hosted this meeting. Each chapter benefited from the careful review of experts in each field. We are most grateful to the following reviewers: John Ahearne, Sigma Xi and Public Policy, Duke University; David Applegate, U.S. Geological Survey, Reston, Virginia; George Apostalakis, Nuclear Engineering Department, MIT; David Bish, Department of Geological Sciences, Indiana University; G. S. Bodvarsson, Earth Sciences Division, Lawrence Berkeley

Laboratory; John Bredehoeft, retired, U.S. Geological Survey; Luther Carter, retired science journalist and book author, Washington, DC; Tom Cotton, JK Research Associates, Washington, DC; Gustavo Cragnolino, Center for Nuclear Waste Analyses, Southwest Research Institute, San Antonio, Texas; Bruce Crowe, Apogen Technologies, Nevada Site Office; Kieran Downes, Science, Technology, and Society, MIT; Jerry Fairley, Department of Geological Sciences, University of Idaho; Mostafa Fayek, Department of Earth and Planetary Sciences, University of Tennessee; Randy Fedors, Center for Nuclear Waste Analyses, Southwest Research Institute, San Antonio, Texas; Robert Finch, Office of Civilian and Radioactive Waste Management, DOE; Alan Flint, U.S. Geological Survey, Sacramento, California; Steve Frishman, Agency for Nuclear Projects, state of Nevada; Bernd Grambow, Subatec, Ecole des Mines, Nantes, France; Mickey Gunther, Department of Geological Sciences, University of Idaho; Tom Hanks, U.S. Geological Survey, Menlo Park, California; Kate Helean, Sandia National Laboratories; Brittain Hill, Center for Nuclear Waste Analyses, Southwest Research Institute, San Antonio, Texas; Dan Holm, Department of Geology, Kent State University; Lawrence Johnson, National Cooperative for the Disposal of Radioactive Waste, Wettingen, Switzerland; Kim Kearfott, Department of Nuclear Engineering, University of Michigan; Lenny Konikow, U.S. Geological Survey, Reston, Virginia; Werner Lutze, Vitreous State Laboratory, Catholic University of America; Daniel Metlay, staff, Nuclear Waste Technical Review Board; Dade Moeller, Dade Moeller and Associates; Sam Mukasa, Department of Geological Sciences, University of Michigan; William Murphy, Department of Geological and Environmental Sciences, California State University, Chico; Heino Nitsche, Department of Chemistry, University of California, Berkeley; Joe Payer, Department of Materials Science and Engineering, Case Western Reserve University; Jay Quade, Department of Geosciences, University of Arizona; Rob Rechard, Sandia National Laboratories; Robert Roback, Chemistry Division, Los Alamos National Laboratory; Dave Shoesmith, Department of Chemistry, University of Western Ontario; Eugene Smith, Department of Geosciences, University of Nevada, Las Vegas; David Stahl, Office of Repository Development, DOE; Etienne Vernaz, Etablistement de la Vallée du Rhône, Commissariat à l'Énergie Atomique, Marcoule, France; William Weber, Pacific Northwest National Laboratory; Nicholas Wilson, Energy and Environment, Geological Survey of Canada; Susan Wiltshire, JK Research Associates, Washington, DC; Man-Sung Yim, Department of Nuclear Engineering, North Carolina State University; and Yanni Yortsos, School of Engineering, University of Southern California.

We thank the DOE for allowing us to reproduce many of the figures used in the chapters. We especially want to acknowledge Jean Younker, senior technical staff, and Paul Meacham at Bechtel-SAIC, the main contractor for the DOE on Yucca Mountain, for responding to our requests and providing assistance in securing figures from DOE publications.

Allison would like to thank Ted Postol and MIT's Technical Working Group, MIT's Security Studies Program, and the Center for International Studies for providing her with a congenial work environment throughout this process, and George Lewis, Alexander Glaser, Lisbeth Gronlund, David Wright, Marvin Miller, Heidi La Bash, Danielle Mancini, Anthony Duggins, Robert Murray, John Holdren, Matthew Bunn, Graham Allison, Jennifer Weeks, and Jim Walsh for offering valuable support and input along the way. Allison owes much appreciation to Kieran Downes and especially Anne Pollock for helping the book make it to press.

Finally, Allison owes a debt of gratitude to her family, including her mom, Pamela Macfarlane, for helping out with Graham; Graham for improving her perspective on life; and most especially her husband, Hugh, for keeping her on track, reading her chapters, and being there when she needed him.

# Contributors

**David Applegate**  is a senior science adviser for earthquake and geologic hazards at the U.S. Geological Survey. He also teaches in the environmental sciences and policy master's program at Johns Hopkins University, and is an adjunct faculty member in the Department of Geology and Geophysics at the University of Utah. He holds a PhD in geology from MIT.

**David L. Bish**  is a professor of geologic sciences at Indiana University. His research interests include the structures, properties, and origins of fine-grained minerals, such as clay minerals and natural zeolites; X-ray and neutron powder diffraction; and the behavior of minerals under controlled-temperature and water-pressure conditions. He received his PhD from Pennsylvania State University.

**Paul K. Black**  is a principal and founding member of the environmental consulting firm Neptune and Company. He received his PhD in statistics from Carnegie Mellon University. His current areas of interest include the probabilistic modeling of radioactive waste management sites at the Nevada Test Site and the building of decision support tools for the redevelopment of brownfields.

**G. S. Bodvarsson**  is the director of the Earth Sciences Division of the Lawrence Berkeley National Laboratory, and has managed the Nuclear Waste Program and led the development of 3-D modeling of Yucca Mountain. He holds a master's degree in civil engineering from North Carolina State and a PhD in hydrogeology from the University of California, Berkeley.

**William L. Bourcier**  is a research chemist at the Lawrence Livermore National Laboratory, where he has worked on the geochemical aspects of radioactive waste disposal, waste form design, and the modeling of rock-water interactions at a waste repository. He received his PhD in geochemistry and mineralogy at Penn State University, and currently works on innovative water treatment technologies.

**Jordi Bruno**  is the managing director of Enviros Spain Ltd. He is originally a chemist, and holds an MBA from Stanford University and a PhD in inorganic chemistry from the Royal Institute of Technology in Stockholm. For the past twenty-five years, he has worked in research and development programs for high level nuclear waste management. He also is head of the Enresa-Enviros Sustainability and Waste Management Chair.

**Gilles Bussod** is a member of the Science Advisory Board's Research Advisory Committee and a chief scientist with New England Research, Inc. He was the principal investigator for the Busted Butte Underground Transport Test at Yucca Mountain, and his current research is in hydrogeophysics. He holds a PhD in geology and earth and space sciences from the University of California, Los Angeles, and a PhD in geophysics from the University of Paris.

**J. William Carey** is a researcher at the Los Alamos National Laboratory. He received his MS in geology at the University of Oregon and PhD in earth science at Harvard University. His research interests include thermodynamic and kinetic studies of mineral reactions, intracrystalline and surface sorption phenomena, and the reactivity of concrete systems.

**Luther J. Carter** is an independent journalist who has written extensively about Yucca Mountain. His book *Nuclear Imperatives and Public Trust: Dealing with Radioactive Waste* (Resources of the Future, 1987) was influential in policymaking in the late 1980s. A graduate in history from Duke University, Carter is a fellow of the American Association for the Advancement of Science and was a writer for *Science* magazine for fifteen years.

**Esther Cera** is the deputy director of Enviros Spain Ltd. A chemist with a PhD from the Universitat Autonoma de Barcelona, she also holds an MS in environmental management from the Open University. Her work has focused on the geochemical modeling of trace elements mobilization in natural systems, thermodynamic and kinetic modeling at the laboratory scale, and the development of source term models to interpret nuclear waste behavior under repository conditions.

**Steve J. Chipera** is a researcher in the Earth and Environmental Sciences Division of the Los Alamos National Laboratory. He holds an MS in geology from the University of North Dakota, Grand Forks. His research interests include the mineralogy, petrology, and geochemistry of igneous and metamorphic rocks and their alteration assemblages; quantitative phase analysis by X-ray powder diffraction; the thermodynamic modeling of mineral stability; and the interaction of water with hydrous phases.

**Jean S. Cline** is a professor of geology at the University of Nevada Las Vegas. She received her PhD from the Virginia Polytechnic Institute and State University. Her research interests include the physical and chemical evolution of hydrothermal ore systems, fluid evolution in magmatic systems, and the use of fluid inclusions to constrain ore deposit evolution.

**Thomas A. Cotton** is vice president of JK Research Associates, where he works on radioactive waste management policy analysis and strategic planning. He currently supports the DOE's high-level radioactive waste program. Before joining JK Research Associates, he dealt with energy policy and radioactive waste management issues at the congressional Office of Technology Assessment. He received a BS in electrical engineering from Stanford University, an MA in philosophy, politics, and economics from Oxford University, and a PhD in engineering-economic systems from Stanford University.

**Bruce M. Crowe** is a geologist affiliated with Apogen Technologies Inc. and the science adviser for the National Nuclear Security Administration's Nevada Operations Office. He received a PhD from the University of California, Santa Barbara. The primary focus of his current work is probabilistic modeling and risk assessment for environmental management programs at the Nevada Test Site.

**Rodney C. Ewing**   is a professor in the Department of Geological Sciences at the University of Michigan, where he also holds appointments in the Departments of Materials Science and Engineering and Nuclear Engineering and Radiological Sciences. He received his MS and PhD in mineralogy from Stanford University. His research interests are mineralogy, materials science, radiation effects, and nuclear waste management.

**June Fabryka-Martin**   is a hydrologist with the Hydrology, Geochemistry, and Geology Group at Los Alamos National Laboratory. She received her BA in geography at the University of Delaware, and her MS and PhD in hydrology and water resources at the University of Arizona. Her research has focused on the geochemical and isotopic composition of groundwater as indicators of its sources, flow paths, and residence times.

**Alan Flint**   is a research hydrologist for the U.S. Geological Survey, Water Resources Discipline. He has worked on measurements and models to quantify infiltration and recharge for Yucca Mountain as it relates to radioactive waste disposal. Currently, he is working on quantifying recharge on a regional scale throughout the desert Southwest. He received his PhD in soil physics from Oregon State University.

**Lynn W. Gelhar**   is a professor emeritus of civil and environmental engineering at MIT. His areas of specialization are groundwater hydrology, stochastic methods, contaminant transport, and the hydrologic aspects of waste disposal. He received his PhD from the University of Wisconsin.

**Annie B. Kersting**   is a geochemist at the Lawrence Livermore National Laboratory, in the Chemistry and Material Sciences Directorate. Her PhD is in geology, and currently her research focuses on the transport and fate of actinides in the subsurface. She is the director of the Glenn T. Seaborg Institute, supporting research and education programs in nuclear science.

**Werner Lutze**   is a professor emeritus of chemical and nuclear engineering at the University of New Mexico. He is now a research scientist at the Vitreous State Laboratory of the Catholic University of America, Washington, DC. He received his PhD from Technische Universität, Berlin. His research interests include glass, ceramics, radioactive waste management, and environmental remediation.

**Allison M. Macfarlane**   is a research associate at MIT's Program in Science, Technology, and Society. She was most recently an associate professor of international affairs and earth and atmospheric science at Georgia Tech in Atlanta. Her PhD is in geology from MIT, though her research focuses on the international security and environmental policy issues associated with nuclear weapons and nuclear energy.

**Arend Meijer**   received a PhD in geochemistry from the University of California, Santa Barbara. His current research interest is in the redox conditions in groundwater systems particularly as they apply to the transport of redox-sensitive species.

**William M. Murphy**   is an associate professor of geologic and environmental sciences at California State University, Chico. He holds an MS in geology from the University of Oregon and a PhD in geology from the University of California, Berkeley. His research interests are geochemistry, environmental science, water-rock interactions, and the geologic disposal of nuclear waste.

**MaryLynn Musgrove** is a research fellow with the Environmental Science Institute at the University of Texas, Austin, and a research affiliate with the Kennedy School of Government's Energy Technology Innovation Project at Harvard University. She holds a PhD in geologic sciences from the University of Texas, Austin. Her research interests include geochemical and isotopic applications to hydrogeology and paleoclimatology, groundwater and aquifer response to climate change, water resource issues, and the interface between environmental science and science policy.

**Frank V. Perry** leads the Environmental Geology and Spatial Analysis Group at Los Alamos National Laboratory. He received his PhD in geology at the University of California, Los Angeles. Perry's work in volcanic hazards has contributed to programs in radioactive waste disposal for both the United States and Japan, and his research has focused on the geology and geochemistry of volcanic rocks in the western United States.

**Daniel P. Schrag** is a professor of earth and planetary sciences at Harvard University, and director of the Harvard University Center for the Environment. He received a PhD in geology from the University of California, Berkeley. Named a MacArthur Fellow in 2000, Schrag studies how climate has changed through Earth's history.

**David W. Shoesmith** is a professor in the Department of Chemistry at the University of Western Ontario (London, Canada) and a consultant to Bechtel/SAIC, the management and operations company responsible for the Yucca Mountain Project. He received his PhD from Newcastle upon Tyne, and specializes in research on the electrochemistry of materials and corrosion science.

**Kurt Sickafus** is an experimental researcher in materials science at the Los Alamos National Laboratory. He received his PhD from Cornell University. His research interests include radiation damage effects, crystallography, and the microstructure of materials.

**Eric R. Siegmann** is an advisory engineer with Areva/Framatome ANP. For nine years, he worked with the Yucca Mountain Project, where he modeled commercial reactor fuel degradation for the Total System Performance Assessment. He has a PhD in nuclear engineering from Northwestern University and extensive experience in reactor accident analysis and probabilistic risk assessment.

**David Stahl** is a senior technical consultant for Framatome ANP and an adjunct professor at the University of Nevada, Las Vegas. He holds a PhD in materials science, with extensive experience in nuclear materials research and development related to nuclear waste management. Before retiring in 2000 after fifteen years with the Yucca Mountain Project, he was manager of the materials department, where he was responsible for all of the waste package and waste form research and development.

**Greg A. Valentine** leads the Hydrology, Geochemistry, and Geology Group at Los Alamos National Laboratory. He received his BS at the New Mexico Institute of Mining and Technology, and his PhD at the University of California, Santa Barbara. His research has focused on field and computational studies of volcanic eruption processes, ranging from caldera-forming scale to small basaltic cones, in the western United States, Italy, and Japan.

**David T. Vaniman** is a researcher at the Los Alamos National Laboratory. He received his PhD in earth sciences from the University of California, Santa Cruz. His research interests include the mineralogy and geochemistry of igneous rocks, metamorphic rocks, and their alteration assemblages.

**Chris Whipple**   is a principal at Environ Corporation, an international consulting company that offers technical analysis on a broad range of scientific disciplines, including engineering, public heath, and regulatory affairs. He holds an MS and a PhD in engineering science from the California Institute of Technology. His expertise and experience includes risk assessments and environmental analyses, regulatory and litigation issues, and risk communication.

**Nicholas S. F. Wilson**   is a research scientist at the Geological Survey of Canada, Calgary. A geologist by training, he received his PhD from Dalhousie University, Halifax, Canada, and completed postdoctoral research at the University of Nevada, Las Vegas. His current research centers on mineral deposit and basinal studies, and approaches these problems by integrating field, petrographic-paragenetic studies with geochemistry and geochronology.

# 1

## Introduction

Allison M. Macfarlane and Rodney C. Ewing

The issue of how to solve the problem of nuclear waste has already generated heated debate among policymarkers, scientists, and the public, and promises to continue into the future. Take, for example, the most recent announcement by the Environmental Protection Agency (EPA) of its newly revised radiation protection standards for the proposed Yucca Mountain, Nevada, nuclear waste repository. The new standards would keep the original 15 millirem per year limit over the first ten thousand-year life of the repository and add a new dose limit from ten thousand to one million years of 350 millirems. The fact that the second part of the standard is almost twenty-five times the first is already creating disagreement among interested parties. What the new standards really highlight—and underscore the importance of—is the substantial uncertainty involved in predicting the future behavior of nuclear waste in a geologic repository, the focus of this book.

Spent nuclear fuel and high-level waste from reprocessing spent nuclear fuel were first generated over half a century ago. The first high-level waste from the reprocessing of spent fuel was from the plutonium production reactors at Hanford, Washington, as part of the Manhattan Project during the Second World War. Since that time, U.S. nuclear power plants and nuclear weapons production have produced enormous amounts of highly radioactive material (approximately thirty billion curies). In 2001, the United States had forty-five thousand metric tons of spent nuclear fuel from commercial nuclear reactors and two thousand metric tons of defense spent nuclear fuel. The international consensus is that the geologic disposal of high-level waste and spent fuel is an appropriate and safe solution to this problem (National Research Council 2001). Permanent geologic disposal requires the careful selection of geologic formations that can isolate the radioactive waste from the biosphere for long enough that the radioactivity poses no significant risk to the environment.

Although a number of countries (the United States, France, Germany, Sweden, Finland, Switzerland, and Japan) have active geologic repository programs, there is, at present, no operating geologic repository for spent fuel or high-level waste anywhere in the world. The U.S.-proposed repository at Yucca Mountain, which will be the focus of this book, has run into substantial technical and political challenges. In the meantime, highly radioactive waste, in the form of used nuclear reactor fuel and highly radioactive wastes from the reprocessing of spent fuel, continues to accumulate at nuclear power plants and storage facilities around the world. The United States has no reprocessing program, so most of its waste is in the form of used nuclear fuel from commercial power plants. A limited amount of commercial spent nuclear fuel, however, was reprocessed between 1966 to 1972 at West Valley, New York. Most of the high-level waste in the United States, nearly 400,000 cubic meters, is from reprocessing that was part of defense programs. These wastes are stored in huge tanks at nuclear weapons complex sites in Washington State, South Carolina, and Idaho. In 2000, there were approximately 228,300 metric tons of spent fuel in the world, projected to increase to 280,000 metric tons by 2005 (Energy Information Administration 2001). Most of this fuel is still at the 236 nuclear power stations (which together have 433 reactors) where it was originally generated in thirty-six different countries.

Alternatives to geologic disposal have been suggested, but none are as promising given the current state of alternative technologies. Proposals for sending the waste into outer space, for example, entailed enormous costs and risks. Ice cap and subseabed disposal would violate international treaties. Newer schemes, such as the transmutation of the waste into shorter-lived radionuclides, face major technological challenges as well. Partitioning and transmutation proposals, as they are called, separate the actinides and fission products from the waste, and then irradiate them in reactors or with accelerators in order to transmute long-lived radionuclides into shorter-lived ones. Recent studies have pointed out that transmutation would be expensive, with costs 10–50 percent more than the expense of light water reactors with a once-through cycle (Ahlstrom 2004). Moreover, these approaches do not eliminate the production of high-level nuclear waste; the reduced volumes of high-level waste generated would still require disposal in a geologic repository (National Research Council 1996; Ahlstrom 2004). Given the economic costs and the remaining need for a repository, others have suggested that partitioning and transmutation do not provide benefits that compensate for the safety, environmental, and proliferation risks posed by use of this technology (Ansolabehere et al. 2003).

The only other alternative to geologic repositories is long-term storage, ⸏.. is only a temporary solution. Long-term storage can be accomplished at a centra.. ized facility, such as is proposed for the Goshute Indian reservation west of Salt Lake City or reactor sites, and involves the use of dry casks, often steel and concrete canisters designed to passively cool spent fuel assemblies. During extended long-term storage, the radioactivity and heat generated by the spent fuel drop considerably, and this could be used to advantage in designing a geologic repository.

The decision between long-term storage versus disposal in a geologic repository rests partially on ethical grounds (Nuclear Energy Agency 1995). Terrorism has recently presented another motivation both for long-term storage at reactors (transport of spent fuel provides a vulnerable target) and against it (consolidating the spent fuel in one location is safer than multiple reactor sites). The basic argument against geologic repositories is that long-term storage allows the development of better technologies to deal with the waste. That may be so, but there is no guarantee that future generations will have any greater success in solving the problem than the present one. Moreover, the generation that created the waste has an ethical responsibility to deal with it. Without adequate oversight, long-term storage may prematurely release radionuclides to the environment. There is no way to guarantee the presence of that oversight over hundreds of years. How long can we ensure that societies and nations remain stable? A geologic repository, even in the absence of oversight, at least provides, in addition to the waste canister, a large volume of rock to retard the access of radionuclides into the environment. Thus, the solution to the problem of nuclear waste remains as first proposed by the National Academy of Sciences in 1957: mined geologic repositories that isolate the waste from humans and the environment (National Research Council 1957).

Although geologic repositories are the best solution to the nuclear waste problem, there are substantial uncertainties in projecting the performance of a geologic repository far into the future. There is simply no way to guarantee that a repository will not release radionuclides into the environment at some point in the future, and it is difficult to predict when and how such a release will occur. Many technical undertakings—such as space travel, nuclear war planning, and even bridge construction—involve uncertain outcomes, and yet society still attempts them. One way to reduce the uncertainties associated with the performance of geologic repositories is to select a site and an engineering design that reduces uncertainty. Also, it is possible to ensure that a repository could be operated in a reversible fashion. That is, if problems develop with the site, the waste could be retrieved. In the end, our society has created and will continue to create nuclear waste that must be disposed of—most likely in geologic repositories.

The issue of how to deal with nuclear waste has become more pertinent in recent years with the rekindling of interest in nuclear energy (Meserve 2004; National Commission on Energy Policy 2004). Nuclear energy is often cited as a way to produce energy without the emission of the greenhouse gases, particularly $CO_2$, that are responsible for global warming (Houghton et al. 2000). If nuclear power is to play an important role in the reduction of carbon emissions, however, most studies have suggested a need for a three- to tenfold increase in nuclear power production (Fetter 2000; Sailor et al. 2000; Ansolabehere et al. 2003). Such an expansion would require an enormous effort to manage the additional spent fuel and high-level waste that would be generated. A tenfold increase in the number of nuclear power plants over those that existed in 2000 would result in the production of more than seventy thousand metric tons of spent fuel each year (Macfarlane 2003). For an open fuel cycle, in which spent fuel is not reprocessed but disposed of directly, this would require the construction and opening of a geologic repository, with the capacity of the one proposed for Yucca Mountain, every year. Such an expansion in nuclear energy production simply cannot move forward without resolving the problem of the safe disposal of nuclear waste.

Although there is an international consensus among nuclear experts that nuclear waste can be safely disposed of in a geologic repository, after many decades of effort there is still no geologic repository licensed to receive spent nuclear fuel or high-level waste. In the United States, work on the Yucca Mountain Project has continued for more than twenty years at a cost of over seven billion dollars. Even once the much-anticipated license application is submitted by the Department of Energy (DOE) to the Nuclear Regulatory Commission (NRC), the review process may take up to four years. Clearly, the technical issues and the accompanying uncertainties related to predicting the long-term behavior of a geologic repository are and will continue to be a challenge (National Research Council 2003; Long and Ewing 2004). In this book, using the experience of developing a geologic repository at Yucca Mountain as an example, we focus on some of the technical issues that must be resolved in order to demonstrate safe disposal over long periods, probably hundreds of thousands of years. Although the Yucca Mountain site is unique in many respects, many of the issues that we highlight here are the same for other geologic repositories.

One of the goals of this book is to emphasize the importance of the disciplines in the geosciences (i.e., climatology, hydrology, geochemistry, and mineralogy) in developing strategies for the long-term disposal of nuclear waste. Unfortunately, the evaluation of the performance of geologic repositories has often focused on the performance of near-field engineered barriers, such as the design of the packages that will contain the waste or the drip shield at Yucca Mountain. Engineered barriers are significant, particularly over the short-term, but it is the geology and hydrology of the repository site that provide the basis for long-term containment of the radioactive wastes. Indeed, the original concept of geologic disposal incorporated multiple natural barriers, including delayed access of water, long groundwater travel times, and sorption of radionuclides along the transport path (National Research Council 1983). In fact, the most robust form of geologic disposal involves not only the properties of the natural features, such as the rock and the hydrology, but the fact that these natural barriers act *in concert* with the engineered barriers, such as the waste package and the waste form. As central as natural systems are to the concept of geologic disposal, one must account for the inevitable and inherent uncertainties in modeling the behavior of geologic systems, particularly over long time spans (many hundreds of thousands of years) and great distances (tens of kilometers). Hence, we have asked the chapter authors to focus on the uncertainty in each of the aspects of the repository system that they describe.

This volume is designed to make these crucial technical issues understandable to a wider audience, including scientists and engineers not directly involved in the Yucca Mountain Project, laypersons, and policymakers. We believe that sound policy is based on a solid understanding of the technical issues. At this time, we must face the problem of high-level nuclear waste disposal and its long-term solutions. While geologic repositories may offer the best solution, we must endeavor to understand the complexity and uncertainty of the multidisciplinary science that is required to support this strategy.

## Characteristics of High-Level Nuclear Waste

Fuels for light water reactors typically consist of uranium dioxide ($UO_2$) in which the isotopic concentration of fissile uranium-235 is approximately 4 atomic percent. The radioactivity of spent fuel one month after discharge is about $10^{17}$ becquerel per metric ton of fuel, which corresponds to $3.35 \times 10^{17}$ becquerel for a single spent fuel assembly. This radioactivity is generated by the fission products that make up 3–4 percent of the waste (e.g., ruthenium-106, iodine-131, cesium-137, and strontium-90), the transuranium elements (e.g., plutonium-239, neptunium-

237, and americium-241), and the activation products of metals, such as cobalt, nickel, and niobium, in the spent fuel assemblies. The penetrating ionizing radiation (beta and gamma) comes mainly from the short-lived fission products (cesium-137 and strontium-90, both with half-lives of about 30 years). The less penetrating alpha radiation comes mainly from the very long-lived actinides (e.g., plutonium-239 and neptunium-237, with half-lives of 24,100 years and 2.1 millions years, respectively). The chemical composition of the spent fuel, when it is initially removed from the reactor, contains hundreds of short-lived radionuclides. The radioactivity decreases over time, declining after 10,000 years to 0.01 percent of the level at one month after removal from the reactor (Hedin 1997).

Spent fuel is highly radioactive. One year after discharge from a reactor the dose rate measured one meter from the fuel assembly is one million millisieverts per hour. A person exposed to this radioactivity from one meter away would absorb a lethal dose in less than a minute, and would incur a 5 percent chance of developing a fatal cancer in less time than that (Hedin 1997). For comparison, the exposure from natural background radioactivity is on the order of three millisieverts per *year*. The dangerously intense radiation from a spent fuel assembly necessitates remote handling either underwater or in shielded hot cells. Currently, storage is accommodated through the use of water pools and shielded casks.

The decay of fission products also produces substantial quantities of heat. Spent fuel that has just been removed from a reactor generates about 2,000 kilowatts per metric ton (Hedin 1997), but after 120 years the energy release drops to 1.47 kilowatts per metric ton (Bodansky 1996). Though the total volume of high-level waste is not large, in comparison to wastes from other energy-producing systems, but due to its high radioactivity and high thermal heat output, it would still require a large volume of rock for storage.

The first ten thousand years after disposal are the most complicated because the spent fuel contains substantial quantities of shorter-lived fission products, and significant amounts of heat are still being generated. After one thousand years, the thermal output of the spent fuel is negligible and the radiotoxicity is dominated by a limited number of long-lived radionuclides: plutonium-239, plutonium-240, neptunium-237, uranium-233, americium-241, technicium-99, iodine-129, and selenium-79, as well as the remaining uranium-235 and uranium-238. The calculated exposure of a person to radiation essentially depends on the time of release from the waste package, and the retardation and dilution provided by the geologic barriers between the repository and the exposed person.

## Geologic Setting of Yucca Mountain

Yucca Mountain is located in the Great Basin Desert, 145 kilometers northwest of Las Vegas (figure 1.1), and receives only 170 millimeters of rainfall annually (OCRWM 1998). The mountain is not an imposing landform as compared with others in the region; its crest reaches elevations of 1,500 to 1,930 meters (OCRWM 2002). The rocks at Yucca Mountain formed between 11.6 and 13.5 million years ago, after multiple episodes of explosive volcanism associated with the Timber Mountain caldera complex (formed by subsiding volcanoes) to the north and uplift associated with extensional faulting (Carr et al. 1986; Sawyer et al. 1994). Yucca Mountain falls in the southwestern Nevada volcanic field, which comprises six calderas (figure 1.2), formed 7.5 to 14 million years ago (OCRWM 2001).

Yucca Mountain consists of alternating layered deposits of welded and nonwelded tuffs that dip ten degrees to the east (figure 1.3). Tuff is a fine-grained rock composed of fragmented volcanic ash and pulverized rock. A specific type of tuff that forms much of Yucca Mountain was deposited as ash flows, and can be dense if its grains were fused or welded by residual magmatic heat on deposition. The uppermost unit, the Paintbrush Group, contains three subunits: the welded Tiva Canyon tuff, the nonwelded Yucca Mountain and Pah Canyon tuffs; and the Topopah Spring tuff, the unit into which the repository would be sited (figure 1.4). Below the Paintbrush Group lies the Calico Hills Formation, consisting of glassy nonwelded tuffs and lavas that contain secondary alteration minerals called zeolites (OCRWM 2002), and the Crater Flat Group, another unit of welded and nonwelded tuffs. Underlying these volcanic units, which formed in the Miocene epoch (five to twenty-four million years ago), are Paleozoic and Precambrian carbonate and clastic (formed from sediments) rocks that form the major framework rocks for regional groundwater flow. In general, the welded tuff units tend to be fractured and the nonwelded units are less fractured.

The Topopah Spring tuff was formed 12.8 million years ago and reaches thicknesses of 375 meters at Yucca Mountain (OCRWM 2002). The Topopah Spring unit is subdivided into an upper member, rich in crystal content, and a lower crystal-poor member, both of which contain lithophysal-rich and lithophysal-poor subunits (figure 1.4). Lithophysae are cavities of varying size formed by gas bubbles that were produced after the rocks were deposited and during the process of welding and cooling. The repository would be located in the densely welded, lower, crystal-poor member, containing both lithophysal-rich and lithophysal-poor subunits (OCRWM 2001).

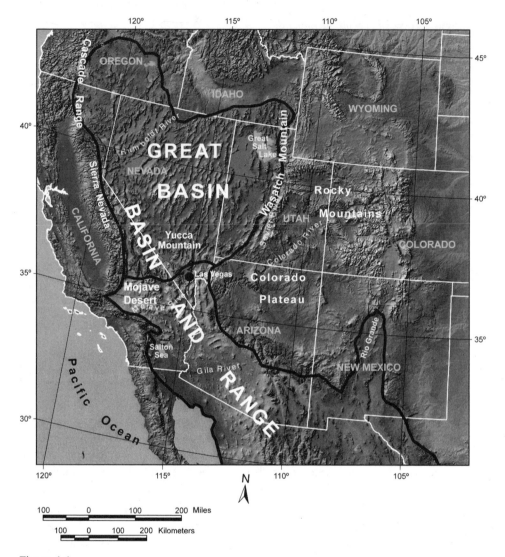

**Figure 1.1**
Location map of Yucca Mountain in the Great Basin and the Basin and Range of the western United States. From Office of Civilian Radioactive Waste Management 2001, figure 1.7.

**Figure 1.2** (*opposite*)
Volcanic rocks in the Yucca Mountain region. Note the position of the four circular $Q_{bo}$ cones southwest of Yucca Mountain (1 million years), the 3.7 million-year-old basalts also southwest of Yucca Mountain, and the Lathrop Wells cone at the southern end of Yucca Mountain (0.08 Ma). From Office of Civilian Radioactive Waste Management 2001, figure 1.8.

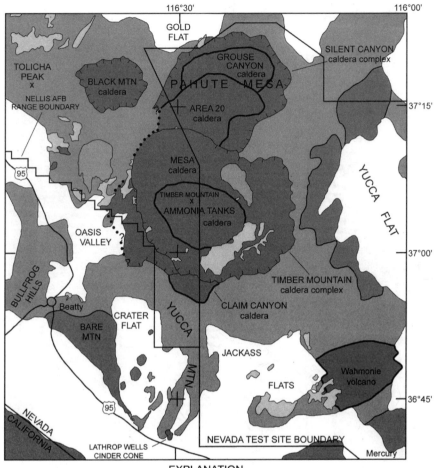

116°30'
116°00'

GOLD
FLAT

SILENT CANYON
caldera complex

GROUSE
CANYON
caldera

TOLICHA
PEAK
x

BLACK MTN
caldera

PAHUTE MESA

NELLIS AFB
RANGE BOUNDARY

AREA 20
caldera

37°15'

MESA
caldera

YUCCA FLAT

TIMBER MOUNTAIN
x
AMMONIA TANKS
caldera

OASIS
VALLEY
?

37°00'

TIMBER MOUNTAIN
caldera complex

BULLFROG
HILLS

Beatty

CLAIM CANYON
caldera

BARE
MTN

CRATER
FLAT

YUCCA

JACKASS

Wahmonie
volcano

MTN

FLATS

36°45'

NEVADA
CALIFORNIA

95

LATHROP WELLS
CINDER CONE

NEVADA TEST SITE BOUNDARY

Mercury

### EXPLANATION

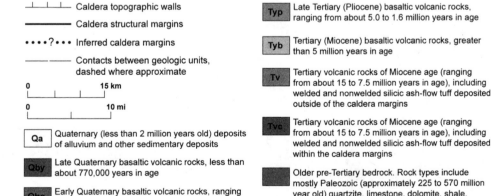

⊥⊥⊥⊥ Caldera topographic walls

▬▬▬ Caldera structural margins

••••?••• Inferred caldera margins

—————— Contacts between geologic units,
dashed where approximate

0                    15 km

0              10 mi

**Qa** Quaternary (less than 2 million years old) deposits
of alluvium and other sedimentary deposits

**Qby** Late Quaternary basaltic volcanic rocks, less than
about 770,000 years in age

**Qbo** Early Quaternary basaltic volcanic rocks, ranging
from about 1.6 to 0.77 million years in age

**Typ** Late Tertiary (Pliocene) basaltic volcanic rocks,
ranging from about 5.0 to 1.6 million years in age

**Tyb** Tertiary (Miocene) basaltic volcanic rocks, greater
than 5 million years in age

**Tv** Tertiary volcanic rocks of Miocene age (ranging
from about 15 to 7.5 million years in age), including
welded and nonwelded silicic ash-flow tuff deposited
outside of the caldera margins

**Tvc** Tertiary volcanic rocks of Miocene age (ranging
from about 15 to 7.5 million years in age), including
welded and nonwelded silicic ash-flow tuff deposited
within the caldera margins

Older pre-Tertiary bedrock. Rock types include
mostly Paleozoic (approximately 225 to 570 million
year old) quartzite, limestone, dolomite, shale,
siltstone, and sandstone.

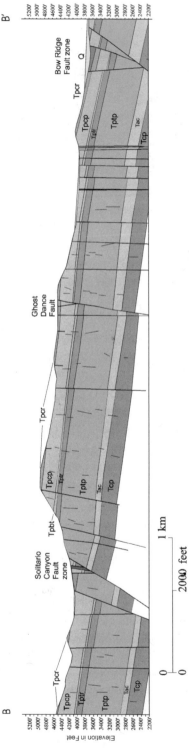

**Figure 1.3**

Cross-section from west (B) to east (B′) across Yucca Mountain. From Office of Civilian Radioactive Waste Management 1999, figure 1.10.

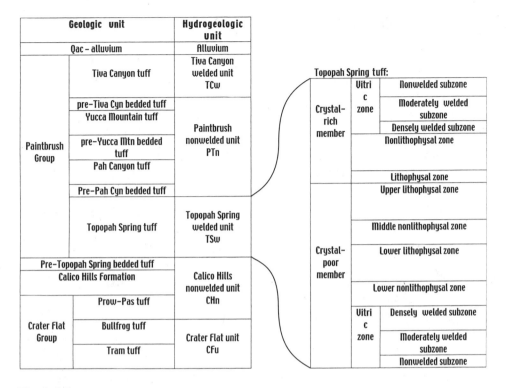

**Figure 1.4**
Stratigraphic column of rocks at Yucca Mountain. On right is detail of the Topopah Spring's welded tuff.

Due to its location in the extensional Basin and Range province, the Yucca Mountain region remains tectonically active, exhibiting both seismic and volcanic activity. These processes are important to characterize and understand because of their possible effects on the future behavior of the repository. Faulting can affect water movement through the mountain by both creating and destroying fast pathways for transport; affecting the stability of the waste-containing drifts, potentially damaging the waste packages; and leading to more faulting (Ferrill and Morris 2001). Volcanism is more significant to repository behavior than faulting. An eruption beneath the repository could lead to the dispersion of radionuclides into the atmosphere (explosive eruption) or the more rapid dispersal of radionuclides into groundwater. The latter could occur due to corrosive conditions affecting waste packages from the infiltration of magma and the associated acidic gases into the drifts that contain the waste packages.

The Yucca Mountain region consists of fault-bounded blocks, and Yucca Mountain itself is bounded by the Solitario Canyon fault located on its west side (figure 1.5). Most of the block-bounding faults experienced movement in the last two million years, are oriented generally north-south, and dip fifty to eighty degrees to the west (OCRWM 2001). The repository footprint is bounded by two smaller-scale faults: the Sundance fault on the west and the Ghost Dance fault to the east (figure 1.5). The Sundance fault experienced eleven meters of movement prior to two million years ago (OCRWM 2001). The local area surrounding Yucca Mountain remains seismically active. For instance, in 1992 an unexposed fault on Little Skull Mountain, located less than twenty kilometers south of Yucca Mountain (figure 1.6), caused a magnitude 5.6 earthquake with thousands of aftershocks (von Seggern and Brune 2000). More recently, in June 2002, the same fault system produced a magnitude 4.4 earthquake further east on the related Rock Valley fault (Seismological Laboratory 2004).

The fault and fracture orientations at Yucca Mountain originate from a stress field that was oriented east-west prior to ten million years ago and then shifted to the west-northwest (Zoback et al. 1981). Thus, the majority of faults are oriented north-south and northeast-southwest (figure 1.5). Some faults and fractures were created by thermoelastic contraction during the cooling of the volcanic rocks that make up the mountain, resulting in faults oriented in other directions (Sweetkind and Williams-Stroud 1996). The faults that will most likely experience slip in the future are those oriented north-northeast to south-southwest (0–N55°E) and dipping forty to sixty-five degrees (Ferrill et al. 1999).

The area surrounding Yucca Mountain is dotted with small cinder cones, but the probability of future volcanic activity at the repository site continues to be debated. Three episodes of basaltic volcanism have recently affected the region (figure 1.2). The oldest occurred about 3.7 million years ago, resulting in a number of cinder cones along a north-south trend adjacent to the west flank of Yucca Mountain (Fleck et al. 1996). This episode was followed by the formation of four scoria (a dark, frothy-type rock with many tiny cavities) cones and lava flows just west and southwest of the repository location one million years ago (Crowe et al. 1995). The most

**Figure 1.5** (*opposite*)
Geologic map of faults in the vicinity of Yucca Mountain. The repository access tunnel is shown by the dark solid line and the outline of the repository itself is shown by the dark dotted line. From Office of Civilian Radioactive Waste Management 2001, figure 1.14, p. 1-35.

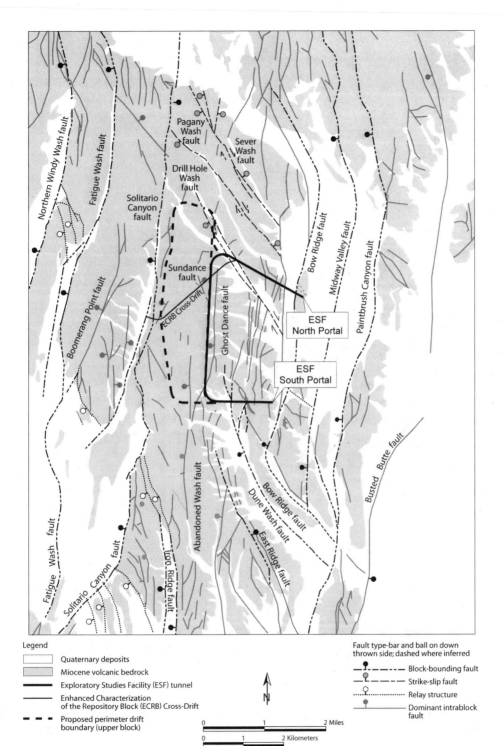

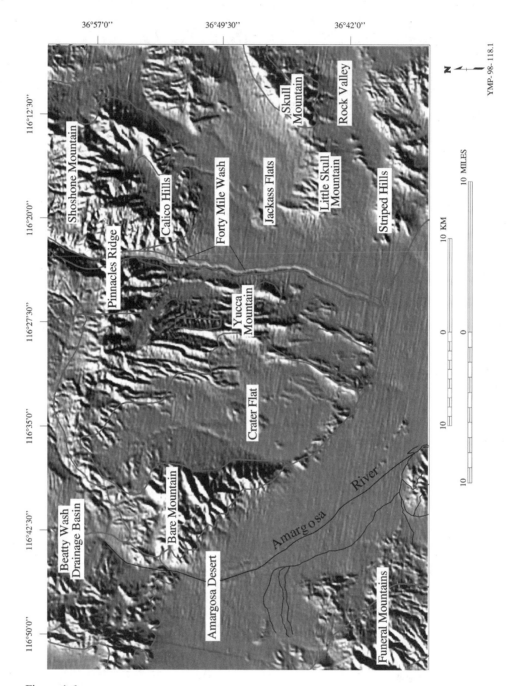

**Figure 1.6**
Map showing locations of physiographic features near Yucca Mountain, including Little Skull Mountain in the southeast corner. From Office of Civilian Radioactive Waste Management 1999, figure 1.2.

recent episode eighty thousand years ago created the Lathrop Wells cone, located at the southern tip of Yucca Mountain (Heizler et al. 1999).

Water is the medium by which radioactive elements would be transported to the biosphere. Thus, understanding the geohydrology of Yucca Mountain is essential to understanding the performance of the geologic repository. The hydrologic system includes the infiltration of rainwater at the surface into the rock units below, transport through the unsaturated zone (the rock units from the surface to the water table) along the fractures and pores of the rock, seepage into the tunnels containing the waste packages, the corrosion of waste packages and waste forms, the release from the waste package and transport to the water table, and finally lateral transport in the saturated zone, below the water table, tens of kilometers to the south to Amargosa Valley.

Current analysis by the DOE uses an average infiltration rate in the unsaturated zone on the order of five millimeters per year (OCRWM 1999). The infiltration rates partly depend on predictions of climate change over the next tens of thousands of years at Yucca Mountain. The DOE uses U.S. Geological Survey (2000) estimates of variations in climate over the last four hundred thousand years to predict future climates. The models suggest that Yucca Mountain could be dominated by a cooler and wetter glacial transition climate over part of this time period. Water infiltration is predicted to increase during a transition in climate within the next ten thousand years due to higher precipitation, colder temperatures, and less evapotranspiration (OCRWM 2001).

Studies of the hydrogeology of the unsaturated zone are still under way at the DOE. Water in the unsaturated zone exists either in matrix pore spaces in the bulk rock, in fractures, or as perched water bodies—that is, small lenses of water that lie above the water table (figure 1.7) (OCRWM 1999). According to the DOE's conceptual model, water transport in the unsaturated zone is controlled by: the fracture flow in the Tiva Canyon welded tuff, with permeabilities of approximately $1 \times 10^{-18}\,\mathrm{m^2}$; the matrix flow in the Paintbrush nonwelded unit (corresponding to the Yucca Mountain and Pah Canyon tuffs), with permeabilities of $1 \times 10^{-13}\,\mathrm{m^2}$; and the fracture flow in the Topopah Spring tuff, with permeabilities on the order of $1 \times 10^{-17}\,\mathrm{m^2}$ (Flint et al. 2001; Campbell et al. 2003).

The unsaturated zone appears to contain a number of perched water bodies, as suggested by the saturated areas in cores samples (OCRWM 1999). All perched water identified so far is located below the repository horizon, most of it near the contact between the Topopah Spring welded unit and the Calico Hills nonwelded unit (figure 1.7). Perched water has a different composition from pore matrix water,

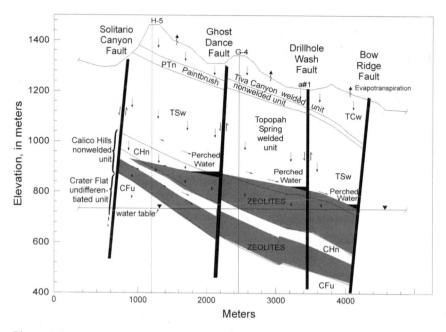

**Figure 1.7**
Schematic cross-section of Yucca Mountain showing the hydrogeologic units, the location of the water table, the perched water zones, and the depth of zeolite-rich horizons. From Office of Civilian Radioactive Waste Management 1999, figure 2.4.

and its younger age indicates that it is recharged via rapid pathways (OCRWM 1999).

The water table at Yucca Mountain is located at least five hundred meters beneath the surface in the volcanic rocks of the Crater Flat Group (figure 1.7). In general, the flow of groundwater is from north to south, following the topographic gradient, which is higher in the north (OCRWM 1999). The tuffs that make up the upper local aquifer interfinger with alluvial (terrestrial, unconsolidated) sediments to the south, and groundwater reaches the surface in the Amargosa Desert (OCRWM 1999). A deeper regional aquifer lies in the Paleozoic and Precambrian carbonate rock units. Recent studies suggest that groundwater infiltrated Yucca Mountain and the adjacent areas between twelve and sixteen thousand years ago, and local infiltration rates at that time ranged between eight and ten millimeters per year (Flint et al. 2002; Kwicklis et al. 2003). One unusual and poorly studied feature of the hydrology near the repository site is an apparent steep hydraulic gradient located just to the north of the proposed repository (OCRWM 2002).

**Engineering Design**

The repository design is still not finalized; the DOE maintains that it can carry a range of design options through the licensing process. (The following description represents the best available knowledge as of August 2005.) These include "hot" versus "less hot" designs where the spacing between the waste packages in various combinations with drip shields, backfill, and different closure dates determines the temperature distribution. The repository may remain "open" for hundreds of years after the waste is emplaced in order to ventilate the heat generated by radioactive decay. The repository itself will be located directly below the ridge of Yucca Mountain, two to three hundred meters beneath the surface and two to three hundred meters above the water table, in the Topopah Spring unit of the Paintbrush Group (figure 1.3). The location is designed to minimize intersections with faults, important pathways for downward-percolating water. The repository design currently consists of fifty-eight emplacement drifts, which will each be 5.5 meters in diameter (figure 1.8). The drifts will be spaced eighty-one meters apart, on a center-to-center basis (OCRWM 2001). The total length of the drifts will amount to 56.2 kilometers, equivalent to 1,150 acres of area, for a total mass loading of fifty-six metric tons per acre (OCRWM 2001).

The DOE will load canisters of high-level waste into the drifts, the vast majority of which will be spent nuclear fuel from commercial nuclear power plants. In the Nuclear Waste Policy Act of 1982, Congress authorized seventy thousand metric tons of waste to be loaded into the repository, with the understanding that the bulk of this (sixty-three thousand metric tons) will originate from electricity production and the rest from the nuclear weapons complex. The waste package itself will consist of a cask that has an outer layer 2.0- to 2.5-centimeter thick of Alloy-22, a nickel-chromium-molybdenum alloy, and a 5-centimeter thick inner layer of stainless steel (figure 1.9). The casks themselves will have different dimensions, depending on the waste form they enclose. Spent fuel, the most abundant waste form, will be in a variety of lengths due to different reactor types. Waste forms from the nuclear weapons complex include vitrified high-level waste and spent or partially used fuels. The DOE may also use titanium drip shields, designed to protect the waste package from falling rocks and dripping water (figure 1.10).

As of early 2005, the DOE had not yet decided a number of the repository design issues, the most significant being the thermal conditions in the repository. Until recently, the DOE had planned to operate a "hot" repository, where thermally hot

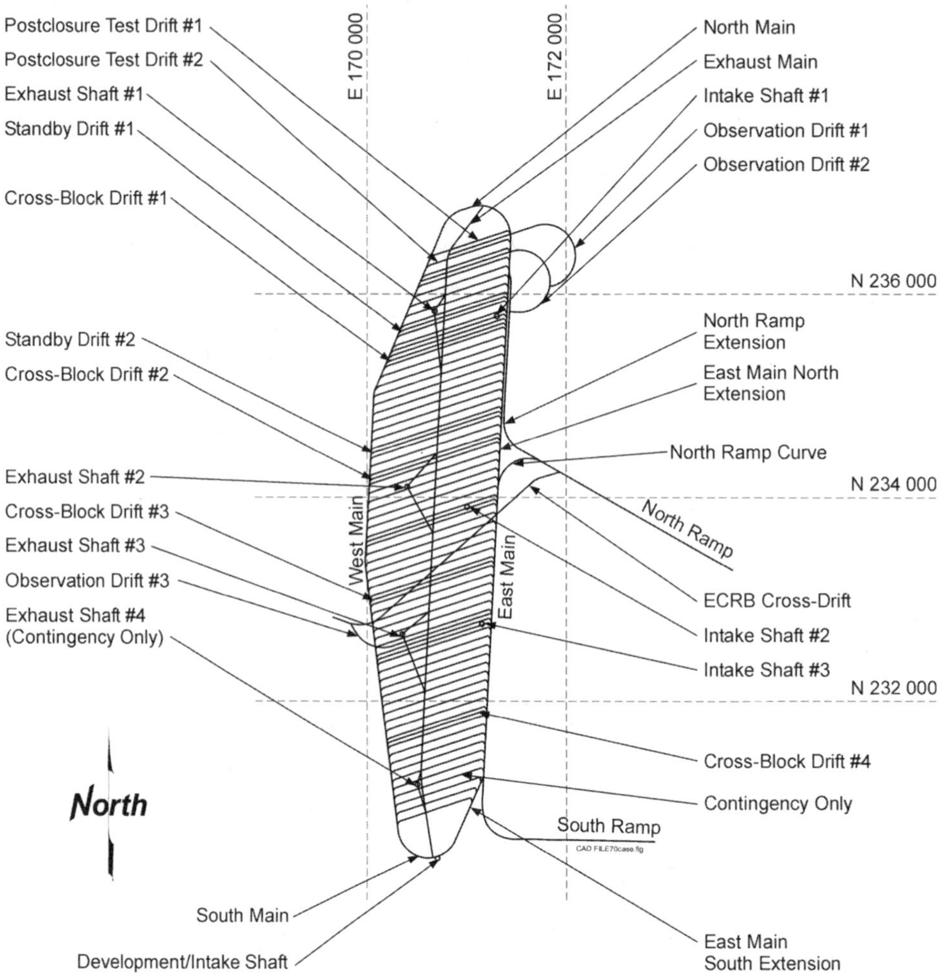

**Figure 1.8**
Schematic map of the plan for the repository drift locations for emplacement of seventy thousand metric tons of waste. From Office of Civilian Radioactive Waste Management 2001, figure 2.38.

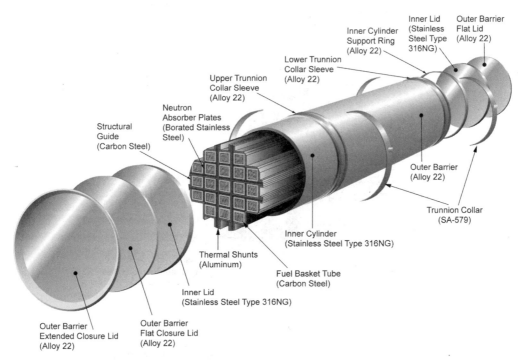

**Figure 1.9**
Schematic diagram of a waste canister showing the positioning of Alloy-22 and stainless steel.
From Office of Civilian Radioactive Waste Management 2001, figure 3.2.

radioactive waste would heat the surrounding repository rocks to temperatures greater than 100°C (OCRWM 1998). In order to decrease the uncertainties associated with repository performance, however, the DOE is considering operating the repository at temperatures no greater than 85°C (OCRWM 2001). To keep the repository that cool would require many years of forced ventilation (using fans), smaller waste packages, more tunnel excavation or emplacement area, and/or a lower linear thermal loading. Alternatively, waste could be cooled aboveground for many tens of years prior to emplacement in the repository.

### Technical Issues

In attempting to understand the technical issues at Yucca Mountain, we first set the stage by discussing issues related to the policy choices in high-level nuclear waste

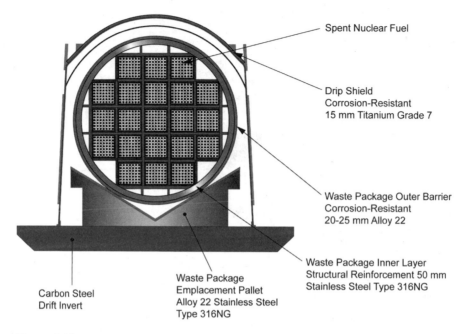

Spent Nuclear Fuel

Drip Shield
Corrosion-Resistant
15 mm Titanium Grade 7

Waste Package Outer Barrier
Corrosion-Resistant
20-25 mm Alloy 22

Waste Package Inner Layer
Structural Reinforcement 50 mm
Stainless Steel Type 316NG

Waste Package
Emplacement Pallet
Alloy 22 Stainless Steel
Type 316NG

Carbon Steel
Drift Invert

**Figure 1.10**
Cross-section of an emplacement drift showing the position of the waste package, the drip shield, and the waste package pallet. From Office of Civilian Radioactive Waste Management 2001, figure 3.1.

disposal. These include the history of the policy process, the regulatory process, the method of determining site suitability, and finally, the policy process itself. Next, we investigate the detailed technical issues related to nuclear waste disposal at Yucca Mountain. In doing so, we examine external factors such as climate change and volcanism that could affect repository performance, and then turn to the repository itself. Here we consider hydrologic issues, such as the potential for water at the repository level and the transport of water and radionuclides both above and below the water table. We then discuss the conditions that the waste in the repository might encounter, and the predicted behavior of the different kinds of waste and waste package materials.

In human terms, the potential repository at Yucca Mountain has already had a long history. In February 2002, the DOE and President Bush declared the Yucca Mountain site to be suitable for a high-level waste geologic repository. The governor of Nevada vetoed the use of the Yucca site for nuclear waste disposal, but Congress overrode this objection in July 2002. Thomas Cotton recounts the story of high-level nuclear waste disposal in the United States in chapter 2.

The next step in the process is the license application that the DOE is preparing for the NRC's evaluation. Although federal law required that the license application be ready within ninety days of congressional approval of the site, at the earliest the DOE will not submit a license application until 2006. Once the application is submitted, the NRC has up to four years to determine whether the site will meet the standards set by the EPA. William Murphy explains the details of regulating the Yucca Mountain site in chapter 3.

The main tool that the DOE and the NRC used to evaluate the Yucca Mountain site is probabilistic performance assessment, an engineering analysis that combines the results of hundreds of subsystem models into a metamodel, which produces a result that can be used to determine compliance with EPA regulations. In chapter 4, Chris Whipple explains the process used in probabilistic performance assessment and clarifies why it is used in the determination of site suitability at Yucca Mountain. Rodney Ewing, in chapter 5, analyzes the pros and cons of probabilistic performance assessment and suggests alternative approaches to site suitability assessment.

The nuclear waste story in the United States is still in flux. In July 2004, the U.S. Court of Appeals for the District of Columbia Circuit ruled that part of the EPA standard was inconsistent with a congressional directive. In the 1992 Energy Policy Act, Congress directed the EPA to base its standard on the recommendation of the National Academy of Sciences (National Research Council 1995). In particular, the EPA failed to heed the recommendation of the academy on the time over which compliance with the standard would be determined. The academy suggested evaluating repository performance at the time of peak dose, expected to occur hundreds of thousands of years after the waste is put into the repository. The EPA instead selected a ten-thousand-year time of compliance—the same time period used in the initial standard for disposal of high-level and transuranic waste promulgated in 1985 (part 191 of chapter 40 of the Code of Federal Regulations, or hereafter 40 CFR 191). The EPA is now in the process of finalizing another new standard. The new standard will require additional analyses from DOE in its license application. In chapter 6, Allison Macfarlane examines how nuclear waste policies evolved in the United States and discusses the interaction between science and policy that have occurred over that time.

In evaluating the technical issues that could affect site suitability, we asked chapter authors to address a specific set of questions. In particular, in order to understand the uncertainties associated with each aspect of the performance assessment, we wanted to know the following:

• What was the state of knowledge of the subject
• What was not known
• Whether what was not known could affect the safety of the repository
• How long it would take to develop the knowledge necessary to address the issues for which present knowledge is lacking

The answers to these questions are essential to understand the technical issues and determine what will be required to have a reasonable expectation that the repository is safe.

We begin our discussion of the technical issues at Yucca Mountain with an overview of some of the more significant geologic issues by David Applegate in chapter 7. One of the largest uncertainties at the site is the potential for future volcanic activity near the repository. Bruce Crowe, Greg Valentine, Frank Perry, and Paul Black explore this issue in chapter 8. An important factor in predicting the performance of a repository, especially a "dry" one like Yucca Mountain, is the rate of water infiltration into the mountain. MaryLynn Musgrove and Daniel Schrag in chapter 9 take a crucial look at the effects of future climate change at Yucca Mountain on the infiltration rate of water into the repository.

Hydrologic issues dominate the predicted performance of a repository at Yucca Mountain. One question that the DOE and even the National Academy of Sciences has grappled with for years is that of the potential for water to rise to the level of the repository at some time in the future. Nicholas Wilson and Jean Cline provide the definitive analysis of this issue in chapter 10. Their chapter illustrates how a focused scientific effort can successfully address an important but much debated technical issue.

The potential for cask corrosion and the consequent transport of radionuclides depends on the amount of water that reaches the repository. This depends in large part on how the water travels in the unsaturated zone, the region above the water table. June Fabryka-Martin, Alan Flint, Arend Meijer, and Gilles Bussod discuss this issue in chapter 11, and explain what is not yet known about transport in the unsaturated zone.

Once a waste package is breached and radioactive material is exposed, a number of issues will affect the ability of radionuclides to travel to the water table. For radionuclides such as plutonium, perhaps the most important technical issue is the ability of materials called colloids to transport them. Annie Kersting evaluates the existing knowledge on this issue as it applies to Yucca Mountain in chapter 12. David Bish, William Carey, Steve Chipera, and David Vaniman extend the discussion of radionuclide transport by examining the interaction of radionuclides with minerals in Yucca Mountain rocks in chapter 13. Once radionuclides reach the water table, they will be transported in the saturated zone. In chapter 14, Lynn Gelhar describes where they might be transported and the gaps in knowledge about water flow and contaminant transport in the saturated zone at Yucca Mountain.

The environmental conditions in a repository full of radioactive waste are expected to be complex, in large part due to the interactions of the thermally hot and highly radioactively waste package with the surrounding rock and any water the rock contains. The result will be the coupled thermal, mechanical, and hydrologic processes that G. S. Bodvarsson discusses in chapter 15. William Murphy explores the local or near-field geochemical environmental conditions in the repository in chapter 16.

How the waste itself will degrade and how the waste packages will corrode has been the focus of much of the characterization work accomplished by the DOE. In chapter 17, David Shoesmith discusses how the waste packages themselves, formed from a highly corrosion resistant metal, might degrade. The DOE plans to provide backup to the waste packages through the use of drip shields that will protect the waste canisters from rockfall and water seepage. David Stahl evaluates their behavior and utility in chapter 18.

Most of the waste to be stored in Yucca Mountain is spent fuel from nuclear reactors. This is in the form of uranium dioxide pellets that fill a long thin tube, the cladding, of zirconium-alloy metal. In chapter 19, Eric Siegmann looks at the ability of the zircaloy cladding to resist corrosion. Once the waste package is breached and the cladding is corroded, the spent fuel itself will be exposed. Jordi Bruno and Esther Cera examine the predicted behavior of spent fuel in chapter 20.

Waste that originates from the U.S. nuclear weapons complex will arrive in the form of glass or crystalline ceramic. The DOE has already vitrified a substantial quantity of high-level waste at the Savannah River site in Aiken, South Carolina, and all of the high-level waste at West Valley, New York. The DOE also plans to vitrify wastes at the Hanford facility in Richland, Washington. Werner Lutze discusses the issues associated with disposing of glass in a Yucca Mountain repository in chapter 21. In addition, some of the U.S. government's excess weapons plutonium may be immobilized into a crystalline ceramic. In chapter 22, William Bourcier and Kurt Sikafus explore the expected performance of a ceramic waste form in the Yucca Mountain repository.

We close the book with two views on nuclear waste disposal. In chapter 23, Luther Carter, a longtime observer of the nuclear waste program in the United States (Carter 1987), presents his outlook on the future and suggests ways to improve current U.S. nuclear waste disposal policy. Allison Macfarlane offers her own closing thoughts in chapter 24. The authors of the chapters in this book represent a range of disciplines and views on the suitability of the Yucca Mountain site as a nuclear waste repository. Indeed, besides academic researchers, the author list includes scientists at the U.S. national laboratories, contractors, think tanks, and DOE staff with the Yucca Mountain Project. They all agree, however, that geologic disposal is the best path forward for the safe disposal of spent nuclear fuel and high-level nuclear waste.

## References

Ahlstrom, P.E. (2004) *Partitioning and Transmutation Current Developments, 2004.* SKB, Stockholm. Technical Report TR–04–15.

Ansolabehere, S., Deutch, J., Driscoll, M., Gray, P.E., Holdren, J.P., Joskow, P.L., Lester, R.K., Moniz, J., and Todreas, N.E. (2003) *The Future of Nuclear Power.* An Interdisciplinary MIT Study. http://web.mit.edu/nuclearpower/.

Bodansky, D. (1996) *Nuclear Energy: Principles, Practices, and Prospects.* Woodbury, NY: American Institute of Physics, 396 p.

Campbell, K., Wolfsberg, A., Fabryka-Martin, J., and Sweetkind, D. (2003) Chlorine-36 Data at Yucca Mountain: Statistical Tests of Conceptual Models for Unsaturated-Zone Flow. *Journal of Contaminant Hydrology* 62–63, pp. 43–61.

Carr, W.J., Byers F.M., and Orkild, P.P. (1986) *Stratigraphic and Volcano-Tectonic Relations of the Crater Flat Tuff and Some Older Volcanic Units, Nye County, Nevada.* U.S. Geological Survey. Professional Paper 1323.

Carter, L. (1987) *Nuclear Imperatives and Public Trust.* Washington, DC: Resources for the Future, 473 p.

Crowe, B.M., Perry, F.V., Geisman, J., McFadden, L., Wells, S., Marrell, S., Potts, J., Valentine, G.A., Bowker, L., and Finnegan, K. (1995) *Status of Volcanism Studies for the Yucca Mountain Site Characterization Project.* Los Alamos National Laboratory. LA–12908–MS.

Energy Information Administration (2001) World Cumulative Spent Fuel Projections. May. http://www.eia.doe.gov/cneaf/nuclear/page/forecast/cumfuel.html.

Ferrill, D.A., and Morris, A.P. (2001) Displacement Gradient and Deformation in Normal Fault Systems. *Journal of Structural Geology* 23, pp. 619–638.

Ferrill, D.A., Winterle, J., Wittmeyer, G., Sims, D., Colton, S., Armstrong, A., and Morris, A.P. (1999) Stressed Rock Strains Groundwater at Yucca Mountain, Nevada. *GSA Today* 9, pp. 1–8.

Fetter, S. (2000) Energy 2050. *Bulletin of the Atomic Scientists*, July/August, pp. 28–38.

Fleck, R.J., Turrin, B.D., Sawyer, D.A., Warren, R.G., Champion, D.E., Hudson, M.R., and Minor, S.A. (1996) Age and Character of Basaltic Rocks of the Yucca Mountain Region, Southern Nevada. *Journal of Geophysical Research* 101, pp. 8205–8227.

Flint, A., Flint, L.E., Bodvarsson, G.S., Kwicklis, E.M., and Fabryka-Martin, J. (2001) Evolution of the Conceptual Model of the Unsaturated Zone at Yucca Mountain. *Journal of Hydrology* 247, pp. 1–30.

Flint, A., Flint, L.E., Kwicklis, E.M., Fabryka-Martin, J., and Bodvarsson, G.S. (2002) Estimating Recharge at Yucca Mountain, Nevada: Comparison of Methods. *Hydrogeology Journal* 10, pp. 180–203.

Hedin, A. (1997) *Spent Nuclear Fuel: How Dangerous Is It?* SKB, Stockholm. Technical Report 97–13.

Heizler, M.T., Perry, F.V., Crowe, B.M., Peters, L., and Appelt, R. (1999) The Age of Lathrop Wells Volcanic Center: An $^{40}Ar/^{39}Ar$ Dating Investigation. *Journal of Geophysical Research* 104, pp. 767–804.

Houghton, J. (2000) *Global Warming: The Complete Briefing.* Cambridge: Cambridge University Press, 251 p.

Kwicklis, E., Meijer, A., and Fabryka-Martin, J. (2003) Geochemical Inverse Modeling of Groundwater Mixing and Chemical Evolution in the Yucca Mountain Area. In *Proceedings of the 10th International High-Level Radioactive Waste Management Conference.* Las Vegas. LaGrange Park, IL, American Nuclear Society, LA–UR–02–5858.

Long, J., and Ewing, R.C. (2004) Yucca Mountain: Earth-Science Issues at a Geologic Repository for High-Level Nuclear Waste. *Annual Reviews of Earth and Planetary Science* 32, pp. 363–401.

Macfarlane, A. (2003) Will Nuclear Energy Drive the Twenty-first Century? In *Fueling the Future*, ed. A. Heintzman and E. Solomon. Toronto: Anansi Press, pp. 130–151.

Meserve, R.A. (2004) Global warming and nuclear power. *Science* 303, p. 433.

National Commission on Energy Policy (2004) *Ending the Energy Stalemate: A Bipartisan Strategy to Meet America's Energy Challenges.* December. http://www.energycommission.org/ewebeditpro/items/O82F4682.pdf.

National Research Council (1957) *The Disposal of Radioactive Waste on Land.* Washington, DC: National Academy Press, 146 p.

National Research Council (1983) *A Study of the Isolation System for Geological Disposal of Radioactive Wastes.* Washington, DC: National Academy Press, 345 p.

National Research Council (1995) *Technical Bases for Yucca Mountain Standards.* Washington, DC: National Academy Press, 205 p.

National Research Council (1996) *Nuclear Wastes: Technologies for Separation and Transmutation.* Washington, DC: National Academy Press, 592 p.

National Research Council (2001) *Disposition of High-Level Waste and Spent Nuclear Fuel.* Washington, DC: National Academy Press, 198 p.

National Research Council (2003) *One Step at a Time: The Staged Development of Geologic Repositories for High-Level Radioactive Waste.* Washington, DC: National Academy Press, 216 p.

Nuclear Energy Agency (1995) *The Environmental and Ethical Basis of Geological Disposal of Long-Lived Radioactive Wastes.* Paris: Organization for Economic Cooperation and Development, 18 p.

Office of Civilian Radioactive Waste Management (OCRWM) (1998) *Viability Assessment of a Repository at Yucca Mountain.* Department of Energy. DOE/RW–0508.

Office of Civilian Radioactive Waste Management (OCRWM) (1999) *Geology/Hydrology Environmental Baseline File.* Department of Energy. 01717–5700–00027 REV 01 DCN 1.

Office of Civilian Radioactive Waste Management (OCRWM) (2001) *Yucca Mountain Science and Engineering Report.* Department of Energy. DOE/RW–0539.

Office of Civilian Radioactive Waste Management (OCRWM) (2002) *Yucca Mountain Site Suitability Evaluation.* Department of Energy. DOE/RW–0549.

Sailor, W.C., Bodansky, D., Braun, C., Fetter, S., and van der Zwann, R. (2000) A Nuclear Solution to Climate Change. *Science* 288, pp. 1177–1178.

Sawyer, D.A., Fleck, R.J., Lanphere, M.A., Warren, R.G., and Broxton, D.E. (1994) Episodic Volcanism in the Miocene Southwest Nevada Volcanic Field: Stratigraphic Revisions, 40Ar/39Ar Geochronologic Framework, and Implications for Magmatic Evolution. *Geological Society of America Bulletin* 106, pp. 1304–1318.

Seismological Laboratory (2004) *Preliminary Report on the M 4.4 Earthquake near Little Skull Mountain, Southern Nevada, [on June 14, 2002].* http://www.seismo.unr.edu/htdocs/monitoring/06142002/06142002_eq.html.

Sweetkind, D.S., and Williams-Stroud, S.C. (1996) *Characteristics of Fractures at Yucca Mountain, Nevada: Synthesis Report.* U.S. Geological Survey. Administrative Report.

U.S. Geological Survey (2000) *Future Climate Analysis.* U.S. Geological Survey. ANL–NBS–GS–000008–Rev 0 00.

von Seggern, D.H., and Brune, J.N. (2000) Seismicity in the Southern Great Basin, 1868–1992. In *Geologic and Geophysical Characterization Studies of Yucca Mountain, Nevada: A Potential High-Level Radioactive-Waste Repository,* ed. Whitney, J.W., and Keefer, W.R. U.S. Geological Survey. Digital Data Series 58, Revision 1.0, Chapter J.

Zoback, M.L., Anderson, R.E., and Thompson, G.A. (1981) Cenozoic Evolution of the State of Stress and Style of Tectonism of the Basin and Range Province of the Western United States. *Royal Society of London Philosophical Transactions* A300, pp. 407–434.

# Part I
## Policy

# 2

# Nuclear Waste Story: Setting the Stage

Thomas A. Cotton

The United States is approaching a key milestone in a decades-long decision process for developing a geologic repository for high-level radioactive waste and spent nuclear fuel. That milestone is whether the U.S. Nuclear Regulatory Commission (NRC) will authorize the construction of a repository at the candidate repository site proposed by the U.S. Department of Energy (DOE) at Yucca Mountain.

The Nuclear Waste Policy Act of 1982 established a clear mandate and direction for the federal government's lagging program on high-level radioactive waste management. The act, as amended by the Nuclear Waste Policy Amendments Act of 1987, laid out a detailed process and schedule for the DOE to evaluate candidate repository sites, and to design, construct, license, operate, and close geologic repositories for spent nuclear fuel and high-level radioactive waste.

This process consists of several key steps:

- Selecting a strategy for the long-term isolation of the wastes
- Determining safety standards for this isolation
- Selecting candidate sites
- Characterizing the sites
- Approving a site for the development of a repository
- Authorizing the construction of facilities
- Licensing the receipt and emplacement of waste
- Operating the repository and confirming its performance
- Authorizing facility closure, decommissioning, and license termination
- Postclosure monitoring and oversight

This chapter places the pending decision about Yucca Mountain in the context of the historical steps leading up to the decision, and the steps that will follow if the site is approved.

The U.S. repository development program can be broken down into five broad phases. The process began with the original conception of a national program for managing high-level nuclear waste. Next came the initial implementation of the Nuclear Waste Policy Act, followed by the redirection of the effort in the Nuclear Waste Policy Amendments Act. The next phase was the characterization of the candidate repository sites, which led to recommendations. The fifth phase will be the actual construction and operation of the repository.

## Phase I: Developing a National Program for High-level Waste Management

Highly radioactive wastes were first generated in the United States as a by-product of producing plutonium for nuclear weapons. Because of national security priorities and severe time pressures, the liquid wastes were stored in steel tanks pending the later development of suitable methods for long-term management. At the request of the Atomic Energy Commission, a committee of the National Academy of Sciences considered the concept of the permanent disposal of highly radioactive wastes in deep geologic formations. In 1957, the committee recommended that salt formations be evaluated as disposal sites (National Academy of Sciences 1957). In 1969, the Atomic Energy Commission conducted experiments involving the retrievable emplacement of spent reactor fuel in an abandoned salt mine in Lyons, Kansas, but a premature recommendation to use this mine as a repository failed for both technical and political reasons (Office of Technology Assessment 1985).

The Atomic Energy Commission then proposed to develop a long-term Retrievable Surface Storage Facility (RSSF) for high-level waste. This approach would have allowed the federal government to meet its obligation to accept commercial high-level waste while allowing more time for work on a repository (Office of Technology Assessment 1985). The RSSF proposal was subsequently dropped by the Atomic Energy Commission's successor, the Energy Research and Development Administration, after environmental groups and the Environmental Protection Agency (EPA) objected that economic factors might turn the storage facility into a permanent repository (Office of Technology Assessment 1985). This objection has plagued all subsequent efforts to develop a federal storage facility before a permanent disposal system is available (Bunn et al. 2001).

After abandoning the RSSF proposal, the Energy Research and Development Administration turned again to the development of a permanent disposal system. The agency initiated a process to find sites for geologic repositories. In parallel, other disposal concepts were evaluated.

In a related development, the expectations of the technical community about the form of wastes requiring disposal changed. It was originally anticipated that spent reactor fuel would be reprocessed to recover plutonium and uranium for reuse, leaving only the residual reprocessing wastes for disposal. During the late 1970s, however, reprocessing efforts in the United States were halted for economic and policy reasons—most important, the concern that the separation of plutonium could promote the proliferation of nuclear weapons. This left no clear disposition path for the spent fuel that was accumulating in storage pools at reactor sites.

Several actions focused attention on the waste disposal problem. In 1976, California prohibited the siting of any new reactors until the federal government demonstrated a means of disposal for high-level waste. Soon thereafter, the NRC (1977) stated that "it would not continue to license reactors if it did not have reasonable confidence that the waste can and will in due course be disposed of safely," and later conducted a generic proceeding (the "Waste Confidence" proceeding) to develop a formal position about the prospects for disposal (NRC 1984).

Against this backdrop of growing concerns, a national discussion of radioactive waste policy was precipitated by President Carter's establishment in 1978 of the Interagency Review Group on Nuclear Waste Management. Composed of representatives from fourteen federal agencies, the group solicited the views of a wide range of interested parties during its deliberations. Public meetings addressing the DOE's (1980) *Final Environmental Impact Statement for Management of Commercially Generated Radioactive Waste*, deliberations during the NRC's Waste Confidence proceeding, and meetings of several consensus-building groups contributed to the national policy discussion. The Interagency Review Group (1979) found that a consensus had emerged on a number of fundamental elements of policy. These included:

• The responsibility for resolving military and civilian waste management problems should not be deferred to future generations.

• The most promising technology for the permanent disposal of high-level waste and spent fuel was geologic disposal, and efforts to develop geologic repositories should not be delayed pending the further evolution of alternative disposal technologies (such as space disposal).

• The search for where to locate a repository should consider a number of sites in a variety of geologic environments leading to the option of having at least two repositories, preferably in different regions of the country.

• The interim storage of waste should not be a substitute for continuing progress toward opening the first repository.

The Nuclear Waste Policy Act of 1982 endorsed and codified these broad conclusions. The act resolved a number of important issues concerning long-term waste isolation, as described below.

### Approach to Long-term Waste Management

The act resolved the question of how strong a commitment to make to the development of disposal facilities that provide long-term isolation through the use of passive engineered and natural barriers rather than through the continued human control and maintenance required by storage. Legislation favoring long-term storage in a manner that would require monitoring and that would allow retrieval of the waste was rejected in favor of the act's focus on permanent disposal.

### Choice of Disposal Technology

The Interagency Review Group report and the waste management environmental impact statement assessed the technical status of a range of technologies for disposal in various media—including sediments under the seabed, deep boreholes, polar ice caps, and even outer space. The conclusion: geologic disposal was the most promising and most advanced technology. The report also evaluated a range of waste management strategies, from proceeding to site and develop geologic repositories, to waiting to choose a disposal technology until alternatives had been further developed. This process concluded with the DOE's decision (codified by the act) to develop repositories. While the act also provided for research and development on alternative disposal technologies, it exempted the repository from the need to consider in its environmental impact statement any alternatives to geologic storage.

### Wastes to Be Disposed Of

The discussions preceding the act's passage included a consideration of whether the repositories should be only for high-level radioactive waste from reprocessing spent fuel (as originally expected) or for spent fuel as well. Some argued (U.S. Senate

1978) that the reprocessing and recycling of the plutonium was advantageous for safe long-term waste disposal, and that spent fuel should not be disposed of since it was a valuable energy resource. Others contended (U.S. Senate 1978) that the separation and reuse of plutonium heightened the risk of the proliferation of nuclear weapons, and that spent fuel could be disposed of safely without reprocessing. Several studies (International Nuclear Fuel Cycle Evaluation 1980; American Physical Society 1978) had supported the conclusion that the reprocessing of spent fuel is not required for the safe isolation of the waste. Reflecting this conclusion, the act provides for repositories for the disposal of both high-level waste and spent fuel. At the same time, it mirrors the disagreement about the possible future reuse of spent fuel in the fact that it refers only to "such spent fuel as may be disposed of" and requires the retrievability of spent fuel (but not high-level waste) for possible economic reasons.

## Site Selection

The act established a schedule for the DOE to nominate at least five sites in a first round of repository selection and recommend three of them to the president for further characterization before selecting one for the first repository. Provisions were made for state and tribal participation in and review of the site characterization activities. This permitted participation included a right to disapprove of a site recommendation, subject to override by Congress. A similar process for selecting a second repository site was also required. Since most of the first round sites were in the West, it was expected (though not explicitly stated) that this second-round choice would focus on sites in the East to provide geographic equity.

## Federal Role in Interim Spent Fuel Storage

The act gave waste owners the primary responsibility for storing waste until it was accepted by the DOE. Anticipating a repository by 1998, Congress rejected proposals to provide substantial federal funds for a facility providing the interim storage of spent fuel. Instead, Congress authorized a limited amount of user-funded capacity (nineteen hundred metric tons) at existing federal facilities. This storage was made available to utilities that could demonstrate an inability to provide their own storage quickly enough to prevent disruptions to reactor operation. The federal storage option expired unused in 1990. The act deferred the decision about whether larger, longer-term storage (Monitored Retrievable Storage, or MRS) facilities were needed, directing the DOE to recommend a design and site for such a facility so that

Congress could decide whether to authorize its construction and operation. As with the earlier RSSF proposal, the debate was dominated by concerns that the availability of a long-term federal storage facility would reduce the incentive for continued work on a geologic repository, and that storage would become a de facto permanent solution. In response, the act required disposal in repositories to proceed whether or not an MRS facility was provided.

### Paying for Disposal

In 1970, the Atomic Energy Commission established the policy that the federal government would accept high-level waste for long-term disposition, and that the generators of that waste would pay the full costs at the time the waste was delivered to the government. The 1982 act reaffirmed the responsibility of waste generators for the full costs of disposal, but established a new funding mechanism requiring utilities with nuclear reactors to pay an up-front waste fee on nuclear-generated electricity at the time spent fuel was produced. It also required the DOE to enter into contracts with the nuclear utilities providing that the DOE would begin accepting and disposing of waste by January 31, 1998, in exchange for payment of the fee. This mechanism was intended to ensure that the federal government recovered the full costs of disposal, and at the same time insulate the program from the uncertainties of the annual federal budget cycle by providing a large and stable source of funds. (Subsequent decisions about the treatment of the revenues from the fee in the federal budget process have prevented the latter intention from being realized.) The contractual obligation was also expected to give the federal government an incentive for timely implementation of the act. This quid pro quo arrangement and the associated obligation of the federal government have been upheld by federal courts, and now form the basis for litigation filed by utilities over the federal government's failure to fulfill that obligation.

### Safety Standards for Isolation

When the act was being debated, both the EPA and the NRC (1981) were already developing standards and regulations for geologic repositories. The act directed the EPA to promulgate generally applicable standards for the protection of the environment from off-site releases of radioactive material from repositories. It also assigned the NRC the responsibility for implementing the EPA standard through the development and implementation of licensing requirements. The standards and regulations are discussed in more detail in chapter 3 (this volume).

## Phase II: Initial Implementation of the Nuclear Waste Policy Act

The aggressive schedule established by the Nuclear Waste Policy Act meant that the sites already under consideration when the act was passed formed the basis of site screening for the first repository. In 1983, the DOE identified nine potentially acceptable sites—six in the West (four in bedded salt, one in basalt, and one in tuff) and three in the South (all in domed salt). In 1984, the DOE issued draft environmental assessments on all nine sites. This was followed in 1986 by final environmental assessments on five of these sites—three in salt formations, and one each in basalt and tuff. Later in 1986, three sites were designated for characterization: a basalt site in Hanford, Washington; a tuff site at Yucca Mountain; and a salt site in Deaf Smith County, Texas. To comply with the requirement for a regional distribution of repositories, the DOE focused second-round siting efforts (1983–1986) on crystalline rock in the eastern United States. With more time available for this round, the DOE was able to conduct a broad screening process in seventeen states before selecting for further consideration twelve rock bodies in seven states on the East Coast and in the upper Midwest. At the same time that the DOE announced the three candidate first-round sites, however, it also indefinitely deferred the search for a second repository site. While the DOE justified this decision on the grounds that the additional waste disposal capacity was not needed, some saw it as a response to the strong political reaction that had arisen following the designation of the candidate rock bodies in the East (Carter 1987; Easterling and Kunreuther 1995). The cancellation of the second round added to the aggravation of the western states hosting the first-round sites, and was seen as an abandonment of the act's commitment to regional equity (Colglazier 1991).

The DOE's indication in 1986 that it intended to recommend a site near Oak Ridge, Tennessee, for an MRS facility added to the rising political attention to the repository siting decisions. Although the local community of Oak Ridge endorsed this recommendation (subject to conditions), the state government strongly opposed it.

## Phase III: The Nuclear Waste Policy Amendments Act of 1987

In reaction to the growing controversy, appropriations for the nuclear waste program were limited to use on non–site specific work, effectively preventing the characterization of the three candidate repository sites. In addition, a bill was introduced that would have halted the program pending recommendations from a blue-ribbon commission and the enactment of new legislation by Congress (Colglazier 1991). This legislation was rejected in favor of the Nuclear Waste Policy Amendments Act of 1987, which allowed the federal waste management program to continue, but in a significantly modified form.

The amendments act terminated the consideration of the first-round salt and basalt sites, halted all of the second-round activities in crystalline rock, and directed the DOE to study only one site—Yucca Mountain—to determine its suitability as a repository. It also required the DOE to deliver a report on the need for an additional repository between January 1, 2007, and January 1, 2010.

The decision to evaluate only a single site focused resources and energy in one place rather than three. In fact, the escalation of the estimated cost of characterizing sites was a major rationale for the decision to narrow the process to one. Yet the decision had a number of offsetting negative effects as well. First, it produced a strong sense among the citizens of Nevada that they had been treated unfairly. It also created the impression that the siting process was purely political, with no technical basis, even though the information available to Congress at the time suggested that Yucca Mountain was a reasonable choice if only one site was to be characterized (DOE 1986). Focusing on a single site also eliminated the opportunity to compare the projected performance of repositories at three well-characterized sites before making a siting decision. Finally, by eliminating work on potential backup sites, the amendments act increased the uncertainty about when a repository would become available. If the Yucca Mountain site is not approved, new legislation will be required to authorize another approach for the federal government to meet its responsibilities for the management of high-level waste and spent nuclear fuel.

The amendments act also created the Nuclear Waste Technical Review Board to strengthen the independent oversight of the program. The board's members are selected and appointed by the president from candidates recommended by the National Academy of Sciences. The board reviews the technical work of the DOE waste program, and reports its conclusions to the secretary of energy and Congress.

The amendments act also dealt with interim storage by authorizing the construction of an MRS facility, but nullifying the 1986 selection of the MRS site in Oak Ridge. To prevent any MRS facility from becoming a de facto permanent repository, the amendments act linked the steps for developing an MRS facility to the equivalent steps for the repository. The DOE cannot select an MRS site until a repository site is recommended, and cannot construct the facility until the repository receives a construction authorization. The amendments act also set up a voluntary process to find a site for the MRS facility or repository, with an independent nuclear waste negotiator who would seek to develop an agreement between a state or an Indian tribe and the federal government. The proponents of an MRS facility hoped the negotiator could locate a state or tribe willing to accept a facility that could operate earlier than the repository. This voluntary process did not produce a site, however, and was abandoned when the authorizing provisions expired in January 1995.

### Phase IV: Post–Amendments Act Site Characterization and Recommendation

Following the selection of Yucca Mountain for a detailed evaluation, the DOE issued a site characterization plan for review and comment by the state of Nevada, the NRC, and the public, as required by law. Once that was done, the DOE was allowed to begin site characterization, which by law and regulation, had to include testing in an underground facility at the site. It soon became clear that the process would take longer than originally anticipated. The DOE (1989) concluded that the technical and institutional complexities of site characterization made it unlikely that a repository would be available before 2010. As a result of such complexities, an underground exploratory studies facility was not completed at Yucca Mountain until 1996.

In 1996, Congress directed the DOE to prepare a viability assessment of the Yucca Mountain site to inform decisions about the continuing evaluation of the site. This assessment, published in 1998, concluded that Yucca Mountain remained a promising site for a repository (DOE 1998).

As 1998 neared, the pressures on the federal government to provide a storage facility to accept commercial spent nuclear fuel increased. Utilities and states with reactors filed lawsuits to force action, allow fees to be withheld and put into escrow, and win damages against the federal government for its anticipated failure to meet the obligation to begin accepting waste in 1998. Federal courts first concluded that the contracts created an absolute obligation on the DOE to begin accepting fuel in 1998, and after the target date came and passed, then decided that the federal government was in default and the utilities could pursue damage claims in court. The first suit for damages was filed in 1998.

Starting in 1995, both houses of Congress passed various bills to provide a storage facility at the Nevada Test Site near Yucca Mountain as a way to begin accepting spent fuel earlier than the expected date of repository operation. The Clinton administration opposed any efforts to locate a storage facility in Nevada before crucial decisions about the use of the Yucca Mountain site for a repository had been made. President Clinton vetoed a bill passed in 2000 to allow storage at Yucca Mountain after a repository had received a construction authorization. As a result of this action, the ability of the federal government to begin accepting spent fuel as required by contracts depends on the development of a geologic repository.

On February 14, 2002, Secretary of Energy Spencer Abraham recommended the Yucca Mountain site to President Bush as suitable for further development (DOE 2002). Following the process laid out in the Nuclear Waste Policy Act (NWPA), the secretary first held public hearings in the state of Nevada (in fall 2001) and informed the state government of his intention to recommend the site. As required by law, the recommendation to the president was accompanied by preliminary comments from the NRC concerning the adequacy of the available data for inclusion in a license application. Other supplemental documentation for the recommendation included an impact report from the state of Nevada, comments received on the site recommendation, and an environmental impact statement.

On February 15, 2002, immediately following the secretary of energy's recommendation of Yucca Mountain, President Bush recommended the site to Congress as qualified for application for construction authorization of a repository. The state of Nevada vetoed the recommendation in April 2002, as permitted by the Nuclear Waste Policy Act. Congress held hearings to receive testimony from the state of Nevada, the DOE, the NRC, the Nuclear Waste Technical Review Board, and others. Following the procedures laid out in the act, Congress then voted in July 2002 to overrule Nevada and approve the site. The state of Nevada filed a number of lawsuits challenging various aspects of the site recommendation process. On July 9, 2004, the U.S. Court of Appeals for the District of Columbia Circuit rejected all of the challenges to the site recommendation. At the same time, the court upheld a contention that the EPA did not follow the advice of the National Academy of Sciences concerning the time period for assessing compliance with safety standards for Yucca Mountain, as required by the Energy Policy Act of 1992, and sent the standard back to the EPA for a reconsideration of the compliance period. This is discussed further in chapter 3.

With congressional approval of the site, the DOE is required by the act to submit a license application for a repository to the NRC. The NRC has three years to decide the license application once it receives it from the DOE, with the possibility of a one-year extension. If the NRC decides not to grant the license, Congress would have to approve new legislation directing another course of action, since current law does not allow a search for other repository sites or the construction of a storage facility before the approval of construction authorization for a repository.

## Phase V: Potential Post–Site Approval Repository Development

Approval of the site triggered a lengthy evaluation of the proposed repository's safety that will occur in the NRC licensing process. The act specifies a three-stage approval process by the NRC: authorization to construct a repository, approval of the receipt and possession of waste in the repository, and authorization to close and decommission a repository. These steps, laid out in the NRC regulations (10 CFR 63) discussed in chapter 3, provide the framework for examining what happens after site recommendation.

### Authorizing Construction

The NRC would first review the DOE's application for the authorization to construct repository facilities. NRC approval of construction would not allow the DOE to bring any waste to the site but rather only to construct surface and underground facilities.

The construction of the repository can itself be divided into stages. The DOE (2001) already plans to construct the underground repository in a stepwise manner; it is also considering the construction of the surface facilities in modules spread over a period of time, rather than all at once. At the DOE's request, the National Research Council (2003) explored the concept of staged repository development and recommended one particular form, "adaptive staging," as "a promising approach to successful repository development."

### Licensing the Receipt and Emplacement of Waste

After completing the construction of the facilities needed for initial operation, the DOE may apply for a license to "receive and possess" waste that would allow the DOE to begin emplacing waste in the repository. The operating license may include licensing conditions (e.g., on the amount of waste emplaced per unit volume of the repository) that the NRC deems warranted. Any substantial changes in design or operating procedures from those specified in the license would require a license amendment.

## Operating the Repository and Confirming Its Performance

During repository operation, the DOE must conduct a performance confirmation program to verify the assumptions, data, and analyses used in the performance assessment as well as any related findings that permitted the construction of the repository. Key physical parameters would be monitored to detect any significant changes that could affect compliance with the performance objective.

## Optional Retrieval of Waste

The repository design must allow waste retrieval within a reasonable period starting at any time up to fifty years after the start of waste emplacement. The NRC could approve closure sooner than fifty years after emplacement, but a license amendment would be required for any action that would make the waste difficult to retrieve. The DOE is designing the repository to allow normal operation for more than one hundred years, and with maintenance, for as long as three hundred years. This will give future generations ample opportunities to make their own decisions about repository closure.

## Authorizing Facility Closure, Decommissioning, and License Termination

The permanent closure of the repository requires a license amendment. This will require an update of the assessment of the postclosure performance of the repository, and a consideration of the site data along with the results of tests and experiments obtained during repository operation. It is only at this point that the emplacement of waste in the repository in a retrievable way changes from retrievable storage to true "disposal." The extended period of retrievability before closure makes the development of a repository quite different from that of a nuclear reactor. With a reactor, the principal risks being regulated begin when the reactor starts operating and end before decommissioning. With a repository, the principal risk being regulated—the very long-term risk posed to future populations by possible releases from buried waste—is committed to only at closure. This would occur after at least fifty to perhaps as many as three hundred years of evolution in science relevant to waste isolation, and the evaluation of data on the behavior and effects of a fully loaded repository.

## Postclosure Monitoring and Oversight

A repository at Yucca Mountain would not simply be abandoned after closure. The National Energy Policy Act of 1992 requires the DOE to continue to oversee the Yucca Mountain site to prevent activities that could adversely affect the repository, and the NRC regulations require that the application to close the repository include a plan for postclosure monitoring and continued oversight.

### Perspective on the Site Recommendation Decision

Geologic disposal is a complex and extended process, not a single action that is approved in toto in one decision and executed swiftly. The remaining steps in the process from congressional approval of the site through termination of the license are expected to take at least seventy-five years and could take decades more. This will allow ample opportunities to review progress and change course.

The decision about whether to continue work on the Yucca Mountain site at each stage of the process involves value judgments about acceptable risk and equity that "can be informed, but not decided, by science and technology" (Dreyfus 1999). A key judgment concerns the acceptability of the irreducible uncertainties that are inevitable with long-term radioactive waste management. The issue is whether the level of knowledge is adequate to inform the current stage of repository development (National Research Council 2001; Nuclear Waste Technical Review Board 2002).

The policy judgment about the adequacy of the knowledge base to support decisions must consider both what is known and what can be known. The National Research Council (1990, 2001) has repeatedly warned against unrealistic and unachievable expectations about the degree of certainty that can be obtained in predicting the long-term performance of a repository. The standard of proof of repository performance developed by the EPA (EPA 2001) and adopted by the NRC is based on a "reasonable expectation." This standard recognizes that it is not possible to "prove" the extremely long-term performance of a repository in the normal sense of the term, and that a reasonable level of confidence in performance is all that can be expected.

Far more is known about the Yucca Mountain site than was originally expected to be available before a license application. The secretary of energy, the president, and Congress have concluded that this level of knowledge is sufficient to support a decision to enter into the next stage of the repository development process: licensing. During the site approval debates, Congress emphasized the sequential nature of the repository decision process, pointing out that approval did not allow the DOE to construct a repository but only to take the next step by applying to the NRC for the authorization to do so (U.S. Senate 2002; U.S. House of Representatives 2002).

In transmitting the resolution of approval to the full Senate, the Committee on Energy and Natural Resources both recognized the long-term nature of the process of developing a geologic repository and reaffirmed the national commitment to that process:

We agree with the Governor [of Nevada] that, even if the repository can be licensed and built in accordance with the Secretary's schedule, it will take years to begin, and decades to complete the shipment of high-level radioactive waste and spent nuclear fuel to it. But we do not see that as a reason to abandon the Nation's commitment to the permanent disposal of these wastes in an underground repository. We agree with the National Academy of Sciences that "geological disposal remains the only scientifically credible long-term solution available to meet the needs for safety without reliance on active management." We see no reason to abandon the commitment to geological disposal Congress made twenty years ago.(U.S. Senate 2002)

Congressional approval of the site recommendation represents a major transition for the DOE high-level waste program. Through the end of 2007 at the earliest, the attention of the program and its overseers will be focused on the development and intensive review of the license application along with the science, engineering, and analysis supporting it. It will be up to the NRC to determine whether the Yucca Mountain site will be allowed to move into the next stage of development: the construction of a repository.

## References

American Physical Society (1978) Report to the American Physical Society by the Study Group on Nuclear Fuel Cycles and Waste Management. *Reviews of Modern Physics 50*, pp. S107–S142.

Bunn, M., Holdren, J.P., Macfarlane, A.M., Pickett, S.E., Suzuki, A., Suzuki, T., and Weeks, J. (2001) *Interim Storage of Spent Nuclear Fuel: A Safe, Flexible, and Cost-Effective Near-Term Approach to Spent Fuel Management.* A Joint Report from the Harvard University Project on Managing the Atom and the University of Tokyo Project on Sociotechnics of Nuclear Energy. June.

Carter, L.J. (1987) *Nuclear Imperatives and Public Trust: Dealing with Radioactive Waste.* Washington, DC: Resources for the Future, 473 p.

Colglazier, E.W. (1991) Evidential, Ethical, and Policy Disputes: Admissible Evidence in Radioactive Waste Management. In *Acceptable Evidence: Science and Values in Hazard Management*, ed. Mayo, D.G., and Hollander, R.D. New York: Oxford University Press. pp. 137–159.

Department of Energy (DOE) (1980) *Final Environmental Impact Statement for Management of Commercially Generated Radioactive Waste*. DOE/EIS–0046–D. October.

Department of Energy (DOE) (1986) *A Multiattribute Utility Analysis of Sites Nominated for Characterization for the First Radioactive-Waste Repository: A Decision-Aiding Methodology*. DOE/RW–0074. May.

Department of Energy (DOE) (1989) *Report to Congress on Reassessment of the Civilian Radioactive Waste Management Program*. Department of Energy. DOE/RW–0247. November.

Department of Energy (DOE) (1998) *Viability Assessment of a Repository at Yucca Mountain*. Department of Energy. DOE/RW–0508.

Department of Energy (DOE) (2001) *Supplement to the Draft Environmental Impact Statement for a Geologic Repository for the Disposal of Spent Nuclear Fuel and High-Level Radioactive Waste at Yucca Mountain, Nye County, Nevada*. DOE/EIS–0250D–S. May.

Department of Energy (DOE) (2002) Recommendation by the Secretary of Energy regarding the Suitability of the Yucca Mountain Site for a Repository under the Nuclear Waste Policy Act of 1982. February. http://www.ymp.gov/new/sar.pdf.

Dreyfus, D.A. (1999) Comments on the Status of U.S. Policy for the Long-Term Management of High-Level Radioactive Waste. Paper presented at the Workshop on Disposition of High-Level Radioactive Waste through Geological Isolation, National Academy of Sciences, Irvine, California, November 4–5. Cited in National Research Council 2001.

Easterling, D., and Kunreuther, H. (1995) *The Dilemma of Siting a High-Level Nuclear Waste Repository*. Boston: Kluwer Academic Publishers, 286 p.

Environmental Protection Agency (EPA) (2001) 40 CFR Part 197: Public Health and Environmental Radiation Protection Standards for Yucca Mountain, Nevada. http://www.epa.gov/radiation/yucca/docs/yucca_mtn_standards_060501.pdf.

Interagency Review Group (1979) *Report to the President by the Interagency Review Group on Nuclear Waste Management*. NTIS Report. TID–29442. March.

International Nuclear Fuel Cycle Evaluation (1980) *Waste Management and Disposal: Report of INFCE Working Group Seven*. Vienna: International Atomic Energy Agency.

National Academy of Sciences (1957) *The Disposal of Radioactive Waste on Land*. NAS–NRC Publication 519. Washington DC: National Academy Press, 142 p.

National Research Council (1990) *Rethinking High-Level Radioactive Waste Disposal: A Position Statement of the Board on Radioactive Waste Management*. Washington, DC: National Academy Press, 34 p.

National Research Council (1996) *Nuclear Wastes: Technologies for Separations and Transmutation*. Washington, DC: National Academy Press, 592 p.

National Research Council (2001) *Disposition of High-Level Waste and Spent Nuclear Fuel: The Continuing Societal and Technical Challenges*. Washington, DC: National Academy Press, 198 p.

National Research Council (2003) *One Step at a Time: The Staged Development of Geologic Repositories for High-Level Radioactive Waste*. Washington, DC: National Academy Press, 201 p.

Nuclear Regulatory Commission (NRC) (1997) *NRDC Petition for Rulemaking on Waste Management.* SECY–77–48B, Attachment 1, June 1. Cited in Office of Technology Assessment 1985.

Nuclear Regulatory Commission (NRC) (1981) Disposal of High-Level Radioactive Waste in Repositories: Licensing Procedures. *Federal Register* 46, p. 13973.

Nuclear Regulatory Commission (NRC) (1984) Waste Confidence Decision. *Federal Register* 49, pp. 34658–34688.

Nuclear Waste Technical Review Board (2002) Letter to the Honorable Dennis Hastert, the Honorable Robert C. Byrd, and the Honorable Spencer Abraham. January 24.

Office of Technology Assessment (1985) *Managing the Nation's Commercial High-Level Radioactive Waste.* U.S. Congress Office of Technology Assessment. OTA–O–171. March.

U.S. House of Representatives (2002) *Approval of Yucca Mountain Site: Report to Accompany H.J. Res. 87.* Committee on Energy and Commerce, 107th Cong., 2nd sess., May 1. Washington, DC: US Government Printing Office, serial no. 107–425.

U.S. Senate (1978) *Nuclear Waste Disposal: Hearings before the Subcommittee on Science, Technology, and Space of the Committee on Commerce, Science, and Transportation.* 95th Cong., 2nd sess., August 9–10, 16, part 1. Washington, DC: U.S. Government Printing Office, serial no. 95–136.

U.S. Senate (2002) *Approval of Yucca Mountain Site: Report to Accompany S.J. Res. 34.* Committee on Energy and Natural Resources, 107th Cong., 2nd sess., May 1. Washington, DC: US Government Printing Office serial no. 107–159.

# 3

## Regulating the Geologic Disposal of High-Level Nuclear Waste at Yucca Mountain

William M. Murphy

The geologic disposal of nuclear waste is conceptually simple and logically sound, but the processes of repository development and regulation are technically as well as socially complicated. Although most countries judge geologic disposal to be the best solution to the problem of high-level nuclear waste, no regulated geologic repository for high-level waste presently exists. In the United States, regulations governing the geologic disposal of high-level waste have evolved over decades as laws were enacted, site selection occurred, scientific understanding of candidate sites advanced, repository designs evolved, and methods of safety assessment improved.

In February 2002, President Bush notified Congress that the administration considers Yucca Mountain qualified for the Department of Energy (DOE) to submit an application for repository construction authorization. The governor of the state of Nevada issued a notice of disapproval in April 2002, and Congress overrode the governor's objection in July 2002. The next major step is the submission of the license application for repository construction by the DOE and the evaluation of this application by the Nuclear Regulatory Commission (NRC). Examining the evolution of this process provides insight on regulating the geologic disposal of high-level waste at Yucca Mountain.

### Brief Regulatory History

In 1970, the Atomic Energy Commission announced plans for a permanent high-level waste repository at an abandoned salt mine near Lyons, Kansas. The ensuing controversy and rejection of the project promoted the creation of separate development and regulatory agencies for the civilian uses of nuclear energy. Soon afterward, the Energy Reorganization Act of 1974 established the NRC. The Nuclear Waste

Policy Act, passed by Congress in 1982 and signed by President Reagan in 1983, set the national framework for the geologic disposal of high-level radioactive waste.

The Nuclear Waste Policy Act specified the roles of the principal agencies charged with managing high-level waste disposal. The DOE determines site suitability and then develops, builds, and operates the facility. The Environmental Protection Agency (EPA) sets standards to assure the protection of the public health and the environment. The NRC regulates the process by establishing licensing requirements, implementing standards, and licensing the repository.

In 1983, the NRC promulgated high-level waste disposal regulations in Part 60 of chapter 10 of the Code of Federal Regulations (10 CFR 60), which invoked the anticipated EPA standards. In 1985, the EPA initially set generally applicable environmental standards for high-level waste disposal in 40 CFR 191. The NRC then proposed modified regulations to conform to the EPA standard, but these modifications were never enacted. In 1987, a federal court remanded the EPA standard, ordering regulators to modify it to ensure consistency with existing environmental laws. The EPA accordingly revised and reissued its high-level waste standard in 1993. In the meantime, Congress amended the Nuclear Waste Policy Act in 1987 specifying Yucca Maintain as the only site to be characterized, and the Energy Policy Act of 1992 mandated the EPA to prepare a standard specifically for Yucca Mountain. Also in 1992, the Waste Isolation Pilot Project Land Withdrawal Act exempted Yucca Mountain from provisions of the original EPA standard.

The Energy Policy Act of 1992 also provided for the National Academy of Sciences to make recommendations to the EPA on the scientific basis for the health and safety standards for Yucca Mountain. The Energy Policy Act directed the EPA to develop radiation protection standards consistent with National Academy of Sciences recommendations. On behalf of the National Academy of Sciences, the National Research Council Committee on Technical Bases for Yucca Mountain Standards published its recommendations in 1995 (National Research Council 1995). In 2001, the EPA presented its Yucca Mountain high-level waste standard (40 CFR 197). Although the recommendations of the committee were considered, the EPA standard for Yucca Mountain diverged on several points from the National Academy of Sciences position, as discussed below.

The NRC's 10 CFR 60 persists as a general regulation for the geologic disposal of high-level waste. In 2001, however, the NRC issued substantially modified regulations, 10 CFR 63, applicable specifically to high-level waste disposal at Yucca

**Table 3.1**
Key events in regulation of geologic disposal of high-level nuclear waste, 1974–2002

| | |
|---|---|
| Establishment of the NRC | 1974 |
| Nuclear Waste Policy Act | 1982 |
| NRC high-level waste regulations, 10 CFR 60 | 1983 |
| DOE site selection guidelines, 10 CFR 960 | 1984 |
| EPA standard, 40 CFR 191 | 1985 |
| Nuclear Waste Policy Act Amendments | 1987 |
| Energy Policy Act of 1992 | 1992 |
| National Academy of Sciences Technical Bases for Yucca Mountain Standards | 1995 |
| EPA standard for Yucca Mountain, 40 CFR 197 | 2001 |
| NRC regulation for Yucca Mountain, 10 CFR 63 | 2001 |
| DOE Yucca Mountain suitability criteria, 10 CFR 963 | 2001 |
| Presidential recommendation of Yucca Mountain site | 2002 |

Mountain and consistent with the EPA standard for Yucca Mountain. According to this revised regulation, the repository at Yucca Mountain is exempt from prior regulatory criteria in 10 CFR 60.

The DOE followed a similar path with its guidelines. In 1984, the DOE issued general guidelines for comparing and selecting high-level waste disposal sites—10 CFR 960. In 2001, the DOE amended these policies in 10 CFR 963 to focus on the criteria for evaluating the suitability of the Yucca Mountain site.

In each case, specific rules for Yucca Mountain superseded the general rules that the EPA, the NRC, and the DOE had developed for high-level nuclear waste disposal. Table 3.1 summarizes the regulatory history of the geologic disposal of high-level waste in the United States.

## Standards and Regulations

Regulation of high-level waste disposal requires judgments on three essential issues. First, what is the appropriate time scale for regulations to apply to the geologic disposal of high-level waste? Second, how are the effects of the potential failure of the isolation of radioactivity to be quantified, and how is safety to be measured? Third, over what spatial domain is safety to be evaluated? Although regulations pertain to many other aspects of high-level waste management, this chapter focuses on regulations regarding time, effect, and space for the conditions at Yucca Mountain once the repository is closed.

**Time**

The selection of a time scale for the regulation of high-level waste at Yucca Mountain is not a straightforward process. Radioactivity provides a physical basis for evaluating time scales. Short half-life fission products (e.g., cesium-137 and strontium-90) dominate high-level waste's initial radioactivity; these nuclides will decay to low levels within three hundred to one thousand years. Primary heating effects from the waste itself in a geologic repository occur during this period. It is for the long half-life radionuclides that geologic isolation is required. For example, plutonium-239, plutonium-240, iodine-129, neptunium-237, technetium-99, and americium-243 have half-lives between sixty-five hundred and seventeen million years. (The half-life of uranium-238, the predominant component of natural uranium and spent nuclear fuel, is 4.5 billion years—so long that it contributes only slightly to high-level waste radiation.)

In its original high-level waste standard, the EPA established ten thousand years as the time scale for high-level nuclear waste regulations. A rationale offered for a period of time this short relative to the radioactivity of the wastes is that major geologic changes are unlikely in ten thousand years, so predictions are not compromised by geologic uncertainty. The EPA deemed standards for longer time periods to entail highly uncertain projections. The agency judged ten-thousand-year evaluations to be capable of distinguishing good geologic repositories from poor ones, and asserted that a system capable of meeting requirements for ten thousand years would continue to be protective for much longer times. It acknowledged, however, that longer time frames are potentially important, and deferred to the DOE guidelines to require analysis of site performance over a hundred thousand years in comparisons between potential sites.

In 10 CFR 60, the NRC established additional regulatory time scales pertaining to performance requirements for radionuclide containment and groundwater travel time. Although this regulation no longer applies to Yucca Mountain, it provides a precedent for regulatory time scales. The NRC mandated that the engineered barrier system provide substantially complete containment of waste for three hundred to one thousand years, and that the travel time of groundwater from the emplacement site to the accessible environment be at least one thousand years. These times correspond roughly to the decay period of short half-life fission products.

The National Academy of Sciences committee concluded that there is no scientific basis for the ten-thousand-year time limit on high-level waste disposal standards at

Yucca Mountain (National Research Council 1995). The committee judged that it is technically feasible to assess compliance on a time scale comparable to the period of geologic stability, which for Yucca Mountain is about one million years. It argued that within the time limit of geologic stability, a safety evaluation should be made and compliance to regulations should be assessed for the time when the greatest risk to public health occurs. The committee questioned if a ten-thousand-year compliance period would fulfill the principle of intergenerational equity—that is, the precept that future generations should be subjected to no greater risks than those acceptable to the present generation.

Despite the National Academy of Sciences recommendations, the current EPA standard for Yucca Mountain retains a ten-thousand-year limit for compliance to individual protection and groundwater protection standards. The NRC also adopted a compliance time limit of ten thousand years in its regulations specific to Yucca Mountain. The EPA and the NRC made these judgments based on both science and policy. From a technical perspective, the NRC contended that ten thousand years encompasses the time period when the radioactivity of the wastes is greatest, and that ten thousand years is long enough to require the consideration of changes in conditions such as seismicity, faulting, volcanism, and climate change. From a policy perspective, the NRC contended that ten thousand years is consistent with other regulations, notably the EPA's original 40 CFR 191.

In July 2004, the U.S. Court of Appeals for the District of Columbia Circuit vacated the EPA regulation for a ten-thousand-year compliance period on the basis that it differs from the National Academy of Sciences committee findings and recommendations. The court ruled that the EPA must either issue a standard that is based on and consistent with the committee report or have Congress authorize deviation from the committee recommendations.

In its standard for Yucca Mountain, the EPA requires the DOE to include in its environmental impact statement performance assessments for times greater than ten thousand years. The long-term evaluations were specified to be an indicator of future performance, but they have no role in licensing under the NRC's 10 CFR 63. To support its recommendation of Yucca Mountain, the DOE repeatedly conducted performance assessments for time periods exceeding ten thousand years. In the 2002 environmental impact statement for Yucca Mountain (DOE 2002), the greatest calculated doses to exposed individuals were predicted to occur between four hundred and six hundred thousand years after site closure.

**Effect**

The effects from a high-level waste repository are generally regulated on the basis of the uncertain predictions of future conditions that are part of performance assessments. Measures of effects include the concentrations of radioactive materials in groundwater, the rates of release of radioactive materials across a boundary, radioactive doses, and the risks, such as the risks of fatal cancers that are consequences of doses. Concentration or release measures can be related to public health effects by evaluating the corresponding radiation doses or health risks. Dose and risk limits can be evaluated for individuals and populations. Concentration, release, dose, and risk measures of effect are all invoked in the regulations for high-level waste disposal.

The EPA standard of 1985 set containment requirements limiting releases of radioactive species for ten thousand years, individual protection requirements limiting annual exposures (doses) for one thousand years, and groundwater protection requirements limiting radioactivity in groundwater for one thousand years. In 1993, two of these standards were amended. The individual protection standard was made to apply for ten thousand years, while the groundwater protection standard was made to conform to provisions of the Safe Drinking Water Act. The containment requirements were designed to limit premature cancer deaths over ten thousand years to no more than one thousand, a risk judged to be no greater than that from unmined uranium ore.

In its 1983 mandate on high-level waste disposal (10 CFR 60), the NRC regulated effects by setting standards both for the repository's subsystems and its overall performance. The subsystem standards, which no longer apply to Yucca Mountain, included limits on radioactivity release and release rates from the engineered barrier system. The overall performance standard in 10 CFR 60 corresponded to the applicable environmental release limit set by the EPA.

The Energy Policy Act of 1992 revamped the basis for regulating effects by mandating that the EPA standard for Yucca Mountain prescribe a dose limit for individual members of the public rather than a radiation release limit from the repository. In its review and recommendations, the National Academy of Sciences committee concurred that an individual protection standard, rather than a release standard, is appropriate for Yucca Mountain (National Research Council 1995). The committee, however, recommended that this standard be based not on the dose to individuals but on the risk to individuals, noting that a risk-based standard would not require modification if future studies show that the relation between risk and dose differs from present understanding. The committee recommended that measures of effect focus on projected risks to the average individual representing the critical group of people that is expected to receive the highest doses. In its standard for Yucca Mountain, the EPA adopted individual protection standards as recommended by the Energy Policy Act of 1992 and the committee. The EPA also set groundwater protection standards and a human intrusion component of a containment requirement. It imposed no general containment (i.e., release) requirements because of the limited potential for widespread population exposures from wastes disposed at Yucca Mountain. The EPA individual protection standard is an annual dose of fifteen millirems to what it calls the reasonably maximally exposed individual—that is, the hypothetical person who would receive the highest exposure to radiation that could reasonably be expected. This limit corresponds to an annual risk of about 8.5 fatal cancers per million people. For comparison, the National Academy of Sciences committee noted that the risk equivalent of dose limits set by authorities outside the United States is one to ten effects per million people per year. The academy suggested using this risk limit as a starting point for rule making by the EPA.

The EPA also set groundwater protection standards corresponding to maximum contaminant levels for radionuclides (expressed as radioactivity concentrations or doses). These limits were established by the EPA under the Safe Drinking Water Act. The EPA set these limits because it contended that future populations should be afforded the same environmental protection as people who generated the wastes. The groundwater radionuclide limits generally fall within the EPA target range of limiting the lifetime risk of fatal cancer from drinking water to between one in ten thousand and one in a million. Yet in 10 CFR 197, the EPA specified that radionuclides in a projected contaminant plume from Yucca Mountain will be diluted in a representative volume of water to apply groundwater protection standards. The specified volume is 3.7 million cubic meters per year, which is based on the projected farming and human uses of groundwater by a community in the vicinity of Yucca Mountain. The characterization of concentrations in a small volume of water, such as that containing the highest radionuclide concentrations of the contaminant plume, is deemed by the EPA to be technically unrealistic for long times in the future.

The DOE opposed separate groundwater protection standards for Yucca Mountain. In public comments on the EPA's 10 CFR 197, the NRC argued against the imposition of groundwater protection standards, asserting that all-pathway individual protection standards are sufficient to protect public health and safety. Nevertheless, the NRC regulations in 10 CFR 63 conform to the EPA standards with respect to groundwater protection standards and the fifteen-millirem annual dose limit.

## Space

Regulations concerning high-level nuclear waste typically designate a controlled area beyond which releases enter the accessible environment. In the original EPA and NRC regulations, the controlled area extended no more than five or ten kilometers from the primary site of waste emplacement. The evolution from a release standard to an individual dose standard, however, modified the regulatory concept of accessible environment. The location of potentially exposed populations emerged as a paramount spatial constraint.

In its Yucca Mountain standard, the EPA modified requirements for the controlled area to permit its extension along the groundwater flow path from the disposal site at Yucca Mountain. The extension ranged as far south as the southwest boundary of the Nevada Test Site, which is about eighteen kilometers from the repository. This location is based on the present occupation at nearby Lathrop Wells, the plans to develop the area in the vicinity of that location, and the improbable agricultural use of groundwater north of there due to rougher terrain and deeper groundwater. The EPA located its reasonably maximally exposed individual at this location and required the representative volume of water used for compliance to groundwater protection standards to be drawn at this point. The NRC's Yucca Mountain regulation adopted these conditions. An eighteen-kilometer compliance boundary in the regulations specific to Yucca Mountain substantially enlarges the geologic repository, which according to the NRC and the DOE definitions includes that part of the geologic setting that provides isolation of the wastes.

The National Academy of Sciences committee concluded that a standard based on the effects to the average member of the critical group would require no designation of an exclusion zone (such as the controlled area) beyond the footprint of the waste emplacement site. The committee asserted that the critical group at risk would be likely to live near Yucca Mountain and would be exposed to radiation through the use of contaminated groundwater.

A difference between the critical group approach of the National Academy of Sciences committee and the reasonably maximally exposed individual approach of the EPA is the location relative to the path of contaminant migration. The EPA standard uses current population distribution to locate the composite reasonably maximally exposed individual. This individual would live directly above the path of the highest concentration of contaminants in the plume, though. The EPA argued that some members of a geographically dispersed critical group could escape the effects of a localized plume of contaminated groundwater emanating from the repository, which would lower the average dose to the hypothetical critical group relative to that of the reasonably maximally exposed individual. In implementing the EPA fifteen-millirem annual dose standard, however, the NRC specified the contaminant dilution in a representative volume of 3.7 million cubic meters per year, which the NRC noted to be consistent with regulating effects to a critical group.

### Uncertainty in the Regulation of High-Level Waste Disposal at Yucca Mountain

The evolution of regulations governing the geologic disposal of high-level waste in the United States reflects changes in federal law, improved site characterization and other scientific understanding, changes in repository design, and advances in performance assessment. Perhaps most important to the development of the regulations has been the selection of Yucca Mountain as the only candidate site. The original regulations (10 CFR 60 of the NRC, 40 CFR 191 of the EPA, and 10 CFR 960 of the DOE) were conceived to be generally applicable and to accommodate the challenge of evaluating long-term geologic isolation of wastes for a spectrum of potential sites. These regulations required judgments of the relative merits of site characteristics—comparisons that became moot once Yucca Mountain was chosen as the only site to be evaluated.

With only one site to regulate, the standards and regulations were adapted to the unique characteristics of Yucca Mountain. The first change in regulations occurred in 1985, when the NRC modified its regulations to address conditions particular to Yucca Mountain, notably the unsaturated hydrologic system. Then the 1992 Energy Policy Act mandated an individual dose standard, as opposed to a release standard. This change was motivated largely by the potential releases of volatile carbon-14 into the atmosphere from a repository in the unsaturated zone at Yucca Mountain (e.g., Van Konynenberg 1991). The change to an individual dose standard avoids regulatory issues from the global atmospheric releases of carbon-14 because they pose negligible risk.

The NRC originally set siting criteria expressed as favorable and potentially adverse characteristics for evaluating and selecting a repository site, and for providing a reasonable assurance of achieving the performance objectives. The general NRC regulations also included requirements for the performance of subsystems. The EPA included assurance requirements in its original standard, but deferred to the NRC to obtain assurance from its licensees. Also, the EPA required the DOE to provide comparative evaluations of a hundred-thousand-year performance in the site selection process. The DOE originally offered technical screening guidelines that specified factors that would qualify or disqualify a site. The NRC's new regulations specific to Yucca Mountain eliminated siting requirements. Similarly, the DOE eliminated qualifying and disqualifying conditions. The NRC also eliminated in its Yucca Mountain regulations general quantitative requirements for subsystem performance. Yet the NRC does maintain requirements to demonstrate site-specific multiple barriers for the Yucca Mountain repository. The EPA devised a unique approach to compliance with its groundwater protection standards specifically for Yucca Mountain. Protected groundwater consists of a representative volume based on patterns of water consumption in the vicinity of Yucca Mountain.

The generic five- to ten-kilometer compliance boundary for releases specified in the original regulations was enlarged to eighteen kilometers, specific to the groundwater flow system and geography at Yucca Mountain. Regulations based on a critical group or reasonably maximally exposed individual located at this distance are in response to conditions at Yucca Mountain.

Certain features of Yucca Mountain make it uncertain if the site could comply with the original set of generic requirements for high-level waste disposal:

• Gaseous carbon-14 releases could violate release and release rate limits for the engineered barrier system in the NRC's 10 CFR 60, and the cumulative release of carbon-14 could violate the limits of the EPA's 40 CFR 191, which are based on atmospheric release and global population effects.

• The thermodynamic incompatibility of the predominant waste form, spent nuclear fuel, with the oxidizing environment at Yucca Mountain (see chapter 20, this volume) could be problematic for the gradual release rate requirement of 10 CFR 60.

• In the more concentrated part of the eventual contaminant plume, it may be difficult to ensure radionuclide levels below the maximum concentration limits of the EPA groundwater protection standards. This concern is highlighted by certain high radionuclide solubilities in oxidizing groundwaters representative of Yucca Mountain.

• A repository at Yucca Mountain might fail the minimum groundwater travel time requirements of 10 CFR 60; such failure is suggested by the fast transport of bomb-pulse chlorine-36 in the unsaturated zone at Yucca Mountain (see chapter 11, this volume) and by the channelized flow in the saturated zone (e.g., Murphy 1995).

• Fast groundwater flow could also threaten assurances of compliance at a five-kilometer boundary on a ten-thousand-year time frame (see chapter 14, this volume).

• Given the site's geochemical and hydrologic characteristics, longer time frames—for example, hundreds of thousands of years—could be problematic for limiting potential doses even at large distances from Yucca Mountain. This observation is supported by performance assessment results showing relatively high doses (hundreds of millirems/year) at long times (hundreds of thousands of years) and eighteen-kilometer distance (DOE 2002).

It is possible that none of these potential problems would disqualify the site under the current EPA standard or the NRC regulation for Yucca Mountain. In the end, the goal is not compliance to details of historical rules devised for a generic site but rather the reasonable assurance (or reasonable expectation) of human health and safety as well as environmental protection.

## Acknowledgments

I would like to thank Rod Ewing and Allison Macfarlane for inviting this contribution to *Uncertainty Underground*. Support from the French Centre National de la Récherche Scientifique helped sustain me during the preparation of a draft of this chapter. Macfarlane and three anonymous reviewers provided suggestions that contributed to the clarity and accuracy of my presentation.

## References

Department of Energy (DOE) (2002) *Environmental Impact Statement for a Geologic Repository for the Disposal of Spent Nuclear Fuel and High-Level Radioactive Waste at Yucca Mountain, Nye County, Nevada.* DOE/EIS–0250.

Murphy, W.M. (1995) Contributions of Thermodynamic and Mass Transport Modeling to Evaluation of Groundwater Flow and Groundwater Travel Time at Yucca Mountain, Nevada. In *Scientific Basis for Nuclear Waste Management XVIII*, ed. Murakami, T., and Ewing, R.C. Warrendale, PA: Materials Research Society: pp. 419–426.

National Research Council (1995) *Technical Bases for Yucca Mountain Standards*. Washington, DC: National Academy Press, 222 p.

Van Konynenberg, R.A. (1991) Gaseous Release of Carbon-14: Why the High Level Waste Regulations Should Be Changed. In *Proceedings 2nd Annual International Conference on High Level Radioactive Waste Management*. La Grange Park, IL: American Nuclear Society, pp. 313–319.

# 4

# Performance Assessment: What Is It and Why Is It Done?

Chris Whipple

The proposed nuclear waste repository at Yucca Mountain has generated many studies that cover a wide range of geologic and engineering issues. The goal of these analyses is to account for all events and processes that could contribute to future human radiation exposures from the stored wastes—everything from water infiltration and corrosion to volcanoes and earthquakes. The Yucca Mountain Project integrates these various analyses into what it calls the Total System Performance Assessment (TSPA) (Bechtel SAIC 2001). The TSPA defines the Yucca Mountain Project team's understanding of how the repository would perform if built and operated.

## How the TSPA Works: Follow the Water

The main pathways for potential exposures to released radionuclides from Yucca Mountain are those in which water is the principal medium of transport. Except in the case of a volcanic eruption, which could release radioactive material into the air, wastes are not expected to move from where they have been deposited without the presence of water. Much of the analysis in the TSPA is therefore aimed at characterizing the amounts and rates of water moving downward through the mountain, and also the characteristics of the waste—for example, its solubility—that determine how susceptible the wastes are to water-borne transport.

The TSPA analysis considers the possible future changes in climate that could increase the amount of precipitation at the site. A key factor in the assessment is the rate at which water infiltrates through the unsaturated zone of the mountain (that is, the area above the water table and below the surface). The TSPA includes the characterization of flow, both for the undisturbed mountain and the case where the heat generated by radioactive decay affects flow patterns. Environmental conditions in the waste drifts such as temperature, humidity, and water chemistry are projected for the initial operating period—on the order of a hundred years—when the drifts are ventilated, and also at later times, after the repository is closed. These anticipated environmental conditions and analyses of water movement define the environment in which the waste will dwell. In particular, these factors underlie the analyses of the corrosion of the drip shields and waste packages.

The Environmental Protection Agency (EPA) standard for Yucca Mountain, 40 CFR 197 (2001), set the regulatory period for the site at ten thousand years. As discussed below, the U.S. Court of Appeals has remanded this aspect of the standard and the EPA will need to revise the standard. Even with the ten-thousand-year standard, the TSPA calculations have been carried out until a million years, a period sufficient to estimate the peak dose that would occur. The TSPA analyses include the possibility that some waste packages may develop cracks at the final closure welds. Releases through such cracks would lead to radionuclide releases much earlier—within the first ten thousand years—than would general corrosion failure of the waste canisters.

Additional modeling describes the nature of the corrosion of spent fuel and high-level waste once waste packages have been breached. As modeled, the corrosion rate is affected by the chemistry of the water that reaches the repository, but is insensitive to the quantity or flow rate of water. This is because the analysis indicates that after the repository is closed, the humidity in the tunnels will approach 100 percent and a water film will cover all surfaces.

The analysis predicts that the most likely initial breaches of the waste packages will be cracks at the closure welds. As the modeling and analyses have evolved, the physical picture of what happens as the waste packages are breached has become more realistic and less conservative. Earlier iterations assumed that once a hole appeared, the waste package would lose all ability to isolate the waste from the environment. Because early designs did not include a drip shield, it was believed flowing water would come into contact with the waste and spent fuel once corrosion caused an opening in the waste package. In the current version of the TSPA, however, the combination of the drip shield and the fact that initial failures of waste packages are modeled as narrow cracks results in calculations of radionuclide releases by diffusion through the initial cracks in the waste package. In contrast to the advective transport that was calculated in earlier versions of the analysis, the release rates due to diffusion are much lower.

Several processes control the rate at which radionuclides are released from the waste package. For the spent fuel (which would make up the largest amount of the waste), cladding will serve as an added metal barrier. While made of a corrosion-resistant zirconium alloy, the cladding is thin in comparison to the waste package, and has experienced wear from the high temperatures and neutron fluxes in the nuclear reactor. For this reason, some small fraction of the cladding is assumed to have leaks at the time of disposal. A second process controlling the release rate is the long period of time over which radionuclides are released from spent fuel or defense high-level waste (the other component that would be disposed of at Yucca Mountain). Finally, the relatively low solubility of some radionuclides (e.g., plutonium and neptunium) limits how much of these radionuclides can be moved by diffusive transport in a thin water film.

The model considers the movement of radionuclides in two forms: as dissolved materials diffusing out of damaged waste packages through water films, and as soluble radionuclides and colloidal particles that can be swept along in flowing water. The model evaluates the transport of radionuclides from the waste packages through the rock below the drifts to the saturated zone; the model also accounts for the mix of fracture and matrix flow that would occur as well as the degree of geochemical retardation expected for significant radionuclides. Saturated flow and transport over a distance of around twenty kilometers is assessed for those radio-nuclides transported down into the groundwater that underlies the repository.

The model calculates radiation doses based on the assumed use of contaminated groundwater by people living in the Amargosa Valley, about twenty kilometers from the site in the direction of groundwater flow. This location is specified in the EPA and the Nuclear Regulatory Commission (NRC) regulations because it is currently the closest downgradient region in which significant farming is practiced. The analyses indicate that the primary exposure pathways would be by the ingestion of contaminated water and food produced locally that has been irrigated with contaminated water.

In addition to the slow, steady processes that would lead to the corrosion failure of waste packages and the migration of wastes to residents of the Amargosa Valley, analyses are made of infrequent events such as large volcanoes, earthquakes, and inadvertent human intrusion. The analyses indicate that of these events, volcanoes pose the highest risk.

For these issues, the Yucca Mountain Project staff developed more than one hundred analysis model reports (see http://www.ocrwm.doe.gov/technical/amr.

shtml), which provide technical details about how specific aspects of performance are modeled. These analysis model reports are combined into a smaller number of process model reports (http://www.ocrwm.doe.gov/technical/pmr.shtml), which cover such topics as flow and transport (in both the saturated and unsaturated zones), the degradation of waste packages and waste forms, engineered barriers, and disruptive events.

### How the TSPA Scope Is Determined

The Yucca Mountain TSPA implements risk analysis through a five-step process. First, scenarios are developed and screened through a process based on features, events, and processes (FEP). Second, computational models of the relevant processes are developed based on some combination of established equations and empirical data describing physical (e.g., hydrologic, chemical) processes. Third, parameter ranges are estimated, including the associated variabilities and uncertainties that arise from, for example, the heterogeneous character of the rock and the lack of complete knowledge of the site or the process that will occur. Fourth, calculations are made, using an approach that samples from the parameter distributions developed in the third step. This process is often referred to as a "Monte Carlo" analysis. The purpose of this approach is to ensure that the uncertainties and variabilities in various parameters are included.

The final step is the interpretation of results. The calculated dose rates can be compared with those specified in the EPA and the NRC regulations, a limit of a dose rate of fifteen millirems per year within ten thousand years from the time the repository is closed. As noted, the period of application of the EPA and NRC standards was remanded in 2004, and no new standard has been promulgated. The standard was found by the court to be inconsistent with the recommendations of a committee of the National Academy of Sciences, which found that there was no scientific basis to limit the period of applicability of the standard to a period shorter than the time to peak dose. The EPA's ten-thousand-year performance standard was found to be inconsistent with that recommendation.

Although the performance period of the now-remanded standard was ten thousand years, the Department of Energy (DOE) evaluated performance over a much longer time period because it was required for its environmental impact statement. As a result, performance calculations are available for periods as long as a million years. Figure 4.1 provides the results for a calculation to one million years from the

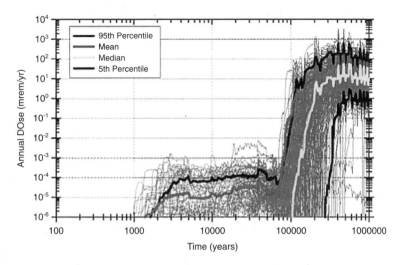

**Figure 4.1**
TSPA results. From Bechtel SAIC 2001, figure 6.5.

TSPA used for the input to the final environmental impact statement and site suitability evaluation. This figure indicates that the calculated dose rate during the first ten thousand years is well below the fifteen-millirem annual limit, but slightly over fifteen millirem per year at the time of peak dose. The models and methods used in this assessment, however, were selected when the standard applied for ten thousand years, and the results for longer time periods come from models that were not refined for compliance assessment.

The list of scenarios considered for evaluation is based on international efforts to identify features, events, and processes (FEPs) relevant to a geologic repository. While no approach can ensure that all possible scenarios have been considered, the FEP list used for scenario screening by the project is from a draft electronic database of 1,261 FEPs developed by the Nuclear Energy Agency based on the input from seven national programs. This list was supplemented by an additional 292 FEPs that are specific to the Yucca Mountain Project (see Freeze et al. 2001).

For the analyses in the TSPA, the FEPs are ignored either if they are sufficiently unlikely or if the consequences of the event are unimportant based on a comparison with regulatory dose limits. The probability threshold, as stated by the EPA standard, is any FEP that the DOE and the NRC estimate to have less than a one in ten thousand chance of occurring during the ten thousand years after disposal (see 40 CFR 197 2001; 10 CFR 63 2001). These aspects of the EPA and NRC standards were not affected by the court's remand.

## Limitations of Our Understanding of Repository Performance

Many of the limitations and weaknesses of the TSPA are inherent to the analysis of a repository system. The analysis is difficult and the uncertainties are large for several reasons:

- The inability to fully characterize the system to be analyzed.
- The lack of sufficiently detailed process knowledge.
- The need for extrapolation for unprecedented time periods.
- The repository system is not fully specified.

These will be addressed in greater detail below.

### The Inability to Fully Characterize the System to Be Analyzed

The Yucca Mountain site is heterogeneous and cannot be fully characterized on an appropriate spatial scale. It would practically require the disassembly of the mountain to discover the details of rock characteristics, including fractures and other discontinuities, for the full repository (on the scale of several kilometers) or full flow path to the Amargosa Valley, about twenty kilometers away. Repository performance would be affected by flow above the repository, by flow and transport in the vadose zone below the repository, and in the groundwater between the repository and the regulatory point of compliance some twenty kilometers downgradient. It is not feasible to fully characterize these large volumes of earth. It is also not feasible to characterize fully the extent and nature of the fast-flow pathways through which flow and transport are thought to primarily occur, except by the use of spatially averaged properties. Although some other geologic settings such as bedded salt or the subseabed would likely be easier to characterize, the requirements for characterization have to be weighed against other advantages such as the low density of local populations, dryness, and public landownership.

### The Lack of Sufficiently Detailed Process Knowledge

Even if the mountain and downgradient saturated zone were perfectly characterized (as would be the case for a fully engineered system), analysis of the future performance would be uncertain due to the lack of a full understanding of the processes that can lead to radionuclide release and migration. Consider, for example, the potential for colloids to act as a vehicle for the movement of plutonium. Colloids may be formed by the corrosion of spent fuel and waste glass containing the residues of reprocessed spent fuel, but the processes are not sufficiently understood for an analysis to be performed from first principles.

### The Need for Extrapolation for Unprecedented Time Periods

The regulatory period for a Yucca Mountain repository was set at ten thousand years following closure, but depending on how the lack of an applicable standard is resolved, it may extend until the time of peak dose. In the performance assessment calculations that have been made, the peak dose typically is found to occur at several hundred thousand years. The ten-thousand-year period and the time to peak dose are short compared to the geologic time scales over which Yucca Mountain was formed, so one would not expect changes in rock properties over such a time period. The volcanic tuff at Yucca Mountain is approximately eleven million years old. A possible exception is that the construction process may induce damage in the rock, and over time, earthquakes may cause rocks to fall from the ceilings of drifts or even cause drifts to collapse. Another exception is that rock that may be affected by the decay heat from the waste may alter in its properties. While rock near the waste may be affected by the initial thermal phase, the properties of the rock not affected by the thermal pulse—in the saturated zone, for example—are not likely to change over the regulatory period.

The performance of some other aspects of the repository system over such time periods is less certain. The analysis attempts to accommodate some of this uncertainty by, for instance, considering climate changes that would affect the precipitation at the site. But our current understanding of climate change allows this to be done only to a limited degree. The extrapolation of performance is least reliable for the engineered components such as the waste canisters and drip shields, where data and testing come from periods short in comparison to the anticipated life of the repository, resulting in the need to extrapolate far into the future based on a relatively short period of experience. While much is known about the basic chemical behavior of the metals proposed for use, the environmental conditions that will prevail at these metal surfaces over such long periods contribute to uncertainties in how these materials will perform. In response to this situation, a variety of tests are being conducted under more extreme water chemistry conditions than are anticipated in order to create conditions under which accelerated aging will occur. But extrapolation from such challenging short-term conditions to more benign but longer-term conditions is uncertain.

### The Repository System Is Not Fully Specified

Many of the details of the design and the operation have yet to be completely defined. One issue that has been the subject of debate and analysis concerns the choice of the repository operating temperature. A cooler design (i.e., where the temperature in the storage drifts is below the boiling point of water) has lower uncertainties in terms of temperature-induced changes in rock properties and chemical conditions near the waste than a hot design (above the boiling point). But the cooler design has drawbacks too. It would require some combination of more kilometers of drifts and a longer ventilation period; it would also be more expensive. The present reference design allows for above-boiling temperatures at the drift walls, but slightly below-boiling temperatures in the rock between drifts. The design is termed "hot," even though it is cooler than the design concept used several years ago. Temperature is not the only issue where the system is not fully specified. The processes for fabricating and inspecting waste canisters and drip shields, for example, have not been set in detail, so estimating the rate of manufacturing defects that could lead to early failures is problematic. Even if these processes were defined, it would be difficult to estimate the rate of defects in canisters or drip shields under full-scale production. Still more uncertainty arises from the lack of knowledge as to what waste goes where; this distribution will determine local thermal loading in the repository. In fact, even the total volume of waste that would go to Yucca Mountain is unclear, given that the current statutory limit of seventy thousand metric tons of heavy metal could be modified in the future—especially if policymakers determine that it would be much less expensive to load more waste into a repository at Yucca Mountain than to locate, characterize, and license a second repository. All of these factors will affect the postclosure conditions within the repository, yet they cannot be determined now from the existing plans and designs.

### Specific Issues with the TSPA

There are thus many factors that limit our ability to project how a repository would perform. Uncertainties in the basic processes, often referred to as conceptual model uncertainty, are considered through a series of sensitivity studies. In addition, the sheer complexity and number of models linked to form the TSPA make it difficult to identify which aspects of the design and the analysis of the system are significant versus which are of secondary importance. While the TSPA can provide insights into the aspects of the system that are crucial to performance, it is not always straightforward to do so. This difficulty is due in part to the fact that the degree of realism in the modeling is not uniform, in that aspects of performance thought to be unimportant may be treated simply through conservative bounding assumptions, while other aspects may be treated more realistically. Moreover, the fact that uncertainties are propagated through the analysis makes it difficult to compare sensitivities in performance to assumptions about parameters and processes.

**Opacity: The Difficulty in Tracking the Calculations**

Given the complexity of the analysis described above, it should not be surprising that the TSPA is a highly complex set of linked computer models. To make matters more complicated, the TSPA is constructed so that the parameter values—for example, permeabilities and radionuclide solubilities—can be represented by probability density functions in order to reflect their underlying variability and uncertainty. When the TSPA model is run, values from each of these many parameter distributions are sampled and the result is computed. This is done repeatedly, so that the distribution of the output measures is obtained (see figure 4.1).

The reason for the use of this method is that except for limited cases, mathematical operations such as the multiplication of probability distribution functions have no closed-form solutions. Historically, an alternative to such uncertainty analyses in risk regulation has been to simply use some conservative point from each distribution (such as the 95th percentile estimate) and combine those points to arrive at a conservative upper-bound estimate. The problem with this approach is that the conservatism can become extreme—that is, the product of the 95th percentile values of several distributions usually gives a value that is above the 99th percentile of the true combined distribution. Because a Monte Carlo method is capable, given a sufficient number of iterations, of accurately propagating uncertainties through the analysis, it is considered to be the best technical practice for such assessments. Indeed, such an approach is required in the NRC's standard for Yucca Mountain, 10 CFR 63.

Even for technically trained individuals, however, it is difficult to follow what is going on in such an analysis. In a deterministic analysis (i.e., one in which point values are used instead of probability functions), one can sometimes trace the intermediate results from each step in the calculation and see how the final answer depends on each intermediate step. With a deterministic analysis, it is also more straightforward to assess the sensitivities of results to changes in input values, taken one at a time. Understanding such sensitivities is much more difficult for a complex probabilistic analysis such as the Yucca Mountain TSPA. But a Monte Carlo model is required to understand how a distribution in uncertainty across many input parameters affects the uncertainty in the overall calculation. While the NRC staff will acquire an understanding of the models and the capability to run them, external parties will find it difficult to assess what goes on in such a complex set of coupled models.

### Difficulties with Assessing Sensitivities and Uncertainties

Uncertainties can arise from three basic sources. First, there are uncertainties about the scenarios that can occur and their likelihood. Second, the process models themselves are uncertain, and competing models may exist. Third, the basis for parameter distributions is often difficult to establish (see, for example, Kaplan and Garrick 1981). One example noted above illustrates this point: What is the likely early failure rate for waste canister welds? This cannot be estimated from actual data.

The intent of the TSPA is to create a computer-based mathematical model of repository performance that is as realistic as possible. While realism may be the goal, all components of the analysis are simplifications of reality, and many of the simplifications are intended to err on the side of overestimating rather than under-estimating risk. To the extent that there are bounding analyses and other conservatisms in the analysis, projections of future performance may overestimate future dose rates and understate the uncertainties associated with these estimates. Because the TSPA is incomplete, however, in that it may not include events and processes that could lead to earlier or larger than anticipated releases, there may be an offsetting effect to the conservative bias in some component models. After all, it is simply not possible to make a fully realistic projection of repository performance for tens or hundreds of thousands of years. But it is important to distinguish between high uncertainties and high risk. The TSPA results suggest that while the uncertainties in predicting performance for the next one hundred centuries are great, even the upper end of the estimated dose rate at ten thousand years is well below the regulatory limit. Nevertheless, given the remand of the standard noted above, the regulatory target is itself a significant uncertainty at the present time.

One consequence of the mix of conservative analyses with more realistic ones is that the sensitivities of repository performance to various input parameters are likely to be misestimated. This creates an inability to distinguish "true" sensitivities in repository performance from model sensitivities, except judgmentally. One issue that has been raised repeatedly by the U.S. Nuclear Waste Technical Review Board, for example, is the repository operating temperature. The TSPA indicates that long-term performance is insensitive to thermal loading, a result that is plausible, given that the temperature profiles for hot and cold designs converge at a time that is relatively early in the regulatory period. But the TSPA models for the corrosion of the waste package and drip shield are not particularly sensitive to temperature, nor do the uncertainties in the corrosion rates increase significantly at higher temperatures. This leaves open the issue of whether performance is insensitive to temperature or whether there is such sensitivity but that the TSPA is incapable of revealing it. In general, the corrosion rates are sensitive to temperature, which suggests that the uncertainties associated with a hot design have not been fully captured in the TSPA.

**What Is TSPA Good For?**

Given the uncertainties and limitations of the TSPA described above, one may ask where the value of the TSPA lies. It is critical to understand that some of these uncertainties did not originate in the TSPA; the TSPA simply reflects them, albeit imperfectly. Indeed, one of the most important purposes served by the TSPA is the disclosure of the assumptions, models, and analytic methods underlying the project staff's understanding of the repository system. Individual specialists can review the process models and data in their areas of expertise, and the broader technical community can participate in evaluating the technical components of the TSPA. This is not a simple or straightforward task, given how the components of the TSPA are constructed and how they fit together in the full analysis.

The TSPA provides insights into system performance, conditional on assumptions and simplified models. This is very useful in that it can reveal which uncertainties are relatively unimportant. Even with the limitations of sensitivity analysis, some performance parameters and models can be shown to have little effect on performance as specified by the EPA and the NRC—that is, on the dose rate within the compliance period. The iterative nature of the Yucca Mountain TSPA has been instrumental in focusing research activities within the Yucca Mountain Project

(TSPA analyses have been issued in 1991, 1993, 1995, 1998, and 2000). With each reevaluation, effort has focused on reducing the major uncertainties as indicated by the previous iteration. In some cases this has been through the collection of field data, in other cases by the refinements in the underlying models, and in still other cases by the redesign of the repository system.

Between the 1995 and 1998 revisions to the TSPA, for instance, significant effort was applied to the modeling of seepage into repository drifts and the behavior of waste packages after an initial corrosion penetration. These factors were seen as important to performance, but not realistically characterized in the earlier analysis. Following the release of the TSPA Viability Assessment in 1998, a design review was conducted aimed at removing or reducing major uncertainties in performance. Among the design changes that were made, two are particularly significant. One major change was made in response to the difficulty of modeling water flow in the unsaturated zone during the initial high-temperature period. The design concept evaluated in the Viability Assessment was a high-temperature one in which the repository would be closed immediately following the completion of the waste emplacement. In that design, the entire repository horizon heated up to above-boiling temperatures, and water infiltrating into the mountain was perched above the repository. This situation was difficult to analyze, and the potential for water flowing into cooler parts of the repository could not be precluded or predicted without detailed, time-consuming modeling; even with such modeling, uncertainties would remain high. A change was made to widen the distance between drifts so that the rock midway between drifts would not reach boiling temperatures. This design change to allow water to drain changed the scale of the thermal-hydraulic analysis from the scale of the mountain to the much more analyzable scale of a single representative drift. An additional change to ventilate drifts for an extended period contributed further to remove heat from the mountain and make the water flow more predictable.

The second change in response to insights from the TSPA Viability Assessment was the addition of drip shields. In the TSPA Viability Assessment, scenarios in which waste was transported advectively (i.e., carried in flowing water) from the waste packages to groundwater produced higher calculated dose rates than did scenarios in which the transport of radionuclides in the waste through cracks in the waste package occurred by diffusion. Provided that either the drip shield or the waste package is intact, the advective transport of radionuclides will not occur. For this reason, the addition of the drip shield provides for a system that is more tolerant of the TSPA misanalysis of the long-term corrosion behavior. As long as one of the two corrosion-resistant materials performs well, advective transport will be averted and dose rates will be well below the regulatory limits.

## What Does TSPA Indicate about Repository Performance?

The TSPA indicates that exposures that are significant to the regulatory dose limits are not expected within the ten-thousand-year period that the EPA had selected as the regulatory period, absent major volcanoes and early waste package failures, especially from human intrusion. Given that the modeling of the redundant engineered barriers estimates that each will last for tens of thousands of years, the lack of early doses from the most studied scenario (infiltration, corrosion, release and transport) is not surprising. At longer time periods, exposures due to radionuclides that transport slowly (e.g., neptunium-237) relative to those that contribute to the initial doses (e.g., technicium-99) begin to occur, and these dose rates are higher than those due to the more mobile radionuclides.

Some aspects of the TSPA, however, are counterintuitive. One such insight is that the corrosion models are essentially independent of infiltration. While it might be intuitively assumed that corrosion would be greater with higher water flow, the corrosion process models indicate that for the entire range of postulated infiltration rates, the relative humidity within the drifts approaches 100 percent, and a water film is expected to form on the metal surfaces. When a water film covers the metal surfaces, the corrosion rate is assessed to be the same as for the case where flowing water contacts the metal. Once the waste package is breached, releases would depend on the rate at which water contacts the waste, but before a breach occurs, the corrosion rate does not depend on the presence of flowing water.

## Summary

The Yucca Mountain TSPA is a highly complex analysis, reflecting the complexity of the system and processes it addresses. The standards for performance assessment from the EPA and the NRC have many elements and requirements, but the philosophical approach of both agencies makes the following clear:

• Absolute proof is impossible to attain for disposal due to the uncertainty of projecting long-term performance.

• A reasonable expectation, on the basis of the record, rather than a worst-case analysis, is the general standard of proof required to establish compliance with the standards.

• Compliance demonstrations should not exclude important parameters from assessments and analyses simply because they are difficult to precisely quantify to a high degree of confidence.

• The performance assessments and analyses should focus on the full range of defensible and reasonable parameter distributions rather than only on extreme physical situations and parameter values.

Whether the Yucca Mountain TSPA measures up to these requirements is a regulatory decision to be made in licensing, when and if a license application is submitted.

## References

Bechtel SAIC Company, LLC (2001) *Total System Performance Assessment: Analyses for Disposal of Commercial and DOE Waste Inventories at Yucca Mountain—Input to Final Environmental Impact Statement and Site Suitability Evaluation.* Department of Energy. REV 00, ICN 02. December. http://ocrwm.doe.gov/documents/sl986m3b/index.htm.

40 CFR 197 (2001) *Public Health and Environmental Radiation Protection Standards for Yucca Mountain, Nevada.* Final Rule.

Freeze, G.A., Brodsky N.S., and Swift, P.N. (2001) *The Development of Information Catalogued in Rev 00 of the YMP FEP Database.* Department of Energy. TDR–WIS–MD–000003 Rev 00 ICN 01. February. http://ocrwm.doe.gov/documents/sl981m3_a/index.htm.

Kaplan, S., and Garrick, B.J. (1981) On the Quantitative Definition of Risk. *Risk Analysis* 1, pp. 11–27.

10 CFR 63 (2001) Disposal of High-Level Radioactive Wastes in a Proposed Geologic Repository at Yucca Mountain, Nevada. Final Rule.

# 5

## Performance Assessments: Are They Necessary or Sufficient?

Rodney C. Ewing

In chapter 4, Whipple described how performance assessments are completed, how the results will be used, and the principal sources of uncertainty in such an analysis. The determination with a "reasonable expectation" that regulatory limits have been met is based, however, on the presumption that the challenges in modeling the important physical, chemical, and biological processes at a repository site have been overcome. The implicit assumption is that the analysis is done well enough so as to ensure that a confident judgment of safety can be made. A total system performance assessment will consist of a series of cascading models that are meant in toto to capture repository performance. For the geologic disposal of high-level nuclear waste, this generally means that models must be capable of calculating radiation exposures to a specified population at distances of tens of kilometers for tens to hundreds of thousands of years into the future. There are numerous sources of uncertainty in these models: scenario, conceptual model, and data uncertainty (Andersson and Grundteknik 1999). These uncertainties will propagate through the analysis, and the uncertainty in the total system analysis must necessarily increase with time. For the highly coupled, nonlinear systems that are characteristic of many of the physical and chemical processes, one may anticipate emergent properties that cannot, in fact, be predicted or even anticipated. In this chapter, I discuss the impact of the uncertainties on the analysis and suggest some different strategies for judging the safety of a repository site.

### Structure of a Performance Assessment

In 1738, Jacques de Vaucanson, a French watchmaker, created a life-size mechanical duck (figure 5.1), one of several of his famous automatons (Glimcher 2003). The mechanical duck was a remarkable creation that could move its head, flap its wings,

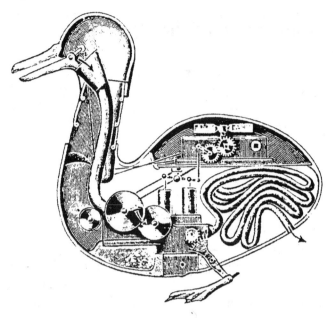

**Figure 5.1**
Vaucanson's mechanical duck. Animated versions of this figure can be found by a quick Web search. From Umberto Eco, *The Picture History of Inventions* (New York: Macmillan, 1963), 360 p.

and even eat from a bowl of grain. The grain was mechanically compacted and finally excreted. To many, this mechanical creation appeared to be a duck—moving and eating as if it were a duck. This mechanical duck relied on the interaction of gears and springs, each designed to capture some aspect of a duck's behavior. Over four hundred parts moved each wing. Taken as a whole, the individual mechanical parts created a machine that mimicked many of the characteristic behaviors of a duck.

Vaucanson's mechanical duck is to a real duck as a performance assessment is to the actual behavior of a geologic repository for nuclear waste. The structure of a performance assessment (figure 5.2) is, in its essence, the same as the structure of Vaucanson's duck. A performance assessment consists of many parts, usually represented by computer codes, that are meant to capture the behavior of a high-level nuclear waste geologic repository over large spatial (tens of kilometers) and temporal scales (tens to hundreds of thousands of years). Each code is an abstraction, a simplified representation, of the important processes that must be evaluated in

# Overview of Model Linkage
## (Total System Performance Assessment - Site Recommendation)

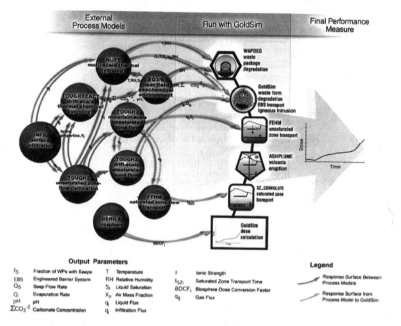

**Figure 5.2**

Diagram showing the linkages between the different subsystem models in the total system performance at the Yucca Mountain repository site. The intricate linkages among the models are meant to simulate the performance of the repository well enough to make a final performance determination. The diagram was provided courtesy of Peter Swift from his presentation to the Advisory Committee on Nuclear Waste on March 25, 2003.

order to determine whether the repository is "safe" (e.g., the hydrology of the site, the geochemistry of the groundwater, waste package corrosion, waste form alteration and dissolution, and the transport of radionuclides through the geosphere to the biosphere). These sophisticated models represented by computer codes are analogous to the wonderful mechanical intricacy of Vaucanson's duck. There are two essential questions:

• How much can one learn about ducks by studying mechanical ducks?
• How closely can a mechanical duck be made to mimic the behavior of a real duck?

These are serious concerns that go directly to the confidence that one can have, or should expect to have, in a performance assessment, such as the one used to evaluate the performance of the Yucca Mountain repository.

• To what extent can we understand the behavior of the actual geologic repository by using a performance assessment?
• To what extent can we expect the performance assessment to actually mimic the real system?

Of course, the answers depend on what one thinks of ducks. If a duck is a *complicated* assemblage of separate systems that taken as a whole is a duck, then the creation of more elaborate mechanical parts or computational simulations will inevitably bring us closer to the real duck and an understanding of the behavior of ducks. If the combination of the separate parts into the larger organism, a duck, or the geologic system, a repository, leads to the *complex* behavior characteristic of highly coupled, nonlinear systems, however, then new properties may arise with the coupling, and the behavior may now have unexpected features. In this case, more elaborate models or finely designed parts may not lead to a clearer understanding of the behavior of ducks or geologic repositories. Certain types of behavior may be unknowable and thus unpredictable.

The struggle to make a compelling safety case for a geologic repository is similar to the challenge of using a mechanical duck to understand the behavior of a real duck. The linked models shown in figure 5.2 represent a mechanical, reductionist approach to capturing the behavior of a geologic system. Yet for a geologic repository, one must address the issue of modeling complexly coupled geologic subsystems (e.g., coupled thermo-mechanical-chemical models). There is also the issue of scaling—that is, the extrapolation of modeled results to larger spatial scales and over long periods. Not only must Vaucanson's mechanical duck behave very much like a real duck, but we want to be able to predict the duck's behavior over an extended period. In most performance assessments of a geologic repository, the peak dose occurs at times greater than ten thousand years, much greater than the five-thousand-year span of recorded history.

Scientists and engineers are increasingly aware of the difficulties of dealing with complex systems (Cipra 2000; Ottino 2004), and the limitations of risk and performance assessments (Ewing et al. 1999; Hoffmann-Riem and Wynne 2002; Ewing et al. 2004). This complexity has led to a variety of approaches in completing performance assessments. The approaches may be broadly grouped into three types: bounding, probabilistic, and realistic calculations. In actual practice, all three approaches may be used in a single performance or risk assessment.

Most commonly, the performance assessment is described as giving a "bounding" or "conservative" result. Such a calculation is certainly a good basis for making a safety assessment. In this case, one is looking for answers that fall below a regulatory standard. But what does a bounding or conservative duck actually look like? Is the analysis so bounding that the behavior of the duck is lost in the mechanical or computational representation? If one makes too many conservative assumptions, then the mechanical duck may stand as a mute, nonmoving representation of a real duck.

A second approach is to use a probabilistic performance assessment. A probability distribution for the function of each part of the duck is used to capture the possible range of parameters and/or bounding conditions. The probabilistic approach may also be used, in principle, to represent conceptual uncertainties—that is, the differences among different models of the same process. A single conceptual model may capture the physics required to describe how ducks fly, but more complicated physical and chemical models will be required to describe how a duck digests its food. With the probabilistic approach, however, the duck becomes a rather strange beast. The duck may have the capability of flight, but only from a probabilistic perspective. And as the probability distributions are widened to capture the full range of behaviors, we may not be able to distinguish between a duck and a mole.

Finally, one may attempt a realistic assessment of the behavior of a geologic repository. A realistic assessment does not require that one actually describes how all parts of the geologic system work but rather usually focuses on the key elements of the analysis that most affect the results of the safety assessment, such as waste package corrosion. In this case, there is an a priori definition of what is important about being a duck. Ducks may be creatures that fly and swim, but is that all there is to being a duck? Have we missed some crucial feature unique to ducks? Similarly, have we missed significant aspects of the workings of a geologic repository? This question goes beyond the usual comprehensive effort to identify all of the important features, events, and processes, which is based on the reduction of the total system into subsystems, but rather asks the question of whether we have captured the performance of the repository with a proper consideration of the highly coupled processes that may have new or emergent properties.

Despite these possible shortcomings, no one can seriously doubt the value of a performance assessment as a systematic and disciplined means of organizing one's understanding of a geologic repository. The results of a performance assessment can be used to identify the major sources of uncertainty and the deficiencies in conceptual models; focus intellectual and financial resources on the key issues that most affect repository performance; and evaluate alternative designs and strategies for nuclear waste containment. Nevertheless, for each of these applications, such as the comparison of different designs, one must reach a certain level of resolution in the analysis that allows one to distinguish between the alternatives.

I am not arguing here against the use of performance or risk assessments to understand how a geologic repository might work. I am also not contending that we should not use models, or that we cannot make reasonable and prudent decisions in the face of large uncertainties. After all, important and costly decisions are routinely made despite a limited understanding of geologic systems, most notably in the exploration for minerals, oil, and gas. Nor are these problems unique to the effort to site and evaluate a geologic repository (Pollack 2003). The same issues of uncertainty in model predictions and the proper course of action are fully exposed in the discussions of climate change (Oreskes 2004). There is, however, a growing body of evidence that the resolution in the analysis of geologic systems may be greatly obscured by the scale of the uncertainties over time. I will argue that the uncertainties in a performance assessment are large and that the resolution of the analysis may fail to answer the essential question, Is the repository safe?

## Is There a Problem?

Much of the faith in geologic disposal is based on a belief that risk can be quantified (Kaplan and Garrick 1981) and that the methods of analysis developed for nuclear reactor safety analysis can be transferred to geologic systems (Garrick and Kaplan

1995). Bredehoeft and colleagues (1978) clearly outlined the challenges to the geosciences in attempting to model the long-term behavior of a geologic repository. Although they acknowledged the value of predictive models in the evaluation and design of geologic repositories, they pointed out that "some components of the models are inherently unpredictable at present and are likely to change at different times. In no sense, therefore, will these models give a single answer to the question of the fate of radioactive waste in geologic repositories." More than a quarter of a century later, the debate over the value and limitations of performance assessments is still ongoing (Ewing et al. 1999, 2004).

Central to the discussion are the clear limitations in the predictive capabilities of models in principle (Oreskes et al. 1994) and practice (Oreskes and Belitz 2001), particularly for geologic systems. The most important difficulties in the use of models within a large-scale performance assessment include the following:

*The models are simplified.* Although there are many physical and chemical processes for which we have a detailed understanding, such as the corrosion of spent nuclear fuel or the nuclear waste glass, the full understanding embodied in the scientifically based model may not be included in the "abstracted" models that are used in the performance assessment. The intention of the analyst is that the abstracted models capture all the significant features of the performance as described in the scientific models, but this is difficult to demonstrate within the context of the fully coupled performance assessment.

*The models may not be properly scaled in time or space.* Most models used in a performance assessment will be based on laboratory- or field-scale tests. The laboratory experiments are generally for short periods and based on just a few grams of material under constant, well-controlled conditions. The field tests are also of relatively short duration and may only extend over spatial scales of just a few tens of meters. In application, the performance assessments model the behavior of tens of thousands of tons of spent nuclear fuel, and track the release and migration of radionuclides over thousands of meters. As one extrapolates to larger temporal and spatial scales, heterogeneities in the system may dominate the performance, such as the fast transport of water along isolated fracture systems. Scaling is an inevitable limitation of studies of geologic systems, and the remedy requires considerable in situ testing and monitoring over long periods.

*Boundary conditions may change over time.* The emplaced nuclear waste in a geologic repository represents a substantial disturbance of the "normal" geologic conditions of the repository. During the first hundreds of years following emplacement, the thermal and radiation fields are transient, and at longer times hydrologic and geochemical conditions may evolve due to changes in climate and water-rock interactions. Thus, at each stage in the future of the repository, one has to anticipate the relevant boundary conditions. In some cases, the shorter-term thermal and radiation fields can be described, but over longer periods one must rely on probabilistic descriptions of changes in climate, seismicity, or volcanic activity.

*The conceptual models may be wrong.* One may select several conceptual models to depict an important process (e.g., the long-term corrosion rate of the borosilicate glass), but there is no assurance that the selected conceptual model is either bounding and conservative or correct. Moreover, the "right" conceptual model may be different for different boundary conditions. Corrosion mechanisms, for example, may be different at elevated temperatures than at lower ones. Thus, there may be profound differences between a hot or cold repository design that are not revealed when one predicts the behavior using the same conceptual models for both thermal regimes.

Each of these four issues is inherent to the difficulties with any performance assessment, and cannot be completely eliminated by additional work or research. A careful site selection process and disposal strategy can, however, reduce the challenge. A repository located below the water table may not be as sensitive to changes in the surface recharge of water due to climate change. For a repository located below the water table fluctuations in the depth of the water table may not affect the conceptual model of repository performance under fully saturated conditions. A repository located in tectonically stable regions may not require an analysis of seismic or volcanic activity.

Of these four issues, the possibility that the conceptual models are wrong stands as probably the greatest source of uncertainty—and also the most difficult to estimate. One approach is to calculate a range of possible outcomes for different conceptual models, but conceptual model uncertainty is difficult to evaluate within the context of a large-scale performance assessment because a bounding assumption in one part of the analysis may obscure the sensitivity of the analysis to the range of outcomes generated by the different conceptual models. As an example, different conceptual models for the corrosion of the high-level waste glass will all give the same answer if they are calculated for a scenario in which the waste package is never breached and water does not reach the waste glass.

The geosciences have made great strides during the past fifty years in describing, modeling, and even predicting the behavior of geologic systems. The actual level of success is rather humbling, though, when compared to the demands placed on the performance assessments of geologic repositories that extend over the next million years. For hydrologic systems, Konikow and Bredehoeft have provided a number of examples of predictive success (or failure) by completing "postaudits" of hydrologic systems—that is, comparing predicted behavior to actual behavior over periods of decades (Konikow 1986, 1992; Konikow and Bredehoeft 1992; de Marsily et al. 1992; Konikow and Ewing 1999). Similar examples may be found for chemical systems. Geochemical models may have nonunique solutions (Bethke 1992), provide greatly varying results for very small changes in fundamental input parameters (Ewing et al. 1999), and results may vary greatly at different spatial scales and for changing boundary conditions (Jensen et al. 2002). One of the ironies of the present situation is that the greater the sophistication of the model, the more difficult it is to test and finally refute the results. Even though hypothesis testing and falsification are the essence of the scientific method, Nordstrom (2004) in a summary of the state of the art for low-temperature geochemical modeling has described the limitations of such models. He emphasizes that models are tools, and that expert judgment and field tests are an essential part of understanding the long-term behavior of geochemical systems.

## Alternative Strategies

Considering the "hard sell" that will be required for a performance assessment when presented to the scientific community, regulators, decision makers, stakeholders, or the general public, what can be done to place the performance assessment in a proper and constructive context?

I suggest the following:

• Analyze and present the uncertainty in the analysis. There is, at present, no standard approach, but it is apparent that one cannot expect a policymaker or members of the public to accept the results of a performance assessment unless there is a clear understanding and presentation of the sources and effects of the uncertainty on the conclusions of the analysis.

• Use multiple criteria in the analysis as part of the decision-making process. The level of uncertainty in the analysis might be in and of itself a criterion. Some sites can be analyzed with lower levels of uncertainty than others.

• Do not present the results only in terms of dose or risk. Although these are critical parameters from a regulator's perspective, it is difficult to place such numbers into context. As an example, one often sees a parameter varied, such as the cladding failure rate or spent fuel corrosion rate, and the effect judged by watching for variations in the exposure to an individual at a distance of twenty kilometers after one thousand years. The actual physical and chemical interactions between the parameters of interest and the impact on exposure is often not evident in this type of sensitivity analysis.

• Analyze multiple barriers, and analyze them as separate units with their own performance specifications. The aggregation of multiple processes—that is, the cascading models—obscures what is supposed to be a "transparent" analysis. Breaking the analysis into smaller pieces, with clearly defined subsystem performance criteria, simplifies the presentation and review.

• Use the natural systems of the repository site as passive or undisturbed barriers to radionuclide release and transport. Then, the models of future performance can be tested against the knowledge of the past behavior of the repository site. Large-scale perturbations of the natural system reduce the ability to use what we know about the site under ambient conditions, and as it has existed for tens of thousands of years, in making the safety case for the repository site.

• Use natural systems to test the performance assessment codes and support, if only in a qualitative manner, the results of the performance assessment.

• Keep it simple. There is a natural tendency to address uncertainties by the use of increasingly sophisticated models, but in practice, this can make the analysis more opaque as well as more difficult to test and review (Nordstrom 2004). An alternative approach is to select designs that simplify the problem. Maintain a low repository temperature where chemical reactions are slower. Use materials, such as copper canisters, for which we have long-term data (native copper samples are available that are hundreds of millions of years old). Maintain, as much as possible, the ambient conditions of the surrounding rock. The less that the construction of the repository and the emplacement of the wastes disturbs the geologic system, the better that past geologic history can serve as an indicator of future behavior.

• The results of the analysis should depend more on the actual properties of the site rather than the assumptions about the bounding conditions (e.g., climate, volcanic activity, and seismic activity). In some cases, a site may be selected so as to reduce the assumptions that are required in the analysis. As an example, a repository constructed in the saturated zone, below the water table, may be less sensitive to climatic variations.

Because of the large uncertainties in performance assessments of geologic repositories, it is misleading to identify a performance or risk assessment as "quantitative" (Apostolakis 2004). A performance assessment is quantitative only in the sense that it provides a numerical result, but the substance of the result is *qualitative*. One danger of treating the results as quantitative is that it causes the license applicant and the regulatory agency to concentrate on numerical compliance with a standard rather than an evaluation of the adequacy of the strategy for the safe disposal of nuclear waste. Still, with all of the limitations fully in mind, performance assessments are a powerful tool in organizing our understanding of the behavior of a geologic repository. Performance assessments remain a necessary part of the political and regulatory process, but a performance assessment may be of limited value in developing and evaluating a strategy for the safe disposal of nuclear waste. A performance assessment is not, by itself, a sufficient basis for determining whether a site is safe.

## Acknowledgments

My use of Vaucanson's duck is directly inspired by the discussion of the same by Paul Glimcher (2003) in his *Decisions, Uncertainty, and the Brain: The Science of Neuroeconomics*. I believe that much can be gained by a broad consideration of the properties and challenges of describing, as well as modeling, complex systems, and Glimcher's book is an exposition on these challenges. I have benefited greatly from conversations with and the writings of Lenny Konikow, John Bredehoeft, and Naomi Oreskes. I have also benefited from numerous discussions and debates with the advocates of performance assessments. I would like to particularly thank John Garrick, Martin Tierny, John Helton, and Peter Swift for their patience and insightful explanations. Although there is an enviable "quantitative" appeal with a performance assessment, my geologic background cautions me against an unqualified acceptance of such results. Much of this chapter first appeared in an invited paper presented at the spring meeting of the Materials Research Society in 2004 at the Scientific Basis for Nuclear Waste Management symposium. (Ewing 2004). Finally, I have only recently become aware of the book by Morgan and Henrion (2003), *Uncertainty*, which is certainly a more thorough exposition of the issues that I have raised in this brief chapter. I recommend *Uncertainty* to those who want to pursue the connection between the uncertainty in risk analysis and policy.

## References

Andersson, J., and Grundteknik, G. (1999) *Data and Data Uncertainties.* SKB Technical Report. TR–99–09, 138 p.

Apostolakis, G.E. (2004) How Useful Is Quantitative Risk Assessment? *Risk Analysis* 24, pp. 515–520.

Bethke, C.M. (1992) The Question of Uniqueness in Geochemical Modeling. *Geochimica et Cosmochimica Acta* 56, pp. 4315–4320.

Bredehoeft, J.D., England, A.W., Stewart, D.B., Trask, N.J., and Winograd, I.J. (1978) Geologic Disposal of High-Level Radioactive Wastes: Earth-Science Perspectives. *Geological Survey Circular* 779, 15 p.

Cipra, B. (2000) Revealing Uncertainties in Computer Models. *Science* 287, pp. 960–961.

de Marsily, G., Combes, P., and Goblet, P. (1992) Comment on "Ground-water Models Cannot Be Validated," by L.F. Konikow and J.D. Bredehoeft. *Advances in Water Resources* 15, p. 367.

Ewing, R.C. (2004) Performance Assessments of Geologic Repositories for High-Level Nuclear Waste: Are They Necessary or Sufficient? In *Scientific Basis for Nuclear Waste Management XXVIII,* ed. Hanchar, J.M., Stroes-Gascoyne, S., and Browning, L. Warrendale, PA: Materials Research Society, pp. 511–520.

Ewing, R.C., Palenik, C.S., and Konikow, L.F. (2004) Comment on "Probabilistic Risk Analysis for a High-Level Radioactive Waste Repository." *Risk Analysis* 24, pp. 1417–1419.

Ewing, R.C., Tierney, M.S., Konikow, L.F., and Rechard, R.P. (1999) Performance Assessments of Nuclear Waste Repositories: A Dialogue on Their Value and Limitations. *Risk Analysis* 19, pp. 933–958.

Garrick, B.J., and Kaplan, S. (1995) Radioactive and Mixed Waste: Risk as a Basis for Waste Classification. *National Council of Radiation Protection and Measurements Symposium Proceedings No. 2.* Bethesda, MD: NCRP, pp. 59–73.

Glimcher, Paul W. (2003) *Decisions, Uncertainty, and the Brain: The Science of Neuro-economics.* Cambridge: MIT Press, 375 p.

Hoffmann-Riem, H., and Wynne, B. (2002). In Risk Assessment, One Has to Admit Ignorance. *Nature* 416, p. 123.

Jensen, K.A., Palenik, C.S., and Ewing, R.C. (2002) $U^{6+}$-Phases in the Weathering Zone of the Bangombe U-Deposit: Observed and Predicted Mineralogy. *Radiochimica Acta* 90, pp. 761–769.

Kaplan, S., and Garrick, B.J. (1981) On the Quantitative Definition of Risk. *Risk Analysis* 1, 11–27.

Konikow, L.F. (1986) Predictive Accuracy of a Ground-water Model: Lessons from a Postaudit. *Ground Water* 24, 173–184.

Konikow, L.F. (1992) Discussion of "The Modeling Process and Model Validation," by Chin-Fu Tsang. *Ground Water* 30, pp. 622–623.

Konikow, L.F., and Bredehoeft, J.D. (1992) Ground-water Models Cannot Be Validated. *Advances in Water Resources* 15, pp. 75–83.

Konikow, L.F., and Ewing, R.C. (1999) Is a Probabilistic Performance Assessment Enough? *Ground Water* 37, pp. 481–482.

Morgan M.G., and Henrion, M. (2003) *Uncertainty: A Guide to Dealing with Uncertainty in Quantitative Risk and Policy Analysis.* Cambridge: Cambridge University Press, 332 p.

Nordstrom, D.K. (2004) Modeling Low-Temperature Geochemical Processes. In *Treatise of Geochemistry, Vol. 5: Surface and Ground Water, Weathering, and Soils,* ed. Holland, H.D., and Turekian, K.K. Amsterdam: Elsevier Pergamon, pp. 37–72.

Oreskes, N. (2004) The Scientific Consensus on Climate Change. *Science* 306, p. 1686.

Oreskes, N., and Beltz, K. (2001) Philosophical Issues in Model Assessment. In *Model Validation: Perspectives in Hydrological Science,* ed. Anderson, M.G., and Bates, P.D. Chichester, UK: John Wiley and Sons, Ltd., pp. 23–41.

Oreskes, N., Shrader-Frechette, K., and Belitz, K. (1994) Verification, Validations, and Confirmation of Numerical Models in the Earth Sciences. *Science* 263, pp. 641–646.

Ottino, J.M. (2004) Engineering Complex Systems. *Nature* 427, p. 399.

Pollack, H.N. (2003) *Uncertain Science ... Uncertain World.* Cambridge: Cambridge University Press, 243 p.

# 6

# Technical Policy Decision Making in Siting a High-Level Nuclear Waste Repository

Allison M. Macfarlane

During his 2000 presidential campaign, then Governor George W. Bush declared in a letter to Nevada Governor Kenny Guinn that "sound science, and not politics," would guide his decision on Yucca Mountain as a site for a nuclear waste repository (Sonner 2000). Two years later, Spencer Abraham (2002), Bush's secretary of energy, repeated this refrain in his recommendation letter to the president by stating, "I have considered whether sound science supports this determination that the Yucca Mountain site is scientifically and technically suitable for the development of a repository. I am convinced that it does." Those who oppose Yucca Mountain see it the opposite way: "Abraham's recommendation has nothing to do with sound science and everything to do with corrupt politics," said Mary Olson, who works with the Nuclear Information and Resource Service's (2002) office in Asheville, North Carolina.

These statements suggest that nuclear waste repository siting is a black-and-white issue—that the policy decision rests entirely on the results of either scientific investigation or political dealings, depending on your point of view. In this chapter, I will argue that neither viewpoint expressed above is correct and that the rationale for the Yucca Mountain site keeps shifting in response to a variety of factors, including political and judicial decisions along with new scientific data. For example, the origin of the use of the performance assessment modeling that forms the basis for the DOE's decision on site suitability is not simply technical but also political, organizational, and cultural. The question I consider in this analysis is whether it is reasonable to expect that technical information will be the sole factor in guiding technical policy decisions.

The situation that is best avoided is one where a veneer of science is used to legitimize technical policy decisions even though other considerations are actually driving the debate. Herrick and Sarewitz point out that public controversy in technical policymaking is often focused on scientific issues, though the real sources of debate are found in "economic interest, ethics, esthetics, equity, ideology, and regional politics" (2000, 320). For example, in developing climate change policy, the U.S. government has concentrated almost exclusively on the "scientific debate" about whether warming is a direct result of human activities, and has given less attention to the potential economic impacts of warming and the mitigation procedures that would have benefits in their own right, such as reducing air pollution. By examining the development of U.S. policy on nuclear waste disposal, we can move beyond the simplistic terms presented by politicians and pro– and anti–Yucca Mountain groups.

## Background

Yucca Mountain was one of three sites identified by the Department of Energy (DOE) as potentially suitable for the disposal of high-level nuclear waste; the other two were in Texas and Washington. These three sites were selected through a site-screening process mandated by the Nuclear Waste Policy Act of 1982. As Cotton describes in chapter 2, Congress truncated the screening process in 1987 by amending the Nuclear Waste Policy Act so as to require the DOE to characterize only the Yucca Mountain site for its suitability as a waste repository.

The DOE maintains that the decision to focus solely on Yucca Mountain was based largely on technical considerations. The four principal technical factors were the site's location in southern Nevada, a region of closed hydrologic basins; the long flow paths between the repository and groundwater discharge points such as springs or rivers; the particular types of rocks at Yucca Mountain, which may slow radionuclide movement; and the aridity of the Yucca Mountain area, resulting in slow groundwater recharge and little water in the unsaturated zone at the repository horizon (OCRWM 1998).

There were also political reasons for the site selection (Colglazier and Langum 1988). Yucca Mountain is on federally owned land shared by the Nevada Test Site, the Nellis Air Force Base, and the Bureau of Land Management. Some of the land about ten to thirty miles north and east of Yucca Mountain was already contaminated with radioactivity from over nine hundred nuclear weapons tests, of which more than eight hundred were underground (DOE 2000). Perhaps most important, in the late 1980s, when Congress made the decision to focus on Yucca Mountain, Nevada was politically weak: with a small population and two recently elected senators, it was outranked by Texas and Washington, which were represented by long-serving congresspeople who had acquired far more power and influence than had Nevada's delegation in the Capitol.

By eliminating two of three sites for consideration as a nuclear waste repository, Congress essentially short-circuited the decision-making process. Moreover, there were only sparse contingency plans in place if Yucca Mountain were to be deemed unsuitable—a decision that, according to Cotton (chapter 2), would require Congress to write entirely new legislation. Thus, strong political pressure exists, amplified by the nuclear industry, whose future depends partially on a resolution to the waste problem, to characterize Yucca Mountain as a suitable site for the disposal of high-level nuclear waste.

## Nuclear Waste Policy Evolution: An Example

To untangle the factors involved in the evolution of the U.S. nuclear waste program, I will use as an example the key distinguishing characteristic of the proposed Yucca Mountain repository: its location above the water table in the so-called unsaturated zone. This area above the water table is often referred to as dry, although in the case of Yucca Mountain, the rocks at the repository level are not in fact dry. My analysis will consider the cascade of effects the selection of a dry repository site has had on the science, policy, and regulations associated with nuclear waste disposal.

In choosing to locate its repository in the unsaturated zone, the United States is almost unique among countries seeking the underground disposal of high-level nuclear waste. Most of these countries—including Canada, Finland, France, Germany, Japan, Sweden, Switzerland, and the United Kingdom—intend to dispose of their waste in the saturated zone, below the water table. Partly, this is a result of geography—few of these countries have large arid regions—but partly this is by design, because the saturated zone confers advantages to the long-term stability of nuclear waste (Ewing et al. 1999). With no free oxygen present, the saturated zone below the water table provides a chemically reducing environment for spent fuel.

The Yucca Mountain site, in contrast, will provide an oxidizing environment for the disposed waste. Spent nuclear fuel contains approximately 95 percent uranium dioxide ($UO_2$) (Janeczek et al. 1996). Because the oxidized form of uranium, U(VI), is more soluble in water than is the reduced form U(IV) in the presence of moisture, spent fuel will corrode faster under the conditions present at Yucca Mountain (see, for example, Shoesmith 2000). As a result, countries considering the saturated zone for a repository have an intrinsic advantage over a site such as Yucca Mountain.

The United States had in fact originally intended to use a saturated site (Metlay 2000). In the mid-1980s, the Nuclear Regulatory Commission's (NRC) draft regulations on the geologic disposal of nuclear waste (10 CFR 60) initially covered *only* saturated zone sites. After preliminary examination of the saturated zone at Yucca Mountain disclosed a high fracture permeability and high water temperature (Stuckless 2003), U.S. Geological Survey geologists suggested that the United States consider the unsaturated zone (Flint et al. 2001a). In response, the NRC changed its regulations to cover unsaturated zones—an example of scientific evidence instigating a policy shift.

Other factors influenced the decision to develop a dry repository. Flint and colleagues (2001a) report that the Energy Research and Development Administration, the precursor to the DOE, was interested in the Nevada Test Site (where Yucca Mountain is located) as a repository site because it was a large piece of federally owned land far from population centers. Furthermore, Yucca Mountain had appeal because the site's rocks were not principally salt; salt sites experienced a number of technical and political problems during the 1970s, and had become politically fraught (Metlay 2000). In addition, the 1982 Nuclear Waste Policy Act required the DOE to investigate three sites in detail that possessed two distinct rock types (Colglazier and Langum 1988).

The Yucca Mountain location was selected because other areas within the Nevada Test Site were deemed unsuitable by the U.S. Geological Survey based on their geology (Dixon and Hoover 1982; Dixon and Glanzman 1982) or by the DOE because of their proximity to future nuclear weapons tests. Because the saturated zone at Yucca Mountain had been previously disqualified, the DOE was left with only the unsaturated zone as a potential site on the Nevada Test Site. Nevertheless, because of the large knowledge base on the geology and hydrology of the Nevada Test Site (Winograd 1972, 1974) and the political reasons noted above, as well as the fact that the Nevada Test Site contained a previously uninvestigated rock type (tuff), the DOE was eager to include the Nevada Test Site as a repository site. In fact, it was easier at that time to change the NRC's draft regulations to include unsaturated zone rocks than to begin a new search for an alternate site. Thus, the scientific, political, and pragmatic factors all supported a decision to consider the unsaturated zone at Yucca Mountain as a potential repository site.

## Implications of a Dry Site

The selection of the unsaturated zone had a number of significant implications for both the science needed to characterize the site and the policies and regulations that guided nuclear waste disposal in the United States. In the early 1980s, the field of hydrogeology, especially the understanding of contaminant transport in the unsaturated zone, was in its infancy. Indeed, the research done at Yucca Mountain has greatly increased our understanding of how transport occurs in the unsaturated zone (National Research Council 2001). By focusing on the unsaturated zone at a time when knowledge of its hydrogeology was underdeveloped, the U.S. government made the job of site characterization, and therefore site selection and approval, that much more difficult.

The second significant implication of the selection of a dry site was that it circumscribed the science to be done in the characterization effort. In particular, the development of the waste canister material was not as straightforward as with a saturated site. In saturated sites, such as the ones Sweden and Finland are considering, the waste canister material could be selected from known natural analogs—in these cases, elemental copper. Copper, as we know from natural deposits, alters slowly in a reducing environment, thus a high degree of certainty can be attached to its predicted performance over geologic time.

There are no equivalent natural analogs for metals that could be used as waste canister materials in oxidizing environments. Instead, the DOE had to seek man-made alloys. The agency ultimately settled on a nickel-chromium-molybdenum alloy called Alloy-22. The DOE Total System Performance Assessment (TSPA) model indicates that the Alloy-22 waste canisters will not fail until about fifty thousand years after repository closure (OCRWM 2001). Other studies show that the waste canisters may be susceptible to corrosion at welds and high temperatures (see chapter 17, this volume; Nuclear Waste Technical Review Board 2003; Cragnolino et al. 2004). In fact, little data exist on the corrosion of Alloy-22 at temperatures above the boiling point of water (Nuclear Waste Technical Review Board 2003). All of the estimates made about the performance of the waste canister are based on laboratory experiments run for durations of six months to two years, and then extrapolated to tens of thousands of years by computer models (OCRWM 2001). It is, however, difficult to project with any certainty the performance of an engineered material over geologic times and in natural, open-system conditions using data based on short times and controlled conditions.

Like spent fuel, most metals will corrode in an oxidizing environment, especially in the presence of water. Thus, the characterization effort at the Yucca Mountain site focused on the behavior of water. The two most important questions were how much water could enter the repository in the future, and how fast could it transport radionuclides to the accessible environment.

Metlay (2000) as well as Flint and colleagues (2001a, 2001b) have documented the evolution of the DOE's understanding of the volume of water that infiltrates repository rock—the so-called percolation flux. After initial, relatively low estimates of 0.5 to 4 millimeters per year (Flint et al. 2001a), estimates were revised to a range of 0 to 80 millimeters per year, with a mean of 4.5 (Flint et al. 2001a). The revision resulted from analyses conducted by Los Alamos National Laboratory scientists (Fabryka-Martin et al. 1994). Their discovery of high chlorine-36 levels in rocks at the repository horizon suggested the existence of fast water transport pathways in the overlying two to three hundred meters of rock, associated with fractures and faults. These high concentrations of chlorine-36 were from atmospheric tests of nuclear weapons conducted over the Pacific Ocean in the 1950s. The chlorine-36 traveled in global atmospheric circulation to the Nevada area, where it was precipitated down to the surface and then traveled through the rock to the repository level in less than fifty years. Because these data are so crucial to the DOE's models, the discovery of bomb-pulse chlorine-36 prompted a number of validation studies to find out whether the Los Alamos results were correct. These studies are under way.

The problem is not so much how fast water can reach the repository as how much water can reach the repository rapidly. The DOE's modeling studies claim that only about 1 percent of percolation flux occurs through fast pathways, so that very little water reaches the repository level (CRWMS M&O 2000). These modeling results depend on a number of assumptions, including that only a few instances of high chlorine-36 levels exist, that no large catchment area is associated with the fast paths, that the high chlorine-36 values are not found in perched water bodies, and that tritium (which like chlorine-36, serves as an indicator of bomb-pulses) is associated with only one high-chlorine-36 sample (CRWMS M&O 2000). Many of these assumptions are subject to question (see chapters 11 and 24, this volume).

The uncertainties about the percolation flux do not end with the proportion of water that flows through fractures but also depend heavily on future climate change. Long and Ewing (2004) point out that because permeability increases nonlinearly with saturation, a relatively small increase in percolation flux caused by climate change (see chapter 9, this volume) could greatly increase the volume of fluid reaching the repository. As the previous discussion suggests, increasing scientific study led to more, not less, uncertainty in the DOE's understanding of how much water would contact the waste package over the repository's lifetime.

Metlay (2000) argued that the changes in the value of percolation flux from the early 1980s to the late 1990s were largely a result of political, psychological, bureaucratic, economic, and regulatory influences. I would contend that the inverse also happened: increasingly complex science resulted in change to policy.

**The Shift to Performance Assessment**

The policy that established how the Yucca Mountain site was to be evaluated was itself a product of multiple factors. These included technical progress, growing awareness of the complex geology of Yucca Mountain, and rivalries between different technical groups—engineers and geoscientists. The evaluation process shifted from ensuring that the site met a set of established criteria to using a predictive computer simulation of future repository behavior. The foundations for this shift were laid in the late 1980s.

In 1987, the First Circuit Court of Appeals in Boston found that the Environmental Protection Agency's (EPA) standards on nuclear waste disposal in a geologic repository, promulgated in 1985, violated the Safe Drinking Water Act and required the EPA to rewrite the standards (Peterson 1987). In 1992, Congress passed the Energy Policy Act, which required the EPA to revise its standards according to advice provided by the National Academy of Sciences. By 1993, the EPA had announced that its new standard would include a ten-thousand-year limit on compliance, in part because modeling of the repository system had improved enough to provide confidence over such a time period (Inside Energy 1993).

This change in the EPA standards obliged the NRC to modify its regulations (10 CFR 63 replaced 10 CFR 60) and the DOE to alter its guidelines (10 CFR 963 replaced 10 CFR 960). In so doing, these agencies moved from a process that evaluated a nuclear waste repository site by looking for favorable or unfavorable conditions and by using system requirements to a process that evaluated the site purely on the basis of the results from a "quantitative" probabilistic performance assessment. The move to sole reliance on performance assessment modeling marked a major shift in the nuclear waste policy process, and it occurred for a combination of reasons.

First, there were numerous objections to the original standards and regulations. The DOE and the NRC complained that the EPA's general standard (which applied to all repositories, not just Yucca Mountain, as the current standards do) "might be impossible or impractical to meet" (Holt 1996). Scientists, in turn, complained about the NRC's regulations. For example, Isaac Winograd, one of the geologists who first suggested using the unsaturated zone for a geologic repository, stated that "the assumption that precise 10,000 year 'predictive' modeling is an essential requirement for determination of the 'performance' of Yucca Mountain, or any other site, is an outgrowth of EPA and U.S. Nuclear Regulatory Commission (NRC) regulations . . . which set discrete upper and lower limits for radionuclide release rates, groundwater travel times, and waste container lifetimes" (1990, p. 1291).

The National Academy of Sciences also criticized the NRC regulations. In its advice to the EPA (as required by the 1992 Energy Policy Act), the academy said:

Because it is the performance of the total system in light of the risk-based standard that is crucial, imposing subsystem performance requirements might result in a suboptimal repository design. Care should be taken to ensure that any subsystem requirements for Yucca Mountain do not foreclose design options that ensure the best long-term repository performance.

For example, in 10 CFR 60, there is a subsystem requirement that "the geologic repository shall be located so that the preemplacement ground water travel time along the fastest path of likely radionuclide travel from the disturbed zone to the accessible environment shall be at least 1,000 years." (National Research Council 1995, pp. 125–126)

The solution to the problem posed by the EPA standards and the NRC regulations was seen by many, including the National Academy of Sciences, as probabilistic performance assessment. The shift to performance assessment may have resolved some problems posed by the original EPA standards, but it created new ones. No longer was site selection held hostage to specific standards, which as research on the site continued and the data became more complex, became more difficult to meet. Performance assessment could better deal with complex data by assigning a weighting to all processes and variables involved in repository performance. As a result, factors that may previously have disqualified the site were now subsumed into a larger model, allowing what were thought to be more significant features and processes to guide the model results.

The idea of using performance assessments to evaluate nuclear waste disposal options was not new; it had been suggested by engineers associated with the issue in the United States since the 1970s. The use of performance assessments for waste disposal was adapted from predictions of nuclear reactor behavior, in which entirely engineered systems operate for three to four decades. Clearly, waste disposal presents a different set of circumstances: geologic repositories are natural systems with engineered features that must function for tens of thousands to millions of years.

Ewing and colleagues (1999) point out that in the 1970s, geologists from the U.S. Geological Survey and the Energy Research and Development Administration questioned the ability of models to predict the performance of geologic repositories. Since then, geologists have continued to warn about the limitations of using models to make geologic predictions (Hodges 1992; Konikow and Bredehoeft 1992; Oreskes et al. 1994; Ewing 1999; Oreskes and Belitz 2001; Bredehoeft 2003), largely because they understood geology to be a science that explains the past but does not predict the future. More important, based on their experience and the theories that guide their work, geologists know that geologic systems are thermodynamically open systems where all the features, events, and processes are neither knowable nor predictable (see chapter 24, this volume). In 1978, for example, U.S. Geological Survey geologists, in discussing the utility of models, said that "some components of the models are inherently unpredictable at present and are likely to change at different times. In no sense, therefore, will these models give a single answer to the question of the fate of radioactive waste in geologic repositories" (Bredehoeft et al. 1978, p. 12).

Another U.S. Geological Survey geologist echoed that sentiment in 1986, stating: "Such predictions—to be generated by complex mathematical models that synthesize knowledge from numerous disciplines are of necessity tenuous because a data base with which to calibrate such models does not now exist, nor is it likely to exist prior to the filling and sealing of the first waste repository" (Winograd 1986, p. 1).

Clearly, the scientists best in a position to judge were unconvinced that mathematical models should be the deciding factor on the suitability of a repository site.

The distinction between the geologists' caution in using models to predict repository performance and the engineers' confidence in this methodology is further emphasized by the following passage from a history of performance assessments: "As discussed here, the method that was conceived and accepted by *the engineering community* in the United States, and by the EPA and NRC as regulators for evaluating the acceptability of a disposal system, was a probabilistic PA [performance assessment]" (Rechard 1999, p. 778–779, emphasis added). Note that Rechard does not mention the geologic community, perhaps indicating that those scientists were not considered participants in the decision-making process.

The engineers won the debate over how to evaluate repository sites: performance assessment became the method accepted by the regulatory agencies. According to the NRC and others, the reason that performance assessment was not used earlier was that it was not well-developed enough to apply to a geologic repository (NRC 1999; Rechard 1999). Probability assessments were first applied to repositories in the late 1980s in the case of the Waste Isolation Pilot Project (Rechard 1999)—the Carlsbad, New Mexico, salt repository that receives transuranic wastes from the nuclear weapons complex.

One technical implication of this shift in the way Yucca Mountain was evaluated was that the performance assessment began to rely more on the performance of the engineered barriers than on the ability of the natural system to prevent radionuclide transport to the biosphere. Metlay (2000) identified this shift from the DOE's 1992 performance assessment model to the 1998 TSPA Viability Assessment model. In a 1999 presentation to the Nuclear Waste Technical Review Board, a DOE representative showed graphs indicating that the waste package was the main contributor to repository performance (OCRWM 1999). More recent DOE documents (Bechtel SAIC 2001) continue to support this new weighting, which shows that nearby residents of Yucca Mountain would receive a five hundred-millirem radiation dose a mere two thousand years after the repository closure. When the waste package (including the canister and spent fuel cladding) is taken into account, doses of that magnitude would be reached only after two hundred thousand years.

Performance assessment is a complicated set of models in which repository behavior over thousands to millions of years is evaluated based on the current understanding of multiple processes and parameters. As such, the weight given to factors can be varied and even used to favor one process over another. The Nuclear Waste Technical Review Board went as far as to accuse the DOE of intending to use performance assessment to cover up problems in its analysis: "The Board [NWTRB] believes that total system performance assessment should not be used to dismiss these corrosion concerns" (2003, p. 2).

What prompted this shift from geology to engineering in performance assessment weighting? Perhaps the results of performance assessment modeling are correct and the engineered features of the repository deserve the weighting they receive. An alternative explanation is that as the DOE studied the natural system in detail, it appeared more complex than expected and therefore the DOE had less control over predicting its behavior; by contrast, the DOE can to some degree prescribe and control the engineered barriers. An example of the apparently increasing complexity of the geologic barriers was the discovery of the elevated chlorine-36 values and the ensuing controversy—a finding that obscured the technical analysis of the natural barriers. I suspect that technical staff and managers at the DOE and the NRC put more trust in the engineered barriers—features, events, and processes that they can control by design—than in the natural barriers, which are less knowable, more dynamic, and thus less predictable.

The problem with such thinking is that the engineered barriers must perform in a natural environment over geologic time periods so that in a sense, *the engineered barriers become part of the natural system*, and therefore are just as difficult to know and predict. Thus, the performance of the engineered barriers depends, foremost, on the geologic environment near the waste canisters. The evolution and behavior of the environment near the waste packages, however, is not easily predictable (see chapters 12, 13, 15–22, this volume). To overcome these uncertainties, the DOE will have to make bounding assumptions about this environment, but the question remains as to whether the DOE can do this with enough certainty to make accurate predictions.

## Conclusions

What does this history and analysis of the policy and technical evolution of U.S. nuclear waste disposal tell us? The initial selection of the unsaturated zone at Yucca Mountain was driven by political issues as well as technical ones: the location itself was for practical reasons desirable, and the DOE managers grew convinced of the technical arguments to use the unsaturated zone. But the decision to focus on a dry site brought with it a set of implications absent from the saturated zone: the need to develop an understanding of the hydrology of the unsaturated zone, an entirely new field of science.

A second consequence of the selection of a dry site was the focus on water transport and the ensuing complexification of an already little-understood area of science. Finally, there was the decision to rely almost exclusively on probabilistic performance assessment to determine whether Yucca Mountain was suitable as a geologic repository, made in part as a response to the complexity of the geology.

Because the science is so new, it would be difficult to determine the data needed on the features, events, and processes to run the performance assessment models with a high degree of certainty. This situation contributed to long delays in the evaluation of the repository. At the same time, the increased uncertainty associated with natural system behavior resulted in a dependence on the engineered barriers to carry the weight of the evaluation on the site. The dependence almost entirely on engineered features removes part of the defense-in-depth strategy inherent in geologic repositories: the multiple barrier concept says that if one barrier fails (engineered or natural), then the other is there to back it up. In the Yucca Mountain case, the DOE is essentially saying—before the repository even begins operation—that it does not trust the geologic barrier. Is this the way to establish a repository?

Perhaps the answer lies in the geology and hydrology of the Yucca Mountain site. With less emphasis on the results of a performance assessment model, and more emphasis on geoscientists' understanding of the natural features and processes at the site and elsewhere, perhaps we can come to a more reliable understanding of Yucca Mountain's ability to perform as a repository. Such an understanding will take time, though, and will never be without uncertainty. That is perhaps a more honest evaluation than presenting a number that meets the standard but may hold little reliability over the long run.

## References

Abraham, Spencer (2002) Letter to the President. February 14, 2002. http://www.ocrwm.gov/ymp/sr/salp.pdf.

Bechtel SAIC (2001) *FY01 Supplemental Science and Performance Analyses, Volume 1: Scientific Bases and Analyses*. Department of Energy. TDR–MGR–MD–000007 REV 00 ICN 01.

Bredehoeft, J.D. (2003) From Models to Performance Assessment: The Conceptualization Problem. *Ground Water* 41, pp. 571–577.

Bredehoeft, J.D., England, A.W., Stewart, D.B., Trask, N.J., and Winograd, I.J. (1978) Geologic Disposal of High-Level Radioactive Wastes: Earth-Science Perspectives. *U.S. Geological Survey Circular 779*, p. 12.

Colglazier, E.W., and Langum, R.B. (1988) Policy Conflicts in the Process for Siting Nuclear Waste Repositories. *Annual Review of Energy* 13, pp. 317–357.

Cragnolino, G., Dunn, D.S., Brossia, C.S., Pan, Y.-M., Pensado, O., and Yang, L. (2004) Corrosion Behavior of Waste Package and Drip Shield Materials. *Nuclear Technology* 148, pp. 166–173.

CRWMS M&O (2000) *Unsaturated Zone Flow and Transport Model Process Model Report.* Department of Energy. TDR–NBS–HS–000002 REV 00 ICN 02. August.

Department of Energy (2000) *United States Nuclear Tests, July 1945 through September 1992.* Nevada Operations Office. DOE/NV–209–REV 15. December.

Dixon, G.L., and Glanzman, V.M. (1982) *Search for Potential Sites.* U.S. Geological Survey. 82–509. May.

Dixon, G.L., and Hoover, D. (1982) *U.S. Geological Survey Research in Radioactive Waste Disposal: Fiscal Year 1979 Circular.* U.S. Geological Survey. 847.

Ewing, R.C. (1999) Less Geology in the Geological Disposal of Nuclear Waste. *Science* 286, pp. 415–417.

Ewing, R.C., Tierney, M.S., Konikow, L.F., and Rechard, R.P. (1999) Performance Assessments of Nuclear Waste Repositories: A Dialogue on Their Value and Limitations. *Risk Analysis* 19, pp. 933–958.

Fabryka-Martin, J., Wrightman, S.J., Robinson, B.A., and Vestal, E.W. (1994) *Infiltration Processes at Yucca Mountain Inferred from Chloride and Chlorine-36 Distributions.* Los Alamos Laboratory Milestone Report. 4317.

Flint, A., Flint, L., Bodvarsson, G., Kwicklis, E., and Fabryka-Martin, J. (2001a) Development of the Conceptual Model of Unsaturated Zone Hydrology at Yucca Mountain, Nevada. In *Conceptual Models of Flow and Transport in the Fractured Vadose Zone*, ed. National Research Council. Washington, DC: National Academy Press. pp. 49–86.

Flint, A., Flint, L., Bodvarsson, G., Kwicklis, E., and Fabryka-Martin, J. (2001b) Evolution of the Conceptual Model of Unsaturated Zone Hydrology at Yucca Mountain. *Journal of Hydrology* 247, pp. 1–30.

Herrick, C., and Sarewitz, D. (2000) Ex Post Evaluation: A More Effective Role for Scientific Assessments in Environmental Policy. *Science, Technology, and Human Values* 25, pp. 309–331.

Hodges, K.V. (1992) General Tectonics. In *Report of the Peer Review Panel on the Early Site Suitability Evaluation of the Potential Repository Site at Yucca Mountain, Nevada*, ed. SAIC. Las Vegas: Office of Civilian and Radioactive Waste Management, pp. 347–400.

Holt, M. (1996) *Civilian Nuclear Waste Disposal.* Congressional Research Service. 92059. November 21.

Inside Energy (1993) Administration Releases EPA Rule on N-Waste Exposure Standards. *Inside Energy/with Federal Lands.* February 8, p. 2.

Janeczek, J., Ewing, R.C., Oversby, V.M., and Werme, L.O. (1996) Uraninite and $UO_2$ in Spent Nuclear Fuel: A Comparison. *Journal of Nuclear Materials* 238, pp. 121–130.

Konikow, L., and Bredehoeft, J.D. (1992) Ground-water Models Cannot Be Validated. *Advances in Water Resources* 15, pp. 75–83.

Long, J.C.S., and Ewing, R.C. (2004) Yucca Mountain: Earth-Science Issues at a Geologic Repository for High-Level Nuclear Waste. *Annual Review of Earth and Planetary Science* 32, pp. 363–401.

Metlay, D. (2000) From Tin Roof to Torn Wet Blanket: Predicting and Observing Groundwater Movement at a Proposed Nuclear Waste Site. In *Prediction: Science, Decision Making, and the Future of Nature*, ed. Sarewitz, D., Roger Pielke, J., and Radford Byerly, J. Washington, DC: Island Press. pp. 199–228.

Nuclear Information and Resource Service (2002) *DOE Recommends Yucca Mountain for Permanent High-Level Dumpsite*. Press release, January 10, 2002. http://www.nirs.org/press/01-10-2002/1.

National Research Council (1995) *Technical Bases for Yucca Mountain Standards*. Washington, DC: National Academy Press, 222 p.

National Research Council (2001) *Conceptual Models of Flow and Transport in the Vadose Zone*. Washington, DC: National Academy Press, 392 p.

Nuclear Regulatory Commission (1999) Disposal of High-Level Radioactive Wastes in a Proposed Geologic Repository at Yucca Mountain, Nevada. *Federal Register* 64, pp. 8640–8679.

Nuclear Waste Technical Review Board (2003) *An Evaluation of Key Elements in the U.S. Department of Energy's Proposed System for Isolating and Containing Radioactive Waste*. November.

Office of Civilian Radioactive Waste Management (1998) *Viability Assessment of a Repository at Yucca Mountain*. Department of Energy. DOE/RW–0508. December.

Office of Civilian Radioactive Waste Management (1999) Post-closure Defense in Depth in the Design Selection Process, NWTRB Repository Panel meeting, Washington, DC, January 25, 1999.

Office of Civilian Radioactive Waste Management (2001) *Yucca Mountain Science and Engineering Report*. Department of Energy. DOE/RW–0539. May.

Oreskes, N., Shrader-Frechette, K., and Belitz, K. (1994) Verification, Validation, and Confirmation of Numerical Models in the Earth Sciences. *Science* 263, pp. 641–646.

Oreskes, N., and Belitz, K. (2001) Philosophical Issues in Model Assessment. In *Model Validation: Perspectives in Hydrological Science*, ed. Anderson, M.G., and Bates, P.D. New York: John Wiley and Sons, pp. 23–41.

Peterson, C. (1987) EPA's Nuclear Disposal Rules Voided. *Washington Post*, July 27, p. A19.

Rechard, R. (1999) Historical Relationship between Performance Assessment for Radioactive Waste Disposal and other Types of Risk Assessment. *Risk Analysis* 19, pp. 763–807.

Shoesmith, D.W. (2000) Fuel Corrosion Processes under Waste Disposal Conditions. *Journal of Nuclear Materials* 282, pp. 1–31.

Sonner, S. (2000) GOP Delegates Say Nuclear Waste Is a Hot Issue in Bush's Bid for Nevada. Philadelphia: Associated Press State and Local Wire. August 1.

Stuckless, J. (2003) The Road to Yucca Mountain: A Brief History. *Geological Society of America Abstracts with Programs* 35, p. 353.

Winograd, I.J. (1972) Near Surface Storage of Solidified High-Level Radioactive Waste in Thick (400–2,000 Foot) Unsaturated Zones in the Southwest. *Geological Society of America Abstracts with Programs* 4, p. 708.

Winograd, I.J. (1974) Radioactive Waste Storage in the Arid Zone. *EOS* 55, pp. 884–894.

Winograd, I.J. (1986) Archaeology and Public Perception of a Transscientific Problem: Disposal of Toxic Wastes in the Unsaturated Zone. *U.S. Geological Survey Circular* 990, p. 1.

Winograd, I.J. (1990) The Yucca Mountain Project: Another Perspective. *Environmental Science and Technology* 24, pp. 1291–1293.

# Part II
## Science and Technology

# I
# Earth Science

# 7

# The Mountain Matters

David Applegate

The advantages of Yucca Mountain as a nuclear waste disposal site are clear enough. Its remote desert location is federally owned land adjacent to the Nevada Test Site, with its legacy of nuclear testing. What little water falls on the arid mountain mostly evaporates or is absorbed by the sparse vegetation rather than percolating down into the rocks below. The water table is one of the deepest in the world, making it possible to build a repository hundreds of meters below the surface that is still hundreds more meters above the aquifer where water flows from the mountain toward areas of human habitation. The region is sparsely populated. Potential exposure for distant populations is limited by the region's internal drainage, whereby neither the surface water nor the groundwater flows into a major river system. Placement in the unsaturated, or vadose, zone also may make it easier to monitor and—were it to become necessary—retrieve the waste. The rocks also contribute to waste isolation: abundant minerals in the rock layers beneath the repository horizon, known as zeolites, can absorb certain radionuclides, thus slowing the progress of waste away from the repository. These assets were some of the major selling points that led the federal government to select the site for the world's first high-level nuclear waste repository (well summarized in Groat 2001).

But the site also has drawbacks, particularly when considering that the repository is being judged on its ability to minimize human exposure over the next ten thousand years or more. Fractured rock provides fast pathways that can carry water quickly from the surface to the repository and on down to the water table. Although zeolites may absorb some radionuclides, others may attach to tiny particles suspended in the groundwater, accelerating their transport through the subsurface. The mountain sits quietly today in splendid isolation, but it has experienced recent earthquakes and is located in the southwestern Nevada volcanic field. Changes in climate are likely to increase the amount of moisture that percolates down from the surface. And a maze of faults and fractures beneath the mountain makes it difficult to model flow pathways.

This chapter provides an earth science perspective on the challenge of storing nuclear waste over a geologically significant period of time—a repository for the ages.

Six aspects of Yucca Mountain and its environs pertain to the site's suitability. First, the rocks that make up the mountain facilitate or impede the flow of water above the repository as well as the flow of water and mobilized radionuclide particles below the repository. Second, future volcanic activity could release radionuclides to the surface or accelerate their transport to human-accessible groundwater. Third, earthquakes could both weaken the repository's engineered barriers and alter the water flow pathways through the mountain. Fourth, changes in the surface environment due to shifting climatic conditions and erosion rates could increase the infiltration of water into the mountain, thus reducing the repository's isolation. Fifth, the presence of natural resources may encourage future human intrusion. The sixth and final issue is just how much of the mountain the repository will come into contact with—that is, How big a footprint will it have?

During site characterization, the first of these six issues generated the most attention because of the importance of understanding controls on fluid flow through the mountain. During the current licensing phase, the low probability but high impact of volcanic activity is likely to play an increasingly significant role in light of recent concerns raised by scientists working for the Nuclear Regulatory Commission (NRC), which oversees licensing.

The one issue that could be considered a "sleeper" is the last of the six: the size of the repository. Most calculations regarding the repository's suitability assume a footprint that seems likely to be a minimum. As larger areas are considered, additional challenges are likely to arise.

A tremendous amount of effort has gone into understanding what is under this desert ridge and how it has evolved over time. Scientists with the U.S. Geological Survey first identified Yucca Mountain as a potential repository location in the 1970s, initially for a repository beneath the water table and later for one in the thick unsaturated zone above it (see Hanks et al. 1999). Since the Department of Energy (DOE) began site characterization in the mid-1980s, scientists working for the U.S. Geological Survey, the DOE national laboratories, and various contractors have mapped the surface in detail. They have examined the geology and geochemistry of the subsurface along more than eight miles (thirteen kilometers) of trenches and tunnels, and in dozens of borehole cores. Through a variety of geophysical techniques, they have also amassed a mountain of data and generated scores of reports (DOE 2001). The state of Nevada and the NRC have contracted their own studies. Even before scientists began their work related to Yucca Mountain, extensive geologic research was conducted at the adjacent Nevada Test Site in conjunction with underground nuclear tests. Moreover, the region surrounding Yucca Mountain has for many years been of great interest to geoscientists seeking to understand fundamental geologic processes and the evolution of western North America.

And they still have a great deal to learn.

## A Mountain Born of Fire

Rocks in this region record more than one billion years of mountains forming, eroding, and reforming. The modern landscape of Yucca Mountain is largely the result of three geologic processes: volcanism, crustal extension, and climate-driven erosion.

The rocks that form Yucca Mountain were produced during a period of explosive volcanism that began fifteen million years ago. The collapsed remains of several large volcanic centers sit thirty kilometers to the north of the repository. Producing eruptions similar to those that formed the stygian landscape of Yellowstone National Park in Wyoming, these calderas ejected vast quantities of fine-grained ash and coarser rocky-glassy material that blanketed the surrounding area to a depth of nearly two kilometers. By comparison, the 1980 eruption of Mt. St. Helens deposited only a few inches of unconsolidated ash over the adjacent region.

The layers that form the topographic expression of Yucca Mountain were produced in eruptions 12.7 to 12.8 million years ago and then subsequently tilted so that they dip gently to the east (Sawyer et al. 1994). If the temperature of volcanic ash is hot enough when it lands, the ash particles will fuse together as they are compacted. The resulting rock, called a welded tuff, is hard and brittle with few pore spaces. Ash that has cooled by the time it lands or is less compacted will form nonwelded tuffs that have a higher porosity and thus are less dense. They are also softer and less brittle than their welded counterparts, and so have less of a tendency to fracture. Because water preferentially flows through fractures in welded tuff and through the porous matrix in nonwelded tuffs, the different properties of these types of tuff have important implications for waste transport. These thick deposits of ash and coarser material are referred to collectively as the Paintbrush Group.

## Rocks above the Repository: Limiting Water Flow

Within the Paintbrush Group, the uppermost unit is the Tiva Canyon welded tuff, roughly 100-meters thick at Yucca Mountain (figure 7.1). Below that is the 30- to 50-meter-thick Paintbrush nonwelded tuff, and below that is the Topopah Spring welded tuff that has a maximum thickness of 375 meters. The repository would be excavated in the Topopah Spring unit, roughly 300 meters below the surface. The rocks above the repository have a simple role to play in site performance: limit the amount of water that reaches the waste emplacement drifts. Under the current conditions, surface processes—runoff, evaporation, and plant transpiration—remove 95 percent of the 19 centimeters of precipitation that falls on the mountain annually (DOE 2001). During wetter climate periods, surface processes will still play the dominant role in limiting the amount of water that reaches the repository. But some water reaches the subsurface and percolates down through the tuff layers. Thus, one of the biggest technical challenges facing the project is to adequately model the flow of water through the unsaturated zone.

For much of the Yucca Mountain Project's history, the DOE assumed that water percolates down slowly through interconnected pore space between mineral grains in the tuffs (Metlay 2000). Such flow would take thousands to tens of thousands of years to reach the repository. That assumption was challenged, however, after water samples from the newly constructed Exploratory Studies Facility in 1996 contained traces of radionuclides produced by atmospheric nuclear tests in the 1950s. This bomb-pulse signature indicated that at least some water was flowing through fractures and faults to reach the repository level from the surface in less than fifty years (Flint et al. 2001). These fractures formed when the ash first cooled and compacted, and then later in response to external stresses acting on the brittle rocks; faults are fractures in which one side has moved relative to the other.

The DOE's models now recognize the dominant role of fracture flow, particularly for the welded units (Flint et al. 2001). Even in the nonwelded Paintbrush tuff, where most of the water does appear to move down through pores, throughgoing fractures allow some water to bypass the slower matrix-flow regime. But is fracture flow a liability or an asset? Critics view fast pathways as a fundamental flaw of the site and a principal reason why the DOE has shifted to a design that relies almost entirely on engineered barriers to meet licensing criteria (Nevada Agency for Nuclear Projects 1998). Large uncertainties remain as to our understanding of how water flows in fractures (see chapter 14, this volume).

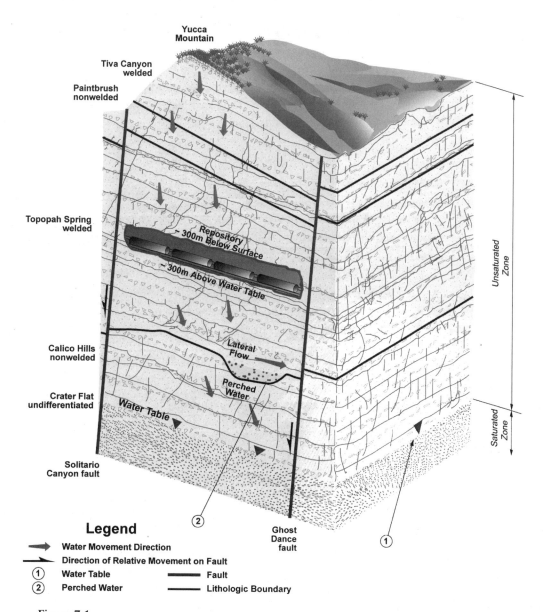

Figure 7.1
Block diagram of repository location and unsaturated zone processes. From DOE 2002, figure 3.23.

The DOE does not see a fundamental flaw; indeed, far from it. The executive summary of the department's massive *Yucca Mountain Science and Engineering Report* states: "The potential repository has been designed to take advantage of the free-draining nature of the repository host rock, which would promote the flow of water past the emplaced waste and limit the amount of water available to contact the waste packages" (DOE 2001). As envisioned by the DOE, if the repository is designed with closely spaced waste containers, the heat from the radioactive waste would keep the walls of the emplacement drifts above the boiling point of water for several thousand years. Down-flowing water would divert around the resulting dry zone into fractures in cooler zones between the drifts (DOE 2001).

In a low-temperature design, or after the dry zone period of the high-temperature design, water will reach the emplacement drifts. But even then, the DOE argues that capillary effects for fractures that intersect an underground opening will limit the flow of water into the drifts. In analog settings such as Spirit Cave in northern Nevada, such a capillary barrier has kept caves dry for thousands of years (DOE 2001). The DOE's view is reinforced by the U.S. Geological Survey, which considers one of the site's natural assets to be "free drainage, through fractures on the floor of the drifts, of any water that enters the tunnels," thus avoiding even transitory flooding of the tunnels (Groat 2001).

### Rocks below the Repository: Retarding Waste Movement

The Topopah Spring welded tuff, which would contain the waste repository, overlies the Calico Hills Formation, which was deposited from another nearby caldera eruption 12.9 million years ago. The Calico Hills consists of mostly nonwelded tuffs and lava flows. On the eastern side of Yucca Mountain, the water table is located within the Calico Hills Formation, roughly six hundred meters below the surface (figure 7.1). Below the water table, all pore spaces and fractures are filled. In contrast to the principally vertical flow direction in the unsaturated zone, flow within the saturated zone is dominantly lateral toward the south. This flow allows material to be transported beyond the repository's boundary.

The rocks below the repository are supposed to help contain its radionuclides. Slowing radionuclide movement allows additional time for radioactive decay, reducing the total amount of radioactive material that will reach the human environment (National Research Council 1995). Because the Calico Hills Formation is mostly nonwelded tuff, travel times through its matrix should be relatively slow. The most notable feature of the Calico Hills tuffs is an abundance of zeolite minerals, which form as a result of the hydrothermal alteration of volcanic glass. Found only in layers that have been below the water table, these minerals are capable of adsorbing significant amounts of soluble radionuclides, thus slowing their release from the mountain (DOE 2001).

The effectiveness of the zeolites, however, is limited on three fronts. As described in chapter 13 (this volume), zeolites are better able to adsorb those radionuclides with a short half-life (Cs-137 and Sr-90) rather than the long-lived radionuclides that pose the greatest long-term risk. Moreover, several of the long-lived radionuclides tend to attach themselves onto tiny particles, known as colloids, suspended in the groundwater (see chapter 12). These colloids may facilitate the transport of radionuclides away from the repository. Finally, the ability of zeolites to retard radionuclides is also compromised by the presence of fast pathways to the water table in the form of throughgoing fractures, and particularly, along the major faults that run through and adjacent to the repository (DOE 2001). Recognizing these limitations, the National Research Council (1995) concluded that radionuclides should be viewed conservatively as a nonsorbing solute, neither taking credit for zeolites or considering the liability of colloid transport.

As much as one thousand meters of additional ash deposits (Crater Flat Group) underlie the Calico Hills Formation. These tuffs are mostly nonwelded and, like much of the Calico Hills, have been altered to zeolites (DOE 2001). They are all products of the same period of explosive volcanism that began fifteen million years ago. The Crater Flat volcanic rocks, however, sit on much older sedimentary rocks that formed during the Paleozoic era (570 to 250 million years ago) and were folded and faulted by mountain-building events in the Mesozoic era (250 to 65 million years ago).

Limestones within this Paleozoic sequence form a deep regional aquifer. This aquifer has a higher hydraulic head than the tuff aquifer, preventing downward flow. Beneath the Amargosa Valley, however, the two aquifers merge downgradient to the south, where the ash deposits thin out and are replaced by unconsolidated sediments eroded from the ranges (DOE 2001). These Paleozoic rocks may be important to site suitability because the water table appears to rise steeply to the north of Yucca Mountain (upgradient), perhaps as high as the repository horizon. Unfortunately, very few boreholes exist in this region to provide enough data to determine whether the gradient is real or represents perched water. As far back as 1992, the National Research Council (1992) argued that a better understanding of that aquifer and the steep gradient was necessary, but the gradient's origin and implications for the repository remain unresolved.

## Potential for Volcanic Disruption

Yucca Mountain is a region with a long history of volcanism. Predicting where the next volcanic eruption will take place is one of the key geologic challenges facing the repository.

Large-scale, explosive volcanism continued to deposit ash over the region until 7.5 million years ago. The explosivity was due to the high silica content in the molten rock, or magma, which made it very viscous. Because it was difficult for gas bubbles to escape the viscous magma, pressure built up and resulted in explosive release. Starting 9 million years ago, another type of volcanism became active. This lower-silica (or basaltic) magma is less explosive, tending to flow out as lava or erupting coarse material to form cinder cones.

The most recent period of basaltic volcanism produced a series of vents and flows in Crater Flat, the valley that lies on the western edge of Yucca Mountain. Four cinder cones are located in Crater Flat, the closest of them less than ten kilometers from Yucca Mountain. They range in age from one million to seven hundred thousand years old (DOE 2001). The most recent volcanic deposit is a cinder cone located sixteen kilometers south of Yucca Mountain at Lathrop Wells. Estimates for the age of this cinder cone range over more than an order of magnitude. But the most widely accepted calculation, based on radiometric dating, places the cone's origin at approximately seventy-five thousand years ago (Heizler et al. 1999).

The young volcanoes in Crater Flat and at Lathrop Wells produced small volumes of lava and other volcanic material, in each case less than one cubic kilometer. Consequently, the DOE refers to the area as one of the "least active basaltic volcanic fields in the western United States," noting that 99.9 percent of the volume of volcanic deposits were produced between 15 and 7.5 million years ago with the remaining 0.1 percent occurring since that time (DOE 2001). But critics point out that it does not take much magmatic material in the right (or wrong) place to cause a disruption of the repository.

In assessing potential threats to the repository, the DOE only considers "disruptive events" if their likelihood exceeds a one in ten thousand probability over the ten-thousand-year compliance period—a chance of one in one hundred million per year. By the DOE's own estimation, the likelihood of a volcanic intrusion was slightly greater than this cutoff and hence needed to be considered. (The risk of a meteorite impact, by contrast, did not make the cut.) This probability was reached through a process of expert elicitation, asking ten volcanologists to evaluate the probability of a repository-piercing eruption at Yucca Mountain (Kerr 1996).

The DOE is considering two volcanic scenarios, both based on a 1.5-meter-thick vertical dike of magma breaching the repository from below. In the first scenario, the dike subsequently reaches the surface, erupting waste-contaminated magma as lava flows and airborne ash. In the second, the dike does not reach the surface but provides a pathway for radionuclide release into the water table (DOE 2001). The DOE's probability-weighted Total System Performance Assessment (TSPA) analysis indicates that neither scenario generates significant human exposure. Yet the DOE's regulators are questioning the assumptions behind the TSPA results.

Volcanism has become a key issue for the repository licensing largely due to issues raised by scientists working for the NRC who argue that the DOE has under-estimated the risk. Those scientists put the risk closer to one in ten million, several times greater than the DOE's estimate (Connor et al. 2000). The higher probability comes from an assumption that the dike would follow preexisting faults and fractures; the DOE, by contrast, assumed that the dike could appear anywhere with equal probability.

The NRC scientists also assert that the DOE is underestimating the fallout that such an eruption would produce as well as the extent to which the magma would disrupt additional waste emplacement drifts (Macilwain 2001). Their models suggest that magma in a dike that intersects the repository drifts will undergo explosive decompression as it encounters the open tunnel (Woods et al. 2002). The resulting shock wave would damage the waste containers. Damage would also result from the intense heat and acidic gases released as magma quickly filled a significant number of drifts. All told, the shock, heat, and corrosive effects of the acidic fluids would compromise many more containers than envisioned in the DOE scenarios.

A separate group of researchers has questioned the DOE's probabilistic estimation of the frequency of volcanic events, contending that a zone of hot material beneath the region may indicate higher recurrence rates (Smith et al. 2002). Clearly, this will be an area of intense scrutiny as the NRC considers the site's license application.

## Potential for Seismic Disruption

Yucca Mountain is located in the central part of a geologically unique region known as the Basin and Range, a broad swath of the western United States and Mexico where Earth's crust has been pulled apart for millions of years. This crustal extension has produced a series of tilted blocks—often compared to dominoes or a row of leaning books on a shelf—typically separated by west-dipping faults. The result is a landscape of north-trending mountain ranges separated by flat valleys. The upthrown side of the blocks are the ranges and the downthrown side are the valley basins, which fill up with sediments eroded from the ranges.

Projecting the repository's future seismic hazard requires an estimate of how much the crust around Yucca Mountain has moved in the past and how fast it is moving today. The more active the system, the more likely that significant earthquake activity will occur during the repository's lifetime. The DOE considers faults to have the potential to move in the future if they have moved in the last two million years, known as the Quaternary period (DOE 2001).

### Measuring Past Movement

Scientists from the U.S. Geological Survey have conducted extensive surveys of past earthquakes in the region. They not only study recorded earthquakes but also dig trenches across faults to look at the age and characteristics of surface deposits and soils affected by faulting, measuring the amount of movement on individual faults as well as the timing. From these studies, they can determine how often earthquakes occur on a given fault and the rate of fault movement. That information is used to generate a probabilistic assessment of the seismic hazard for the region.

Yucca Mountain is itself a fault-bounded crustal block with a steep west face formed by the Solitario Canyon fault (figure 7.2). Like most faults in the vicinity of Yucca Mountain, it experienced most of its movement more than ten million years ago, during the major episode of explosive volcanism (DOE 2001). Over the last million years, it appears to have moved on the order of a few meters (Fridrich et al. 1999). Just to the east of the Exploratory Study Facility's main drift is the steeply dipping Ghost Dance fault, which shows evidence of twenty-seven meters of extensional slip prior to two million years ago.

**Figure 7.2** (*opposite*)
Mapped faults at Yucca Mountain. From Department of Energy 2001, figure 1.14.

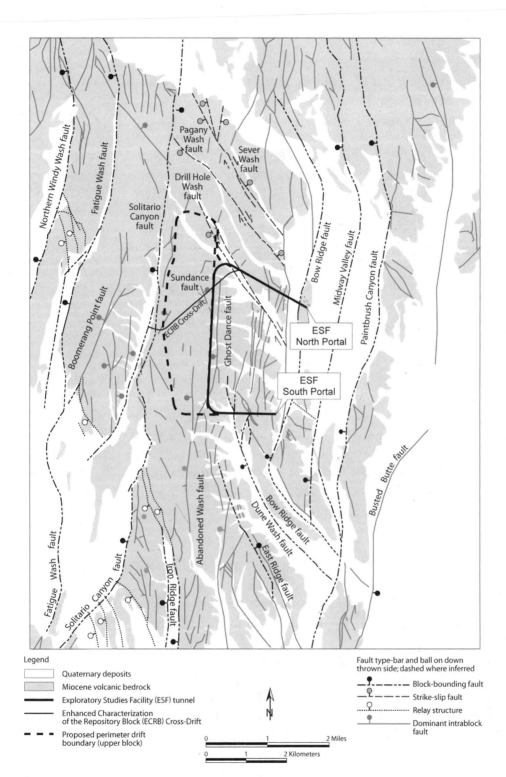

Quaternary deposits

Miocene volcanic bedrock

Exploratory Studies Facility (ESF) tunnel

Enhanced Characterization
of the Repository Block (ECRB) Cross-Drift

Proposed perimeter drift
boundary (upper block)

N

0        1        2 Miles

0        1        2 Kilometers

Fault type-bar and ball on down
thrown side; dashed where inferred

Block-bounding fault

Strike-slip fault

Relay structure

Dominant intrablock
fault

The only fault that intersects the primary repository block is the Sundance fault, which shows eleven meters of extensional slip—all of it occurring at least two million years ago (DOE 2001). A number of faults within ten kilometers of the repository show evidence of movement during the past two million years, with slip rates from one to fifty meters of displacement per million years (0.001 to 0.05 millimeters per year). The average recurrence intervals for these fault movements are between ten thousand and one hundred thousand years (Stepp et al. 2001). Such rates are at or below slip rates for other faults in the Basin and Range (Levich et al. 2000). For example, the Death Valley–Furnace Creek fault system, fifty kilometers to the southwest of Yucca Mountain, is moving at a rate of four to eight kilometers per million years (or four to eight millimeters a year; Stepp et al. 2001).

Recent fault displacements may be small, but the region remains active. Since 1868, there have been twenty-one earthquakes of magnitude 6 or greater within three hundred kilometers of Yucca Mountain (DOE 2001). The largest nearby earthquake struck on June 29, 1992, with a magnitude of 5.4. The quake was centered under Little Skull Mountain on the Nevada Test Site, less than twenty kilometers from the repository site. Although more than an order of magnitude smaller than the devastating Northridge earthquake that struck Los Angeles two years later, the Little Skull Mountain quake nevertheless caused minor structural damage to surface facilities at the repository site. Underground facilities had yet to be built at the site, but tunnels at Little Skull Mountain were unaffected (CRWMS M&O 2000). That's not surprising. Unless one is quite close to the epicenter of an earthquake, most of the shaking is due to seismic waves that travel along Earth's surface. Shaking is more intense in unconsolidated sediment than it is in solid rock. As a result, underground structures typically experience less shaking than ones on the surface.

Several systems of faults in the vicinity of Yucca Mountain are considered the most likely earthquake candidates. The active Bare Mountain fault, twenty kilometers from Yucca Mountain, shows two kilometers of offset, at least some of it in the Quaternary. Faults within or near the repository also show Quaternary offsets. Finally, there are the faults in Jackass Flats east of Yucca Mountain, where the Little Skull Mountain earthquake initiated (Wernicke et al. 1998). Although the Death Valley–Furnace Creek fault system is fifty kilometers away from Yucca Mountain, its large size makes it a major component of the shaking hazard; as a rule, the longer the fault, the greater the magnitude of the earthquake that it can produce. The Death Valley fault system could produce quakes with magnitude 7 or greater (Stepp et al. 2001).

Researchers at the University of Nevada at Reno have postulated that a strike-slip fault similar in size to the Death Valley system may run beneath the Amargosa Valley just to the south of Yucca Mountain (Schweikert and Lahren 1997). Where the fault is exposed to the south, it shows twenty-five kilometers of movement during the last thirteen million years. Such a large fault would be capable of producing large earthquakes, but neither its existence nor its level of activity has been confirmed, and the proposal is controversial. Critics argue that the deformation can be explained by existing normal faults seen at the surface (Stamatakos and Ferrill 1998). The probabilistic seismic hazard analysis prepared for the DOE concludes that "hypo-thesized buried strike-slip faults" are a negligible contribution to the overall seismic hazard facing the repository (Stepp et al. 2001).

**Measuring Current Movement**
In addition to looking at past fault movement, scientists can also study how Earth's crust is deforming at present. Such studies typically have been made through careful measurement over a number of years between fixed benchmarks on the surface. U.S. Geological Survey scientists established such a trilateration network around Yucca Mountain in 1983. The recent advent of the satellite-based Global Positioning System (GPS) has revolutionized these measurements. Taking measurements from a large number of the constellation of GPS satellites (rather than the two or three used in standard GPS navigation) makes it possible to determine the location of a fixed spot with millimeter accuracy.

During the 1990s, GPS networks were established around Yucca Mountain, augmenting the earlier geodetic network. The initial results came from a team of researchers at the California Institute of Technology who made GPS measurements from 1991 to 1997 at a series of five benchmarks along a thirty-four-kilometer-long transect that stretched on either side of the repository parallel to the regional nor-thwest-southeast extension direction. Their results indicated that the crust had moved 1.7 millimeters per year (Wernicke et al. 1998).

To compare numbers from networks of different lengths, the Caltech researchers calculated a strain rate, defined as the measured movement (in millimeters per year) divided by the length of the network (in kilometers). Thus, 1.7 millimeters per year over a thirty-four-kilometer network translates into fifty-one nanostrains per year. With uncertainty expressed at the 90 percent confidence level (one standard devia-tion), that number was reported as 51 ± 9 nanostrains per year, a rate four times the average rate across the Basin and Range, and nearly a quarter of the rate of the

San Andreas fault (Wernicke et al. 1998). Moreover, the rate is a full order of magnitude greater than the rates determined through paleoseismic studies.

Such a rate, if correct, would have significant implications for the earthquake hazard facing the repository. But the U.S. Geological Survey scientists who established the original trilateration network conducted their own GPS investigation between 1993 and 1998 over a fifty kilometers network. They obtained a much lower number: 23 ± 9 nanostrains per year (Savage et al. 2001). They reported that these strain rates were consistent with those obtained from the trilateration network and from rates inferred from the geologic record. When using the same network as in the Caltech study, they came up with only 38 ± 45 nanostrains per year. That estimate's significant uncertainty led the U.S. Geological Survey scientists to propose that the higher rate found by Caltech may have reflected transient crustal relaxation following the 1992 Little Skull Mountain earthquake (Savage et al. 2001). These geodetic networks will continue to operate throughout the licensing phase of the Yucca Mountain Project, and new data are likely to further revise these estimates and interpretations.

### What Is the Seismic Hazard?

Identifying potential earthquake sources and seismic hazards is a means to an end. The question that ultimately matters is: What impact would seismicity have on the repository's ability to contain waste?

In the case of volcanism, the hazard is obvious. For earthquakes, it is not. The threat is probably not a fault rupture within the repository; the DOE (2001) concluded that the annual likelihood of such an event is less than the threshold for consideration. The National Research Council (1995) concluded that the principal hazards were waste canister failure due to rock falls or tunnel collapse and an increase in water flow through existing fractures.

An extreme case of the latter would be the "seismic pumping" of groundwater up into the repository. An example of that phenomena occurred when a magnitude 7 earthquake struck Idaho's Borah Peak in 1983, and water levels in a nearby silver mine rose sixty meters. The Research Council report concluded that such pumping could raise the water table at most a few tens of meters—not enough to reach the repository. The report also concluded that the impact of a seismic event on fracture conductivity is just as likely to be negative as positive; in other words, an earthquake is as likely to impede the flow of water through fractures as to promote it (National Research Council 1995). In its analysis, the DOE concluded that rock falls are "extremely unlikely" to degrade the repository's engineered barriers. The DOE also concluded that hydrologic changes resulting from fault displacement are highly unlikely and will not affect the safety of the repository. The only consideration of earthquakes in the DOE's (2001) base-case TSPA calculations is the effect of seismic vibrations on fuel cladding. The U.S. Geological Survey has concluded that seismic hazards are primarily a problem for surface facilities before the repository is closed (Groat 2001).

Three caveats apply to these assessments. First, the National Research Council (1995) has noted that we need to better understand the impact that seismicity might have on the apparent steep rise in the water table just to the north of the repository. Second, magmatism and extension are linked. If high strain rates exist—if, for example, Caltech's estimates are accurate—they could reflect the increased movement of magma into the crust (Smith et al. 2002). A better understanding of the present strain rates thus has a bearing not only on the seismic but also the volcanic hazards facing the repository. Finally, when the seismic hazard analysis was used to estimate possible peak ground accelerations at very low probabilities (one in one hundred million), it generated results up to eleven times that of gravity. Such extreme ground motions have never been documented in actual earthquakes far larger than any that could be generated by the faults near Yucca Mountain. Research is ongoing to determine whether such motions are physically possible or simply an artifact of extending the models to very low probabilities. Scientists are also looking for indicators such as precipitous talus slopes or precarious rocks that constrain the maximum ground motion experienced in the area (Hanks et al. 2004).

## Climate and Erosion

A third geologic process that has contributed to the present landscape of Yucca Mountain is erosion. Yucca Mountain is located in the southwestern corner of the Great Basin. What rivers there are in this dry region flow into closed basins with no outlet to the sea. The Great Basin includes nearly all of Nevada, large areas of Utah and California, and smaller parts of Idaho and Arizona. Its borders coincide with the northern half of the Basin and Range province.

Because of the region's arid climate, there are few perennial streams or lakes (Utah's Great Salt Lake being the notable exception). But it has not always been this dry. The current interglacial climate is itself an exception; glacial periods account for 80 percent of the past half million years (DOE 2001). The last glacial age, known as the Wisconsin, ended only ten thousand years ago. Although glaciation did not reach this far south, the climate was cooler and wetter. The floors of Death Valley and other closed basins in the region were covered with lakes. Indeed, the Great Salt Lake is a small and shallow version of a mammoth glacial lake that once occupied much of what is now northern Utah.

While changes in climate are inevitable over the next ten thousand years, predicting those changes is notoriously difficult. Several highly respected paleoclimatologists have proposed that current climate conditions could prevail for another fifty thousand years with minor change (Berger and Loutre 2002). Still, the geologic record gives us some sense of what ranges to expect, and those ranges have in turn been incorporated into the DOE's models. The DOE includes in its base-case model of repository performance a transition to glacial-age conditions. Specifically, the model assumes present-day climate with average annual precipitation of nineteen centimeters for four hundred to six hundred years, a so-called monsoon climate with thirty centimeters annual precipitation for nine hundred to one thousand and four hundred years, and then a glacial-transition climate with thirty-two centimeters annual precipitation for the remainder of the ten thousand years. The DOE describes the monsoon climate as warm and slightly wetter than the Yucca Mountain area at present; it would be similar to today's climate in Nogales, Arizona, with precipitation coming in summer storms. The glacial-transition climate would be cooler and wetter, similar to that of the region extending from present-day northern Nevada up to eastern Washington State. Precipitation would come in winter rain and snow, and infiltration rates would be higher because plants would require less uptake of moisture from the ground (DOE 2001).

Assuming climate does change along the lines of the DOE's base-case model, what effect would a cooler, wetter climate have on Yucca Mountain and its suitability as a repository site? The National Research Council (1995) has identified three main impacts: an increase in erosion, in effect bringing the buried waste closer to the surface; a shift in the distribution and activities of the human populations nearby; and higher water fluxes through the mountain leading to a rise in the water table and greater waste release. Because useful predictions about future human societies over this time span are not possible, only the first and third effects are considered here.

As reconstructed by the DOE, the modern appearance of Yucca Mountain and the current drainage systems were established by ten million years ago (DOE 2001). Gradual erosion has been the dominant process ever since, with sediments being shed off the mountains and deposited into the valleys by water and landslides. By dating the age of surface exposure for rocks on Yucca Mountain, the DOE (2001) has determined erosion rates of one to five millimeters per thousand years. In other words, the mountain will look much the same in tens to hundreds of thousands of years as it does today. Surface erosion or down cutting by intermittent streams will thus have a negligible impact on the depth of the repository over its regulatory time frame.

The possibility of large fluctuations in the water table has been a topic of great controversy for more than a decade, and it remains the subject of intensive research efforts funded by the state of Nevada. In 1992, a National Research Council (1992) panel studying this issue reported that "geologic and geochemical evidence does not support the contention that the water table has risen to the proposed repository level in the last 100,000 years." In a letter report to Congress, the Nuclear Waste Technical Review Board (1999) reached a similar conclusion.

With erosion inconsequential, human migration patterns unknowable, and large fluctuations in the water table unlikely, it would appear that the principal concern regarding future climate is the effect that increased water infiltration at the surface has on the amount of water reaching the waste. Although the DOE's (2001) own sensitivity analysis found that the influence of varying infiltration rates was minor over the regulatory time span, that conclusion is based entirely on the assumption that the waste packages remain intact, preventing any release. But we know from earlier DOE performance assessments, which could not assume unbreachable-engineered barriers, that percolation flux is the key parameter for repository performance (Metlay 2000). Climate, just like the water flow pathways through the unsaturated zone, matters.

## Natural Resources: An Attractive Nuisance?

Federal guidelines relating to nuclear waste disposal invariably emphasize that a key criterion for the acceptability of a host site is the lack of natural resources that might provide an incentive for future drilling or mining, as such intrusions would produce a potentially serious exposure pathway. The proximity of natural resources was a significant issue for the DOE's Waste Isolation Pilot Plant in New Mexico. Significant oil deposits were found within a kilometer of this trans-uranic waste repository, and potash is mined directly adjacent to the site (National Research Council 1996). Despite the presence of those resources, the Environmental Protection Agency (EPA) permitted the repository, which received its first waste shipment in 1999.

Recognizing the futility of seeking to predict future human society, the EPA standards for Yucca Mountain do not require the DOE to meet specific exposure guidelines but simply to show that the repository would be resistant to large releases resulting from human intrusion. The DOE (2001) has chosen to meet the standard using a simple case of inadvertent intrusion of the repository by a water well. Although water is certainly an important resource, should other geologic resources such as petroleum and minerals also be considered?

One would expect this issue to be a significant hurdle for a repository located in Nevada—the nation's top producer of gold, and a leading producer of a wide range of metals and industrial minerals (Nevada Bureau of Mines and Geology 2004). The state of Nevada claims that many features common to ore deposits elsewhere in Nevada are found at Yucca Mountain, suggesting "that the potential for valuable mineral resources in the immediate vicinity of Yucca Mountain must be recognized" (Nevada Agency for Nuclear Projects 1998). Before recently shutting down, one of the major gold mines in Nevada was located outside the town of Beatty, less than thirty kilometers from Yucca Mountain. The mine extracted ore from hydrothermally altered ash deposits that are also found in the vicinity of Yucca Mountain (Johnson and Hummel 1991). The state also makes a case for oil and gas potential in the area (Nevada Agency for Nuclear Projects 1998).

In contrast, the DOE's *Yucca Mountain Science and Engineering Report* concludes that there is little of value at Yucca Mountain. In particular, it cites the lack of evidence for hydrothermal alteration or mineralization of the tuff units that might suggest economic quantities of metal ore or uranium. The nearest patented mining claim is sixteen kilometers to the south, although one opportunistic miner did stake out a claim to Yucca Mountain itself right after the DOE announced it as a potential repository site, forcing the federal government to buy him out for a significant sum. The report further concludes that the resources that are found at Yucca Mountain, such as the raw material for concrete or building stone, are widely distributed around the region (DOE 2001).

For its part, the National Research Council (1995) has asserted that natural resources are a nonissue, arguing in one report that our inability to predict the technological capability of future civilizations is matched by an inability to predict which materials will be regarded as valuable resources in the future. The report concluded that resource potential was primarily useful as a criterion for comparing alternative sites rather than assessing the suitability of a particular one.

Given that the resource potential at Yucca Mountain does not appear to come anywhere near the actual resources being developed around the Waste Isolation Pilot Plant at the time of its regulatory approval, it seems unlikely that this issue will play a major role in the licensing process. If, however, deposits are identified—as the state of Nevada contends is likely—this issue could present a challenge to the DOE's approach toward human intrusion. Were that to occur, the value in political capital will surely outweigh the monetary value of whatever resources are found.

### Implications of an Expanded Footprint

When considering the relationship between a mined repository and the mountain that houses it, an obvious question to address is just how much of the mountain will the repository encompass. This question is particularly pertinent for Yucca Mountain because of the complexity and variability of the site's geology. Most of the calculations concerning repository performance assume that the repository footprint will be limited to a single block of drifts off the main tunnel. But it is a distinct possibility, for two very different reasons, that the repository would cover an area twice the size of this base-case scenario, extending into more highly fractured areas of the mountain.

The DOE's base case for the repository calls for the disposal of 70,000 metric tons of heavy metal. Of that total, 63,000 metric tons of heavy metal would be commercial spent fuel, and the remainder would consist of military spent fuel and high-level radioactive waste. The containers would be packed tightly to generate enough heat to keep the emplacement drift wall temperatures above the boiling point of water for thousands of years (DOE 2001). This may delay the time at which liquid water flows through the repository and corrodes the waste containers.

This base case can be accommodated in a single (primary) block of drifts—also known as the "pork chop" because of its shape—extending west from the main access tunnel and covering 1,150 acres (figure 7.2). The pork chop was chosen because it is the least fractured zone in the mountain and is not crossed by a major fault. If the space available in the primary block proves inadequate, however, the DOE (2001) has identified a number of additional fault-bounded blocks within the mountain that would more than double the repository's footprint, to 2,500 acres (figure 7.3).

The amount of material to be stored in the base case represents the statutory limit set by the Nuclear Waste Policy Act of 1982. The total amount of spent fuel that

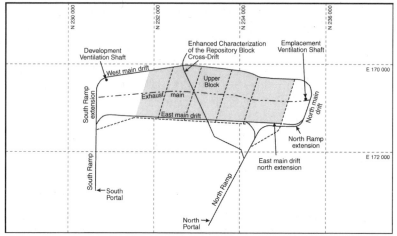

Draft EIS high thermal load repository layout

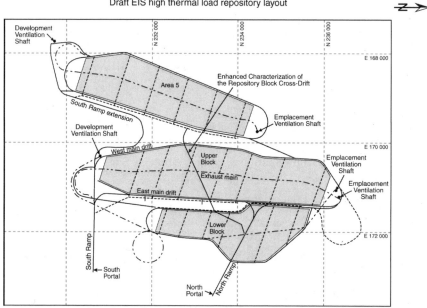

Draft EIS low thermal load repository layout

**Legend**

— · — · —  Ventilation drift

▨  Emplacement drift area

■ ■ ■ ■ ■  Performance confirmation drift

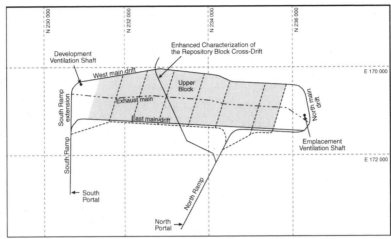

Draft EIS intermediate thermal load repository layout

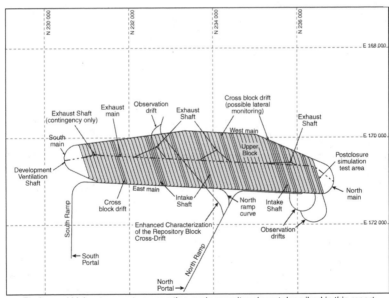

Design and higher-temperature operating mode repository layout described in this report

Note:  The grid system is the Nevada State Plane Coordinate System
converted to metric units.  E = Easting; N = Northing.

**Figure 7.3**
Diagram showing the expansion blocks for repository with draft EIS designs for low, mid, and high T. From Department of Energy 2001, figure 2.7.

will be produced by existing U.S. commercial reactors, however, is likely to exceed that figure by a considerable amount. The exact figure varies anywhere from 80,000 to 140,000 metric tons of heavy metal depending on how many reactors receive twenty-year license extensions from the NRC. Were Congress to remove the statutory cap, the *Science and Engineering Report* states that site design "does not preclude the flexibility to be expanded to accommodate" up to 119,000 metric tons of heavy metal. The report explicitly considers a "full inventory case" of 97,000 metric tons of heavy metal and the expanded footprint required to accommodate such an amount (DOE 2001).

In another modification that would necessitate a larger repository footprint, the DOE is considering a low-temperature design that maintains the emplacement drift temperatures below the boiling point of water. Achieving this lower temperature requires that the heat-generating waste containers be spaced farther apart from one another. The U.S. Geological Survey is particularly vocal in calling for a lower-temperature design "so as not to complicate our understanding of an already complex environment" (Hanks et al. 1999).

A lower-temperature design has several advantages for assessing the response of the mountain's hydrologic, geochemical, and fracture systems to repository conditions. The higher the temperature, the more difficult it is to predict the system's response over the regulatory compliance period. A lower-temperature repository design will also make it easier for the DOE to draw guidance from natural analogs, such as caves that preserve ancient drawings and human remains. The DOE may have another incentive related to one of the key materials in the waste package as well: keeping the waste package surface temperature below 85°C lowers the corrosion susceptibility of Alloy-22, the nickel-chromium-molybdenum-tungsten alloy that is to form the outer barrier of the waste package (DOE 2001). The *Science and Engineering Report* shows the possible layouts of blocks containing additional drifts for a lower-temperature design. The report does not consider, however, the combination of a low-temperature design and a larger statutory limit—that is, the two large-footprint scenarios combined.

From a geologic standpoint, an enlarged repository footprint poses three challenges. The first is the incorporation of sections of the mountain that are much more highly fractured than the primary block, providing greater potential for water flow via fast pathways to the repository and then on down to the aquifer. The second involves the relationship between the repository and the water table. The depth of the water table beneath the mountain is one of the principal advantages of the site, but it has already been noted that this advantage disappears to the north and west of the repository where the water table steeply rises to a level at or above the repository horizon (DOE 2001). Several of the preferred expansion areas shown in the *Science and Engineering Report* would locate waste close to the steep hydrologic gradient. Finally, doubling the repository footprint increases the probability of repository-piercing volcanic eruptions.

The need for an expanded footprint is by no means certain. Even if the DOE adopts a lower-temperature design, there are ways to achieve that end other than the wide spacing of waste packages that would necessitate a larger footprint. Aging the waste prior to emplacement will reduce the heat produced in the repository, and ventilating the repository (either actively or passively) will remove heat after emplacement. If the DOE chooses to maintain the planned schedule of waste emplacement and close the repository in a timely fashion, then wider spacing will be required and with it an expanded footprint. Since the base case for which the DOE is seeking regulatory approval contains only the primary block, there is a very real possibility that the repository being proposed and the repository that is built will be two quite different things.

## Conclusions

From a geologic viewpoint, the Yucca Mountain site has many natural assets and several important liabilities. Fundamental questions remain about the movement of water through the mountain and the potential for volcanic disruption. The DOE has sought to minimize the site's perceived liabilities by adopting a design that relies almost entirely on engineered barriers—fuel cladding, waste containers, and drip shields—to keep water from mobilizing the waste during the repository's regulatory compliance period. Only after those barriers fail does the DOE place any reliance on the mountain. But the NRC will decide the licensing of the repository based on the ability of multiple barriers, both engineered and natural, to contain the waste. Thus, no matter how robust the engineered systems, the behavior of the mountain is critical to the repository's viability.

And that is as it should be. The greatest challenge facing the Yucca Mountain Project has always been the time scale over which it is to operate. Data collected in the present must be extrapolated out to AD 12,000 and beyond. The challenge is big enough for natural systems that can be studied back through time and by way of analogy to other sites. But the engineered systems have no natural analogs that can give us a good understanding of their long-term behavior under shifting conditions. What if the containers fail early? What if the cladding is disrupted by earthquakes or volcanism?

That time scale also challenges the very notion of "normal" operating conditions for the repository. Over geologically significant periods of time, seismic and volcanic events are simply part of the operating mode of a planet that conducts its normal business through such disruptions. As with climate change, they are inevitable but appear to be boundable. That said, this mountain has repeatedly humbled attempts to declare it understood and no doubt has additional surprises in store. This reality of nature's complexity strongly suggests the need for long-term monitoring, and a flexible approach to repository design and operation that encourages the absorption of new information. The site is neither a sure bet nor a fool's paradise, but given the proper approach, it may yet prove more than adequate to the task.

## References

Berger, A., and Loutre, M.F. (2002) An Exceptionally Long Interglacial Ahead? *Science* 297, pp. 1287–1288.

CRWMS M&O (Civilian Radioactive Waste Management System Management and Operating Contractor) (2000) *Yucca Mountain Site Description.* Department of Energy. TDR–CDW–GS–000001 REV 01 ICN 01.

Connor, C.B., Stamatakos, J.A., Ferrill, D.A., Hill, B.E., Ofoegbu, G.I., Conway, F., Sugar, B., and Trapp, J. (2000) Geologic Factors Controlling Patterns of Small-Volume Basaltic Volcanism: Application to a Volcanic Hazard Assessment at Yucca Mountain, Nevada. *Journal of Geophysical Research, B, Solid Earth and Planets* 105, pp. 417–432.

Department of Energy (DOE) (2001) *Yucca Mountain Science and Engineering Report.* Office of Nuclear Waste Management. DOE/RW–0539.

Department of Energy (DOE) (2002) *Yucca Mountain Site Suitability Evaluation.* Office of Civilian Radioactive Waste Management. DOE/RW–0549. February.

Flint, A.L., Flint, L.E., Kwicklis, E.M., Bodvarsson, G.S., and Fabryka-Martin, J. (2001) Hydrology of Yucca Mountain, Nevada. *Reviews of Geophysics* 39, pp. 447–470.

Fridrich, C.J., Whitney, J.W., Hudson, M.R., and Crowe, B.M. (1999) Space-Time Patterns of Late Cenozoic Extension, Vertical Axis Rotation, and Volcanism in the Crater Flat Basin, Southwest Nevada. In *Cenozoic Basins of the Death Valley Region,* ed. Wright, L.A., and Troxel, B.W. Boulder, CO: Geological Society of America Special Paper 333, pp. 197–212.

Groat, C.G. (2001) Letter to Robert G. Card, Undersecretary for Energy, Science, and Environment, U.S. Department of Energy, October 4.

Hanks, T.C., Abrahamson, N.A., Board, M., Boore, D.M., Brune, J.N., and Cornell, C.A. (2004) Observed Ground Motions, Extreme Ground Motions, and Physical Limits to

Ground Motions. Paper presented at the International Workshop on Future Directions in Instrumentation for Strong Motion and Engineering Seismology, Kusadasi, Turkey, May 17–21.

Hanks, T.C., Winograd, I.J., Anderson, R.E., Reilly, T.E., and Weeks, E.P. (1999) Yucca Mountain as a Radioactive Waste Repository. *U.S. Geological Survey Circular* 1184.

Heizler, M.T., Perry, F.V., Crowe, B.M., Peters, L., and Appelt, R. (1999) The Age of Lathrop Wells Volcanic Center: An $^{40}$Ar/$^{39}$Ar Dating Investigation. *Journal of Geophysical Research, B, Solid Earth and Planets* 104, pp. 767–804.

Johnson, C., and Hummel, P. (1991) Yucca Mountain, Nevada. *Geotimes* 36, no. 8 (August), pp. 14–16.

Kerr, R.A. (1996) A New Way to Ask the Experts: Rating Radioactive Waste Risks. *Science* 274, pp. 913–914.

Levich, R.A., Linden, R.M., Patterson, R.L., and Stuckless, J.S. (2000) Hydrological and Geologic Characteristics of the Yucca Mountain Site Relevant to the Performance of a Potential Repository. In *Great Basin and Sierra Nevada*, ed. Lageson, D.R., Peters, S.G., and Lahren, M.M. Boulder, CO: Geological Society of America Field Guide 2, pp. 383–414.

Macilwain, C. (2001) Out of Sight, out of Mind? *Nature* 412, pp. 850–852.

Metlay, D. (2000) From Tin Roof to Torn Wet Blanket. In *Prediction: Science, Decision Making, and the Future of Nature*, ed. Sarewitz, D., Pielke, R.A., Jr., and Byerly, R., Jr. Washington, DC: Island Press, pp. 199–228.

National Research Council (1992) *Ground Water at Yucca Mountain: How High Can It Rise?* Washington, DC: National Academy Press, 242 p.

National Research Council (1995) *Technical Bases for Yucca Mountain Standards.* Washington, DC: National Academy Press, 222 p.

National Research Council (1996) *The Waste Isolation Pilot Plant: A Potential Solution for the Disposal of Transuranic Waste.* Washington, DC: National Academy Press, 184 p.

Nevada Agency for Nuclear Projects (1998) State of Nevada and Related Findings Indicating That the Proposed Yucca Mountain Site Is Not Suitable for Development as a Repository. Reno: State of Nevada.

Nevada Bureau of Mines and Geology (2004) The Nevada Mineral Industry 2003. *Nevada Bureau of Mines and Geology Special Publication MI–2003.*

Nuclear Waste Technical Review Board (1999) Report to the U.S. Congress and the Secretary of Energy, January to December 1998. Arlington, VA: Nuclear Waste Technical Review Board.

Savage, J.C., Svarc, J.L., and Prescott, W.H. (2001) Strain Accumulation near Yucca Mountain, Nevada. *Journal of Geophysical Research, B, Solid Earth and Planets* 106, pp. 16483–16488.

Sawyer, D.A., Fleck, R.J., Lanphere, M.A., Warren, R.G., Broxton, D.E., and Hudson, M.R. (1994) Episodic Caldera Volcanism in the Miocene Southwestern Nevada Volcanic Field: Revised Stratigraphic Framework, $^{40}$Ar/$^{39}$Ar Geochronology, and Implications for Magmatism and Extension. *Geological Society of America Bulletin* 106, pp. 1304–1318.

Schweikert, R.A., and Lahren, M.M. (1997) Strike-slip Fault System in Amargosa Valley and Yucca Mountain, Nevada. *Tectonophysics* 272, pp. 25–41.

Smith, E.I., Keenan, D.L., and Plank, T. (2002) Episodic Volcanism and Hot Mantle: Implications for Volcanic Hazard Studies at the Proposed Nuclear Waste Repository at Yucca Mountain, Nevada. *GSA Today 12*, pp. 4–9.

Stamatakos, J.A., and Ferrill, D.A. (1998) Strike-slip Fault System in Amargosa Valley and Yucca Mountain, Nevada: Comment. *Tectonophysics* 294, pp. 151–160.

Stepp, J.C., Wong, I., Whitney, J., Quittmeyer, R., Abrahamson, N., Toro, G., Youngs, R., Coppersmith, K., Savy, J., Sullivan, T., and Yucca Mountain PSHA Project Members (2001) Probabilistic Seismic Hazard Analyses for Ground Motions and Fault Displacement at Yucca Mountain, Nevada. *Earthquake Spectra* 17, pp. 113–150.

Wernicke, B., Davis, J.L., Bennett, R.A., Elosegui, P., Abolins, M.J., Brady, R.J., House, M.A., Niemi, N.A., and Snow, J.K. (1998) Anomalous Strain Accumulation in the Yucca Mountain Area, Nevada. *Science* 279, pp. 2096–2100.

Woods, A.W., Sparks, S., Bokhove, O., LeJeune, A., Connor, C.B., and Hill, B.E. (2002) Modeling Magma-Drift Interaction at the Proposed High-Level Radioactive Waste Repository at Yucca Mountain, Nevada, USA. *Geophysical Research Letters* 29, 10.1029/2002GL014665.

# 8

# Volcanism: The Continuing Saga

Bruce M. Crowe, Greg A. Valentine, Frank V. Perry, and Paul K. Black

Imagine Yucca Mountain some hundreds to thousands of years from now. Will the site remain an arid mountain range, unchanged by hydrologic, geologic, or volcanic processes, with nuclear waste safely and securely isolated underground? Or will the area be a radioactive wasteland devastated by a volcano that erupted through an underground repository, releasing clouds of radioactive volcanic ash that contaminated hundreds of square kilometers of land? Scientists are faced with the challenge of gathering evidence to determine which of these scenarios is most likely to occur despite their limited ability to predict the future behavior of complex Earth systems (Sarewitz et al. 2000; IPCC, Intergovernmental Panel on Climate Change 2001; Allen 2003). How well can we assess the risk of future volcanism, and can decisions be made with sufficient confidence to support a ten-thousand-year or longer commitment to protect the health and safety of the nearby populations?

The possible recurrence of a volcanic eruption has long been recognized as a tangible risk to a proposed repository at Yucca Mountain. What is the likelihood of such an eruption? How would it affect a repository? These are some of the questions raised in both public and scientific arenas since the inception of volcanism studies for the Yucca Mountain site in the late 1970s.

In the early years of volcanism studies, the U.S. Department of Energy (DOE) concluded that the likelihood of a repository disruption by a volcanic eruption was low, where low was defined arbitrarily as less than one chance in a million, and the estimated probability of a disruption was significantly less than one chance in a million (Crowe and Carr 1980). Most (though not all) investigators and technical reviewers concluded that this low risk of volcanism should not eliminate Yucca Mountain from consideration as a permanent repository (see, for example, National Research Council 1995; Nuclear Waste Technical Review Board 2002). The design decision by the DOE's (1998, 2001) Yucca Mountain Project to place waste canisters in an open tunnel without backfill, however, creates the potential for a low-probability but potentially high-consequence disruptive event consisting of a volcanic eruption *into* a repository. In fact, as of the site recommendation, the volcanic scenario reemerged as the only disruptive event that significantly degraded repository performance during the site's ten-thousand-year regulatory compliance period (DOE 2002). Thus, the effects of future volcanism are currently a leading source of risk to human health, based on the regulatory standards of the U.S. Environmental Protection Agency (EPA 2001).

This chapter provides an overview of the volcanism problem, how it is studied, and its potential impact on an underground repository. It is presented primarily from the perspective of the Los Alamos scientists who conducted, over nearly two decades, the scientific studies for the DOE. Currently, there is not complete agreement over the conclusions of volcanism studies. While recent progress has been made, scientists conducting independent volcanism studies for the state of Nevada and the Nuclear Regulatory Commission (NRC) still disagree with aspects of the DOE volcanism studies. The goal of this chapter is to provide sufficient background on this issue, without descending too far into the technical debate, to enable the reader to draw conclusions about the controversies and the potential impact of future volcanism on the Yucca Mountain site.

The risk of volcanism has two components: the probability that a volcano will occur in the Yucca Mountain region during the repository's life span, and the effects of volcanism on the repository. This chapter emphasizes the former topic because the areas of controversy are well refined through long debate. We also briefly describe the nature and current significance of the volcanic effects on a repository. Yet this topic is technically complex, it is still being studied and debated, and the resulting interpretations and controversies continue to evolve.

## Background

Yucca Mountain is composed of volcanic rocks deposited from a series of large-volume eruptions that occurred just north of the mountain, beginning about fifteen million years ago (figure 8.1; see also chapter 7, this volume). These episodes of eruptive activity ceased about eight million years ago, and scientists agree such episodes are not a future threat to a repository. Of concern, however, are geologically recent episodes of *small*-volume volcanic activity. Evidence of these lesser volcanoes comes from the presence of less than a dozen small scoria cones (scoria is solidified

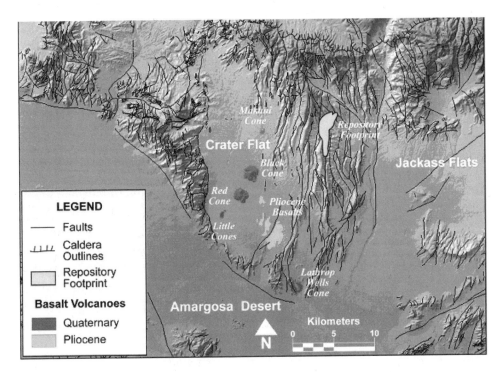

**Figure 8.1**
Pliocene (<5 million years old) and Quaternary (<1.8 million years old) basaltic volcanoes in the Yucca Mountain region. The Pliocene basalt erupted from about five scoria cones that are mostly removed by erosion or covered by basin-fill deposits of Crater Flat. The hachured line marks the southern part of a large caldera complex that was the source of the older large-volume eruptions.

magma with abundant open spaces formed by escaping gases), some well preserved and some highly eroded, in the Crater Flat basin immediately west of the Yucca Mountain site (figure 8.1).

Three distinct episodes of volcanic activity occurred in the Crater Flat basin. Each episode erupted basalt, a low-silica magma generally associated with mildly explosive volcanic eruptions that are comparable to, but smaller in volume than recent eruptions of Mount Etna in Italy. Multiple scoria cones, marking the vent areas of basaltic eruptions, formed along overlapping north-south fissures in southeastern Crater Flat 3.7 ± 0.1 million years ago and extruded lava flows that banked against the southwest flank of Yucca Mountain (the Pliocene basalt of figure 8.1). A cluster of basaltic volcanoes consisting of four scoria cones and the surrounding lava flows (figure 8.1) erupted 1 ± 0.1 million years ago along a 12.5-kilometer, northeast-trending alignment in the Crater Flat basin. The youngest basaltic volcano, the Lathrop Wells volcano, is the major source of scoria-fill material for the more than decade-long construction boom in Las Vegas, and its deposits appear destined for complete removal. This volcano is about seventy-five thousand ± ten thousand years old and is located at the south end of Yucca Mountain, about twenty kilometers from the repository site (figure 8.1). The presence of these geologically young volcanoes has sparked intense study and continuing debate about the potential for volcanism in the future at Yucca Mountain (Crowe and Carr 1980; Crowe et al. 1982, 1995; Smith et al. 1990; Ho 1992; Bradshaw and Smith 1994; Connor and Hill 1995; Fleck et al. 1996; Perry et al. 1998; Connor et al. 2000).

Most of the geologically young volcanoes occur preferentially in the Crater Flat basin (figure 8.1). Tectonic studies show that Yucca Mountain was once a part of the low-standing terrain that demarked the Crater Flat basin, but it was uplifted and now forms a mountain range that bounds the east flank of the basin (Fridrich 1999). Fridrich and colleagues (1999) suggest that based on the geologic record, the most likely location of another basaltic volcano is in the alluvial valley of the modern Crater Flat basin—not the Yucca Mountain site.

But not all the volcanic assessments are as favorable for the safety of the site. Faults and other deep tectonic structures may provide alternative pathways for the ascent of basaltic magma, and some of these structures project from Crater Flat into the Yucca Mountain site (Smith et al. 1990, 2002; Connor et al. 2000). For example, an eleven-million-year-old basaltic dike (subsurface feeder system) intrudes the Solitario Canyon fault at the northwest boundary of the Yucca Mountain site. This dike may be an artifact of the older tectonic setting of Yucca Mountain, or it may be a reminder that basaltic magma has in the past and could in the future penetrate the rocks that form the repository site.

The main question requiring resolution here is whether volcanism is an unacceptable threat to an underground repository. Given the inherent uncertainties, volcanism studies for the Yucca Mountain Project were assessed probabilistically from their inception (Crowe et al. 1982; Crowe 1986). These studies culminated in the mid-1990s with a DOE-convened elicitation of the occurrence probability of volcanism, conducted by a panel of volcanism experts (Geomatrix Consultants 1996). Each expert developed independent models of the probability of a volcanic disruption of a repository at Yucca Mountain and the models were aggregated into a single probability distribution.

## The Volcanism Decision Problem

Evaluations of volcanic hazards at Yucca Mountain are more complicated than traditional volcanic hazard studies that assess the possibility of a future eruption at an existing active volcano. The threat of volcanism for the Yucca Mountain site is, rather, from the birth of a new volcano at an unknown location. There is uncertainty in *when* a new volcano will erupt, *where* it will form, and the *nature* of the volcanic event.

The volcanism issue is a paradox. The small number of past volcanic eruptions is the primary evidence that the probability of a future volcanic event is low. But the rareness of volcanic events also limits the data available for study, thus leading to large uncertainty in quantifying the probability of future events. If there were more volcanic events, there would be more data to assess and project future events and decreased uncertainty. The risk of volcanism would be higher, however.

Volcanism represents a potential disruptive event that could both directly release radioactive waste at the Earth's surface as well as perturb the repository and the hydrogeologic setting of the repository system. Given our imperfect knowledge of the multiple processes associated with the generation, migration, and eruption of magma, we cannot realistically expect to predict the location or timing of future volcanic events. But we can assess the likelihood and effects of volcanism and their associated uncertainty.

If its occurrence probability is sufficiently low, volcanism can be dismissed as an issue of concern. What is the definition of a "sufficiently low" or "very unlikely" event? A target probability was established initially in part B of the EPA (1993) guidance, modified in the revised standard for Yucca Mountain (EPA 2001), and clarified in a proposed rule amendment (NRC 2002). According to this standard, a volcanic event does not require assessment if its occurrence probability is less than one in ten thousand in ten thousand years (i.e., one chance in one hundred million per year). That's roughly the same chance of occurrence as the ultimate low-probability, extremely high-consequence event: global mass extinction from the impact of an asteroid or a comet (Morrison et al. 1994; Chapman 2000).

The DOE (1986) formally decided in the mid-1980s that the estimates of the occurrence probability of volcanism were sufficiently low that the Yucca Mountain site was not disqualified solely by the risk of volcanism. Yet the probability was not sufficiently low (less than one in one hundred million per year) to eliminate volcanism as an issue. This decision continues to apply for the site suitability evaluation

(DOE 2002) and leads to two conclusions. First, the volcanic disruption of a repository is a plausible event during the ten-thousand-year compliance period and the probability of that event must be carefully quantified. Second, studies of the potential radiological releases associated with future volcanic events must be evaluated as a component of the repository performance. And in fact, the performance assessment for the Yucca Mountain site incorporates estimates of the occurrence probability of a repository disruption (CRWMS 1998; DOE 2002).

### Assessing the Probability of a Volcanic Disruption of the Repository

Mathematically, the probability of a volcanic disruption of an underground repository at Yucca Mountain from future volcanic activity ($PR_d$) is a conditional one:

$$Pr_d = Pr(E2 \text{ given } E1)Pr(E1),$$

where E1 is the recurrence rate of a future eruptive or intrusive event in a volcanic zone and E2 is an event that intersects the repository.

Stated simply, the disruption of a repository site requires the following conditions:

• A volcanic event must occur in a volcanic zone (if there is no event, there is no risk to a repository).

• An event must occur at or near the underground repository (distant volcanic events will have limited effects on an underground repository).

A further question that must be addressed is whether a future volcano at the site will in fact disrupt the repository. We can evaluate this by describing how magma might penetrate a repository during an eruptive and/or intrusive volcanic event, and then evaluating the effects of this event on the performance of a repository.

The information needed to assess the probability and consequences of future volcanism is summarized in a logic diagram (figure 8.2). This information includes:

• The definition and number of volcanic events in a specified time period (inputs to E1)

• The spatial, structural, or conceptual models of zones where a volcanic event might occur (input to both E1 and E2)

• The distribution and properties (feeder dike length and orientation) of volcanic events in zones (inputs to E2)

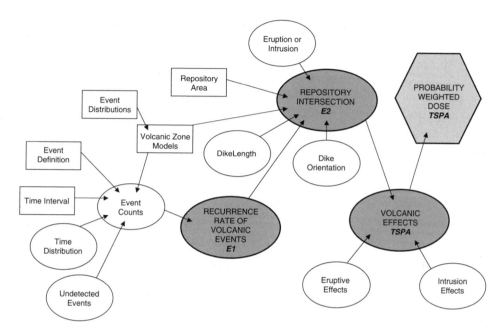

**Figure 8.2**
Logic diagram for the volcanism decision problem. See the accompanying text for descriptions of the diagram elements. TSPA is the system model used to forecast the future performance of the Yucca Mountain site.

- The location and area of an underground repository (input to E2)
- The radiological release and dose associated with volcanic disruption scenarios

The boxes in figure 8.2 represent decision parameters that require subjective judgments of the assumptions and approaches used by both scientists and decision makers to evaluate the probability model for volcanism studies. For example, the assessment of the box labeled *repository area* is determined through decisions by the DOE of the design basis of a repository. Decision variables are treated as decision points, assumptions, or constants in a probabilistic analysis. They represent a form of conceptual model uncertainty, and their representation may be different for a regulator, a scientist, or the DOE.

The ovals in the diagram are probabilistic parameters where uncertainty is represented as a distribution that defines a range of possible values. The small number of past volcanic events in the Yucca Mountain region results in significant uncertainty in these parameters as well as uncertainty in discriminating alternative models. The larger bolded ovals are the E1 and E2 parameters of the conditional probability equation, and are calculated as probability distributions through computer-based simulation models that systematically sample input from the probabilistic and decision parameters.

The output of the simulation model gives the occurrence probability of a future volcanic event disrupting a repository ($Pr_d$). This probability is applied as a weighting factor to the radiological dose calculations established through the performance assessment, and these calculations are compared with the regulatory requirements for licensing a repository. The following sections provide background information, and describe the uncertainties associated with assessing the parameters, variables, and decision logic of figure 8.2.

### Volcanic Event Counts

A volcanic event is defined from the dual perspective of recognizing a surface volcano and understanding how the underpinnings of the volcano could affect an underground repository. The event of concern is the rise of a new pulse of magma that moves through or near a repository and forms a surface volcano. For most probability estimations, a volcano is equated to a volcanic event, where an event includes the length of a subsurface feeder system for the volcano. In practice, individual researchers define volcanic events differently, and this leads to differences in how the resulting data are applied to E1, E2, and the resulting dose calculations.

Volcanoes in the Yucca Mountain region exhibit a range of surface forms. Some consist of a single scoria cone, marking a main eruptive vent. Others consist of multiple, spatially separate cones that may or may not have formed from a single volcanic event (figure 8.1). As a complicating factor, erosional modification of the volcanoes over geologic time makes their remaining deposits difficult to interpret.

The presence of scattered aeromagnetic anomalies identified through geophysical studies further complicates the determination of past volcanic events and requires exploratory drilling to reduce uncertainty. These local anomalies in the magnetic field at the Earth's surface provide evidence of the possible presence of buried basalt volcanoes. Other explanations, such as the faulting of magnetic older volcanic rocks, also may account for some observed anomalies. Multiple aeromagnetic anomalies have been identified in Crater Flat and the Amargosa Valley directly south of Yucca Mountain (Kane and Bracken 1983; Langenheim 1995; Stamatakos et al. 1997; Connor et al. 2000; Blakely et al. 2000; O'Leary et al. 2002). To date, only a few of these anomaly sites have been fully evaluated through exploratory drilling. The DOE, in response to the NRC concerns, has acquired additional high-resolution aeromagnetic data to better identify aeromagnetic anomalies that may be buried basalt volcanoes (Perry et al. 2004). Selected anomalies will be evaluated through drilling, and samples of basalt, if present, will be analyzed to determine the age of the buried volcano. Newly identified buried volcanoes could change estimates of the probability of a repository disruption if the volcanoes are younger than five million years or if they occur in areas where basaltic volcanoes have not been identified (for example, in Jackass Flats east of Yucca Mountain; see figure 8.1).

The next step is to make assumptions about the time between volcanic events in the future (the time-distribution oval in figure 8.2). Here, the limitation from the volcanic paradox reappears; there were too few volcanic events in the past to constrain or discriminate alternative time-distribution models to forecast the timing of future events. A key issue is assessing whether event rates are accelerating, decelerating, or remaining steady.

There is general agreement that the volcanic record of the Yucca Mountain region shows a dramatic decline in the volume of erupted basaltic magma over the last four million years with an associated slight decrease in the average time between eruptions (Crowe et al. 1995). The data remain unclear, however, for selecting time-distribution models for future volcanic events. After many years of debate, the DOE and the NRC have agreed that a steady-state model provides a reasonable representation of the record of past volcanic events despite evidence of decreasing intervals *between* eruptive events as well as observational data on a global scale that volcanism is an episodic process (Smith et al. 2002) and may be unpredictably unique (Shaw 1987).

Choosing a time interval for assessing event counts is yet another requirement for evaluating probability models (the decision box in figure 8.2). Here there is a trade-off. The longer the time period for examining the volcanic record, the greater the number of volcanic events for use in the model (improved statistics). But longer time intervals lower the confidence that the included volcanic record provides a realistic representation of future events that could affect a repository. The best approach is to use a geologic perspective to assess the record of volcanism at time intervals that correspond to natural cycles in the past volcanic record. Most researchers use this approach for developing volcanic probability models (Geomatrix Consultants 1996).

A final issue in establishing recurrence rates is assessing the possibility of undetected volcanic intrusions (the oval in figure 8.2). There are two, somewhat contradictory interpretations. The DOE studies and the expert panel conclude that it is unlikely for magma to migrate to repository depths (three hundred meters) without causing an eruption that forms an easily recognizable surface volcano (Geomatrix Consultants 1996). Undetected intrusions, by the DOE's interpretation, are possible and are included in event counts, but are presumed to be associated in space and time with a surface volcano. This interpretation is based on field examination of deeply dissected volcanoes in the Yucca Mountain region.

The NRC (1999) argues, in contrast, that undetected intrusions need not be associated with a surface volcano, and that there may be as many as two to five times the number of *undetected* intrusions as there are surface volcanoes. This interpretation is based on field studies of eroded volcanoes at locations away from the Yucca Mountain region. The difference in approaches is partly the result of different views of whether shallow intrusions can form at a repository depth (three hundred meters below the surface) without an accompanying eruption. These alternative approaches have not been reconciled, and lead to differences in estimates by the DOE and the NRC of the probability of a repository disruption.

Collectively, the combination of the subjective interpretations of the number of volcanic events, the multiple alternative models of the time distribution of volcanic events, and the differences in opinion over undetected volcanic events leads to significant uncertainty in assessments of the recurrence rate of volcanic events (E1). Estimates of E1 range from $1.5 \times 10^{-6}$ to about $8 \times 10^{-6}$ events per year with a mean of about $4.5 \times 10^{-6}$ events per year (Crowe et al. 1995). These numbers equate to estimated times between volcanic events ranging from as long as 625,000 years to as short as 125,000 years, with a mean estimate of about 250,000 years.

### The Probability of Repository Intersection

Given a probability distribution for E1, the next part of the volcanism problem is to estimate E2—the likelihood that given an event, it intersects the Yucca Mountain site. The disruption ratio, E2, is simply the ratio of the area of the repository to the area of the entire volcanic zone. Other alternative approaches to estimating E2 are possible (see Connor and Hill 1995).

The repository area depends on the repository design. The final design has not been established, but most designs are between five- to eight-square kilometers. The second factor in E2 is the area of relevant volcanic events and particularly the spatial relationship of the volcanic zone to the repository site (the decision box in figure 8.2). This is a complicated and contentious aspect of probability models, and is the second-largest uncertainty component. Multiple approaches are used. One is to establish a probability range representing disruption ratios from *all* viable alternative models of the disruption ratio and to sample this distribution through simulation modeling (Crowe et al. 1995). The DOE established E2 using the results of the expert judgment panel, and the NRC's preference is to evaluate individual models separately with an emphasis on the results of the higher or bounding values of the disruption ratio.

Once again, the limited record of past volcanic events results in insufficient information to discriminate alternative models. Many models have been proposed, none can be proven conclusively, and only a few have been disproved. In practice, virtually every researcher has one or multiple unique models to determine the size of the volcanic zone. The important points of the remaining dispute are whether the Yucca Mountain site should be included in every defined volcanic zone, and how far subsurface feeder dikes can extend from volcanic events.

Compilations of alternative models of the disruption ratio range from about .015 to .0008 (Crowe et al. 1995). Assuming a volcanic event has occurred in the Yucca Mountain region, these ratios translate to about one chance in seventy to one chance in twelve hundred that the event will intersect the repository site.

## The Probability of Volcanic Disruption

The probability distributions of E1 and E2 are combined through simulation to give a probability distribution of volcanic disruption at the Yucca Mountain site (figure 8.2). What are the most current estimates of this probability? The DOE relies on the results of their expert elicitation to establish the probability of the disruptive scenarios. The mean of the expert elicitation is $1.5 \times 10^{-8}$ events per year (1 event in 70 million years) with a 90 percent confidence interval ranging from $5.4 \times 10^{-9}$ (1 event in 180 million years) to $4.9 \times 10^{-8}$ events per year (1 event in 20 million years; see Geomatrix Consultants 1996). These are the DOE's estimates of the probability that a volcanic dike will intersect a subsurface repository. Recent studies have slightly modified these probabilities based on changes in the repository design and the probability of an eruption along a dike that intersects the repository (Perry and Youngs 2000). These modified probability distributions as well as any changes from continuing studies of aeromagnetic anomalies will be used by the DOE in its license application for the Yucca Mountain site. The NRC also assesses repository disruption as a probability distribution, but emphasizes a subset of spatial and structural models for the Yucca Mountain region (Connor and Hill 1995; Connor et al. 2000). The NRC's (1999) approach notes the importance of the bounding nature of its emphasized models and the regulatory need for including conservatism in the probability distribution. The commission concludes that the range of $10^{-7}$ to $10^{-8}$ events per year (1 event in 10 million years to 1 event in 100 million years) adequately represents the uncertainty in estimates of the volcanic disruption of a repository. Further, the NRC argues that there is no basis for distinguishing between values in this probability range because of significant model and scenario uncertainty; it thus uses a conservative                               n 10 million years).

The NRC and the DOE assessm                                      pro- bability estimates are applied to t                               E, in weighting its radiological dose calcι                            ition. Careful examination of the DOE                                hows that the resulting distribution is sk                            pro- bability. Indeed, some values are lο                           bility estimations of disruption of a reμ                         Perry et al. 1998; Connor et al. 2000), aι                       , uses a single, conservative bounding vε                       nance assessment calculations. The NRι                       t-case estimations.

**Consequences of Volcanic Disrupti**

Initial studies of the consequences of the volcanic disruption of a repository focused on the ability of basalt magma to carry radioactive waste from a shallow depth to the surface (Link et al. 1982; Crowe et al. 1983; Valentine and Groves 1996). This made sense as long as the regulatory requirements for high-level waste applied to the cumulative release of radiation at a specified distance from a repository (ten kilometers). But the EPA changed its approach and now requires adherence instead to a standard that limits radiation risk to a controlled group (EPA 2001). This regulatory regime requires an assessment of the dispersal of waste radionuclides in eruptions as well as transport by surface processes. The DOE terminated its volcanism studies in 1996 before completing its planned investigation of the full range of volcanic effects (Valentine et al. 1998). In response to the NRC's continuing work on this topic, however, the DOE reopened its volcanism effects studies in 2002 and they continue today.

The DOE's (2002) performance assessment deals with two disruptive volcanic scenarios:

1. An *intrusion* scenario consisting of a single intrusive event fed by a basalt dike set that degrades waste packages and accelerates the release and transport of radionuclides.

2. An *eruption* scenario in which a feeder dike passes through the repository and erupts at the surface, dispersing radioactive waste. This scenario assumes that any dike reaching the repository level will vent to the surface. Thus the eruptive scenario includes both intrusive and extrusive effects.

The methods for applying these two volcanic scenarios to the repository's regulatory requirements are complicated; the results vary with different versions of the DOE performance assessment and depend on the details of the volcanic occurrence probability. The mean probability-weighted peak dose defined in the DOE (2002) studies from the combined volcanic disruption scenarios is about 0.08 to 0.1 millirems per year. These doses are significantly below the regulatory requirement for the individual protection standard of 15 millirems per year. The NRC has conducted more extensive modeling studies of volcanic effects for eruption scenarios. The NRC's (1999, p. 10–11) estimates for the mean probability-weighted dose range are ten to one hundred times higher than the DOE's estimates, ranging from 1 to 8 millirems per year. While even this dosage would meet the NRC-enforced EPA regulation, the highest values in the range comprise a significant fraction of the individual protection standard of 15 millirems per year. Additionally, the NRC's (1999, p. 12) radiation dose estimates could be higher with the inclusion of contributions from the multiple release pathways associated with the magmatic intrusion of a repository. The DOE has made significant progress in the last 2 years but its result must be viewed cautiously given its somewhat limited studies of eruptive effects. The NRC results must also be viewed cautiously given its propagation of bounding-conservative calculations, a controversial approach for probabilistic-based risk assessment (Morgan and Henrion 1990; Vose 2000).

The volcanism issue is greatly complicated by the decision reflected in the current repository design not to backfill the tunnels following waste emplacement (DOE

2001). Basaltic magma migrating upward beneath the surface moves within sheet-like dikes that have small widths (on the order of a few meters) relative to their extents (typically one to ten kilometers). As long as such a dike is propagating through a backfilled tunnel, magma flowing through it will have limited interaction with waste packages; the effects would be confined mostly to waste packages close to the feeder dike. Without backfilling, however, the risk of magma-waste interaction is higher. When rising magma intersects a void such as a tunnel, there can be a sudden drop in pressure. Mixtures of gas and fragmented magma can expand rapidly into the tunnel, followed by flowing lava or continued fast-moving mixtures of gases and magma fragments. Volcanic gases, which are highly acidic and corrosive, can significantly degrade stainless steel cylinders such as those that will be encasing the repository's radioactive waste. Heat from the magma may also degrade waste package performance. The NRC-sponsored researchers predict that in the event of a volcanic disruption, large numbers of waste packages may be exposed to high temperatures, high pressures, and the high-velocity flow of magma-gas mixtures (Woods et al. 2002). They further argue that multiple pathways can develop for contaminated magma to erupt onto the surface, increasing the potential for higher doses of environmentally released radiation. The DOE convened an igneous consequences peer review panel (Detournay et al. 2003) to evaluate the consequences of the volcanic disruption of a repository. The panel emphasized the uncertainty of consequence studies, disagreed with some of the worst-case aspects of the NRC-sponsored research, and identified a combined program of future experimental and modeling studies now ongoing that could reduce uncertainty in assessing repository disruption by volcanism.

Encouragingly, the NRC and the DOE agreed in late 2001 on the scope of studies needed to assess the potential impact of eruption and intrusion on an open-tunnel (i.e., nonbackfilled) repository. An assessment of the full range of effects of the volcanic disruption of a repository, however, remains unresolved.

**Final Comments**

Despite long-standing debates, both the DOE and the NRC estimates of the probability of the volcanic disruption of the Yucca Mountain site are very low ($10^{-7}$ to $10^{-8}$ disruptive events per year are extremely small numbers). The technical work for volcanism probability studies by the DOE was rated as a program strength and the understanding of the consequences of intrusive activity as a program weakness

(Nuclear Waste Technical Review Board 2002, p. 7). There is, accordingly, scientific agreement that the significance of the volcanism problem is dependent on the consequences of a volcanic event. If the dose/risks are low, as indicated by the current DOE calculations, volcanism is a nonissue. If the dose/risks are significantly higher, as indicated by the higher range of the NRC results, further studies of the effects and uncertainty of repository disruption by volcanism must be conducted. Based on the experience gained from the years of volcanism studies, the actual estimates of volcanic effects probably lie somewhere between the current DOE and NRC result (. . . to be continued).

## Acknowledgments

Finding a balanced perspective for presenting an overview of the volcanism issue is a difficult task. Eugene Smith, University of Nevada, Las Vegas, an anonymous reviewer, Abe Van Luik, Eric Smistad and Jean Younker, of the YMP, Leon Reiter of the NWTRB and the editors of this book improved this chapter substantially through their review and comments.

## References

Allen, M. (2003) Possible or Probable? *Nature* 425, p. 242.

Blakeley, R.J., Langenheim, V.E., Ponce, D.A., and Dixon, G.L. (2000) *Aeromagnetic Survey of the Amargosa Desert, Nevada and California: A Tool for Understanding Near-Surface Geology and Hydrology.* U.S. Geological Survey. Open-File Report 00–188.

Bradshaw, T.K., and Smith, E.I. (1994) Polygenetic Quaternary Volcanism at Crater Flat, Nevada. *Journal of Volcanology and Geothermal Research* 63, pp. 165–182.

Chapman, C.R. (2000) The Asteroid/Comet Impact Hazard: Homo Sapiens as Dinosaur. In *Prediction: Science, Decision Making, and the Future of Nature*, ed. Sarewitz, D., Pielke, R.A., Jr., and Byerly, R., Jr. Washington, DC: Island Press, pp. 107–134.

Connor, C.B., and Hill, B.E. (1995) Three Nonhomogeneous Poisson Models for the Probability of Basaltic Volcanism: Application to the Yucca Mountain Region, Nevada, USA. *Journal of Geophysical Research* 100, pp. 10107–10125.

Connor, C.B., Stamatakos, J.A., Ferrill, D.A., Hill, B.E., Ofoegbu, G.I., Conway, F.M., Sagar, B., and Trapp, J. (2000) Geologic Factors Controlling Patterns of Small-Volume Basaltic Volcanism: Application to a Volcanic Hazards Assessment at Yucca Mountain, Nevada. *Journal of Geophysical Research* 105, pp. 417–432.

Crowe, B.M. (1986) Volcanic Hazard Assessment for Disposal of High-Level Radioactive Waste. In *Recent Tectonics: Impact on Society*. Washington, DC: National Academy Press, pp. 247–260.

Crowe, B.M., and Carr, W.J. (1980) *Preliminary Assessment of the Risk of Volcanism at a Proposed Nuclear Waste Repository in the Southern Great Basin*. U.S. Geological Survey. Open-File Report 80–375.

Crowe, B.M., Johnson, M.E., and Beckman, R.J. (1982) Calculation of the Probability of Volcanic Disruption of a High-Level Radioactive Waste Repository within Southern Nevada, USA. *Radioactive Waste Management* 3, pp. 167–190.

Crowe, B.M., Perry, F., Geissman, J., McFadden, L., Wells, S., Murrell, M., Poths, J., Valentine, G.A., Bowker, L., and Finnegan, K. (1995) *Status of Volcanism Studies for the Yucca Mountain Site Characterization Project*. Los Alamos National Laboratory Report. LA–12908–MS.

Crowe, B.M., Self, S., Vaniman, D., Amos, R., and Perry, F. (1983) Aspects of Potential Magmatic Disruption of a High-Level Radioactive Waste Repository in Southern Nevada. *Journal of Geology* 91, pp. 259–276.

CRWMS M&O (Civilian Radioactive Waste Management System, Management and Operating Contractor) (1998) *Total System Performance Assessment: Viability Assessment (TSPA–VA) Analyses Technical Basis Document*. Las Vegas. ACC:11981008.0001 to .0011.

Department of Energy (DOE) (1986) *Environmental Assessment Yucca Mountain Site, Nevada Research and Development Area, Nevada*. Office of Civilian Radioactive Waste Management. DOE/RW–0073.

Department of Energy (DOE) (1998) *Viability Assessment of a Repository at Yucca Mountain*. Office of Civilian Radioactive Waste Management. DOE/RW–0508.

Department of Energy (DOE) (2001) *Yucca Mountain Science and Engineering Report: Technical Information Supporting Site Recommendation Consideration*. Office of Civilian Radioactive Waste Management. DOE/RW–0539.

Department of Energy (DOE) (2002) *Yucca Mountain Site Suitability Evaluation*. Office of Civilian Radioactive Waste Management. DOE/RW–0549.

Detournay, E., Mastin, L.G., Jr., Pearson, A., Rubin, A.M., and Spera, F.J. (2003) *Final Report of the Igneous Consequences Peer Review Panel*. Bechtel SAIC Company LLC Report.

Environmental Protection Agency (EPA) (1993) 40 CFR Part 191: Environmental Radiation Protection Standards for the Management and Disposal of Spent Nuclear Fuel, High-Level and Transuranic Radioactive Wastes; Final Rule. *Federal Register* 58, pp. 66398–66416.

Environmental Protection Agency (EPA) (2001) 40 CFR Part 197: Public Health and Environmental Radiation Protection Standards for Yucca Mountain Nevada, Final Rule. *Federal Register* 66, pp. 32074–32135.

Fleck, R.J., Turrin, B.D., Sawyer, D.A., Warren, R.G., Champion, D.E., Hudson, M.R., and Minor, S.A. (1996) Age and Character of Basaltic Rocks of the Yucca Mountain Region, Southern Nevada. *Journal of Geophysical Research* 101, pp. 8205–8227.

Fridrich, C.J. (1999) Tectonic Evolution of the Crater Flat Basin, Yucca Mountain Region. In *Cenozoic Basins of the Death Valley Region*, ed. Wright, L.A., and Troxel, B.W. Geological Society of America Special Paper 333, pp. 169–195.

Fridrich, C.J., Witney, J.W., Hudson, M.R., and Crowe, B.M. (1999) Space-Time Patterns of Late Cenozoic Extension, Vertical Axis Rotation, and Volcanism in the Crater Flat Basin, Southwest Nevada. In *Cenozoic Basins of the Death Valley Region*, ed. Wright, L.A., and Troxel, B.W. Geological Society of America Special Paper 333, pp. 197–212.

Geomatrix Consultants (1996) *Probabilistic Volcanic Hazard Analysis for Yucca Mountain, Nevada*. San Francisco. Report BA0000000–1717–220–00082.

Ho, C.-H. (1992) Risk Assessment for the Yucca Mountain High-Level Nuclear Waste Repository Site: Estimation of Volcanic Disruption. *Mathematical Geology* 24, pp. 347–364.

IPCC (Intergovernmental Panel on Climate Change) (2001) *Climate Change 2001: The Scientific Basis; Contribution of Working Group to the Third Assessment Report of the Intergovernmental Panel on Climate Change*. Ed. Houghton, J.T., Ding, Y., Griggs, D.J., Noguer, M., van der Linden, P.J., Dai, X., Maskell, K., and Johnson, C.A. Cambridge: Cambridge University Press, 883 p.

Kane, M.F., and Bracken R.E. (1983) Aeromagnetic Map of Yucca Mountain and Surrounding Regions, Southwest Nevada. U.S. Geological Survey Open-File Report 83–616.

Langenheim, V.E. (1995) *Magnetic and Gravity Studies of Buried Volcanic Centers in the Amargosa Desert and Crater Flat, Southwest Nevada*. U.S. Geological Survey. Open-File Report 95–564.

Link, R.L., Logan, S.E., Ng, H.S., Rokenbach, F.A., and Hong, K.J. (1982) *Parametric Studies of Radiological Consequences of Basaltic Volcanism*. Sandia National Laboratories Report. SAND 81–2375.

Morgan, M.G., and Henrion, M. (1990) *Uncertainty*. Cambridge: Cambridge University Press, 332 p.

Morrison, D., Chapman, C.R., and Slovic, P. (1994) The Impact Hazard. In *Hazards Due to Comets and Asteroids*, ed. Gehrels, T. Tucson: University of Arizona Press, pp. 59–91.

National Research Council (1995) *Technical Basis for Yucca Mountain Standards*. Washington, DC: National Academy Press, 205 p.

Nuclear Regulatory Commission (NRC) (1999) *Issue Resolution Status Report, Key Technical Issue: Igneous Activity, Revision 2*: Washington, DC: U.S. Nuclear Regulatory Commission, Division of Waste Management.

Nuclear Regulatory Commission (NRC) (2002) 10 CFR Part 63: Proposed Rules Specification of a Probability for Unlikely Features, Events, and Processes. *Federal Register* 67, pp. 3628–3631.

Nuclear Waste Technical Review Board (2002) Letter and Attached Report to the U.S. Congress and the U.S. Secretary of Energy on the Board's Evaluation of the Department of Energy's Technical and Scientific Work. January 24.

O'Leary, D.W., Mankinen, E.A., Blakely, R.J., Langenheim, V.E., and Ponce, D.A. (2002) *Aeromagnetic Expression of Buried Basaltic Volcanoes near Yucca Mountain, Nevada*. U.S. Geological Survey. Open-File Report 02–020.

Perry, F.V., Crowe, B.M., Valentine, G.A., and Bowker, L.M. (1998) *Volcanism Studies: Final Report for the Yucca Mountain Project*. Los Alamos National Laboratory Report. LA–13478.

Perry, F., and Youngs, B. (2000) *Characterize Framework for Igneous Activity at Yucca Mountain, Nevada*. Analysis/Model Report. Office of Civilian Radioactive Waste Management. ANL–MGR–GS–000001 Rev 00, ICN 01, 148 p.

Perry, F.V., Cogbill, A.H., and Oliver, R.D. (2004) Reducing Uncertainty in the Hazard of Volcanic Disruption of the Proposed Yucca Mountain Radioactive Repository Using a High Resolution Aeromagnetic Survey to Detect Buried Basalt Volcanoes in Alluvial Filled Basins. Geological Society of America Abstracts with Programs V. 36, p. 33.

Sarewitz, D., Pielke, R.A., Jr., and Byerly, R., Jr. (2000) *Prediction: Science, Decision Making, and the Future of Nature*. Washington, DC: Island Press, 405 p.

Shaw, H.R. (1987) Uniqueness of Volcanic Systems. In *Volcanism in Hawaii, Volume 2*. U.S. Geological Survey Professional Paper 1350, pp. 1357–1394.

Smith, E.I., Feuerbach, D.L., Naumann, T.R., and Faulds, J.E. (1990) The Area of Most Recent Volcanism about Yucca Mountain, Nevada: Implications for Volcanic Risk Assessment. In *High-Level Radioactive Waste Management, Proceedings of the International Topical Meeting*. Las Vegas: American Nuclear Society. pp. 81–90.

Smith, E.I., Keenan, D.L., and Plank, T. (2002) Episodic Volcanism and Hot Mantle: Implications for Volcanic Hazard Studies at the Proposed Nuclear Waste Repository at Yucca Mountain, Nevada. *GSA Today* 12, pp. 4–9.

Stamatakos, J.A., Connor, C.B., and Martin, R.H. (1997) Quaternary Basin Evolution and Basaltic Volcanism of Crater Flat, Nevada, from Detailed Ground Magnetic Surveys of the Little Cones. *Journal of Geology* 105, pp. 319–330.

Valentine, G.A., and Groves, K.R. (1996) Entrainment of Country Rock during Basaltic Eruptions of the Lucero Volcanic Field, New Mexico. *Journal of Geology* 104, pp. 71–90.

Valentine, G.A., WoldeGabriel, G., Rosenberg, N.D., Carter Krogh, K.E., Crowe, B.M., Stauffer, P., Auer, L.H., Gable, C.W., Goff, F., Warren, R., and Perry, F.V. (1998) Physical Processes of Magmatism and Effects on the Potential Repository: Synthesis of Technical Work through Fiscal Year 1995. In *Volcanism Studies: Final Report for the Yucca Mountain Project*, ed. Perry, F.V., Crowe, B.M., Valentine, G.A., and Bowker, L.M. Los Alamos National Laboratory Report. LA–13478.

Vose, D. (2000) *Risk Analysis: A Quantitative Guide*. New York: John Wiley and Sons, 418 p.

Woods, A.W., Sparks, S., Bokhove, O., LeJeune, A.-M., Connor, C.B., and Hill, B.E. (2002) *Modeling the Explosive Eruption of Basaltic Magma into the Proposed High-Level Radioactive Waste Repository at Yucca Mountain, Nevada, U.S.A.* Center for Nuclear Waste Regulatory Analyses, San Antonio, TX. IM 01402.461.155.

# 9

# Climate Change at Yucca Mountain: Lessons from Earth History

MaryLynn Musgrove and Daniel P. Schrag

Yucca Mountain's suitability as a nuclear waste repository stems largely from its very dry climate and deep water table (Bodvarsson et al. 1999). The facility's ultimate goal is to isolate waste materials for more than ten thousand years (Whipple 1996). And there lies a crucial question: How can we tell whether the climate and hydrologic conditions at Yucca Mountain will be stable enough beyond the next ten millennia so that the site remains a safe repository for radioactive materials?

If one were planning for the next decade or two, one might look at instrumental records of precipitation and river discharge from the southwestern United States to determine the range of climate variability in the region. But if one must consider the future climate over thousands of years, the historical record is not as useful; climate fluctuations, such as catastrophic floods, may occur so infrequently that they are not captured in the last century or so of human observations. An alternative method for evaluating future climate and hydrologic conditions is to consider the geologic record of climate change. This record comes from a variety of archives. Sedimentary cements and deposits that form from groundwater in caves tell us about temperature, rainfall, and recharge. Remains of flora and fauna in sedimentary layers record details about moisture and climate. Flood and lake deposits hold information about water levels, shorelines, floods, and droughts. Such geologic and geochemical information on past climates may not be as precise as historical observations, but it has the advantage of extending the time scale of observations back over thousands or even millions of years.

This approach has been applied to the assessment of Yucca Mountain by investigating climate variations over the last few hundred thousand years (U.S. Geological Survey 2000). Proponents of this method of assessment argue that because climate is cyclic, the variability observed over this period of time should provide a good indication of what could happen beyond the next ten thousand years. In this chapter, however, we propose that consideration of the geologic record over just this time period offers an incomplete portrayal of the full range of climate change possible at Yucca Mountain beyond the next ten thousand years. We suggest, based on an expanded view of the geologic record, that there are substantial uncertainties and risks not adequately considered by previous studies.

## Paleohydrology and the Pleistocene

During the Pleistocene, which comprises the last two million years of Earth history, climate has oscillated between cold glacial periods and warm interglacial ones. The timing of these cycles is linked to changes in Earth's orbital parameters, which affect the intensity of sunlight reaching the planet (Hays et al. 1976). For the last ten thousand years (called the Holocene), Earth has been in an interglacial climate. Twenty thousand years ago, Earth was experiencing a maximum glacial climate, commonly called an ice age. Ice sheets covered large parts of North America and Europe, sea level was 130 meters lower than today, and Earth's mean surface temperature was about 5°C (Fairbanks 1989; Bard et al. 1990; Broecker and Hemming 2001). A complete glacial-interglacial climate cycle is approximately one hundred thousand years in duration (Hays et al. 1976). Thus, the U.S. Geological Survey (2000) investigation of the last four hundred thousand years covers the past four cycles.

Hydrologic conditions during previous interglacial periods were generally similar to modern climate. During glacial maxima, however, many independent lines of evidence indicate that the now-arid environment of the southwestern United States was much wetter. For example, the presence of extensive pluvial lakes in basins of the western United States during the last glacial period is well documented (Smith and Street-Perrott 1983; Benson and Thompson 1987). Vegetation remains preserved in packrat middens from the southern Great Basin indicate that at the peak of the last glacial period, the region received as much as 40 percent more annual rainfall than it does now (Spaulding 1985). Uranium-series dating of travertine and cave deposits from the Southwest support wetter conditions during glacial periods for the entire region (Musgrove et al. 2001; Brook et al. 1990; Szabo 1990). In southern Nevada, water table fluctuations during glacial periods reconstructed from subterranean calcite deposits indicate higher water levels and a wetter climate (Szabo et al. 1994). A rigorously dated vein calcite deposit from Devils Hole provides a continuous half-million-year record of the paleohydrologic and paleoclimatic conditions in the vicinity of Yucca Mountain (Ludwig et al. 1992; Edwards et al. 1997). Geochemical variations in this calcite, which formed from groundwater, correlate with global ice and marine records, and are suggestive of cooler and wetter conditions in southern Nevada during glacial periods (Winograd et al. 1988, 1992). Records from playa lake salt deposits in Death Valley and southern Great Basin fossil spring deposits support this view (e.g., Li et al. 1996; Quade et al. 1995).

The U.S. Geological survey (2000) assessment of Yucca Mountain acknowledges wetter climates for the southwestern United States in the geologic past associated with glacial intervals. This assessment concludes that the climate at Yucca Mountain for the next ten thousand years will be dominated by a cooler and wetter glacial-transition climate. It suggests that annual precipitation rates will increase, although not significantly. Water infiltration, however, will increase due to cooler temperatures and decreased evapotranspiration. This potential increase in water infiltration into Yucca Mountain is a critical component in evaluating the site because the infiltration and movement of water through the unsaturated zone (that is, the soil and the rock between the surface and the water table) will determine how long it may take waste canisters to corrode and for water to leach radionuclides from the waste (Whipple 1996). Consequently, there has been a continuing effort to develop and refine a model of unsaturated zone hydrology (Flint et al. 2001a, 2001b; Bodvarsson et al. 1999).

There has been considerable controversy over the origin of carbonate and opal deposits found in fractures of the unsaturated zone (see chapter 10, this volume). If these deposits were formed by upwelling waters from the deep aquifer, then this might mean that the water table fluctuates over time, and could possibly rise again in the future and inundate the waste facility (Szymanski 1989). Following many years of study, the preponderance of evidence indicates that these deposits were formed by surface processes and meteoric waters percolating down through the unsaturated zone (Wilson et al. 2003).

One climate scenario that has been considered and investigated with the unsaturated zone model uses conditions that may have approximated the climate of the last glacial period. This scenario models a fivefold increase in infiltration, which results in an increase of percolation water reaching the potential repository (Ritcey and Wu 1999). This study also examined changes that might occur due to anthropogenic climate change by considering a scenario with a doubling of atmospheric carbon dioxide ($CO_2$), resulting in a twofold increase in infiltration. Infiltration estimates in these scenarios are based on precipitation results from a regional climate model (Ritcey and Wu 1999). These estimates, however, are uncertain due to a variety of modeling complexities. Regardless, given the extremely low infiltration rates estimated for the current climate—from zero to a maximum of forty millimeters per year (Ritcey and Wu 1999)—even a fivefold increase still results in extremely low infiltration.

## Predicting Future Climate Change

The investigation of Pleistocene climates for the Yucca Mountain region provides useful information on how hydrologic conditions may change in response to large changes in global climate and is therefore important for assessing the site's suitability for long-term waste storage. But the climate variability of the Pleistocene is not necessarily a good indicator of what Earth will experience over the next one hundred years—much less the next ten thousand years. In fact, projecting Yucca Mountain's climate variability based on the Pleistocene climate may seriously underestimate the possibility for large changes in the climatic and hydrologic regime.

Atmospheric greenhouse gases, such as carbon dioxide ($CO_2$), absorb infrared radiation and maintain Earth's temperature, and are thus a crucial component of Earth's climate. The $CO_2$ levels during the Pleistocene ranged from a low of approximately 200 parts per million during glacial maxima to as much as approximately 280 parts per million during interglacial periods—a variation that contributed to changing global climate (Petit et al. 1999). Bubbles trapped in the Vostok ice core document that $CO_2$ never exceeded 300 parts per million over at least the last 420,000 years (Petit et al. 1999). Ever since the Industrial Revolution, however, humans have been conducting an extraordinary experiment. The combustion of fossil fuels has driven atmospheric $CO_2$ concentrations above 380 parts per million— a level that continues to rise. Projections for the year 2100 range from a low of 500 parts per million, if we implement extreme reductions in fossil fuel consumption, to over 1,200 parts per million, if the developing world accelerates its industrialization (Houghton et al. 2001). Any serious assessment of the climate at Yucca Mountain over the next several thousand years must surely take such enormous changes into consideration.

What are the expected effects of increased atmospheric $CO_2$ on global climate for the next one hundred years? Based on a wide range of scenarios, by 2100 the global average surface temperature is projected to increase by 1.4°C to 5.8°C, and the average sea level is projected to rise by nine to eighty-eight centimeters (Houghton et al. 2001). These changes will significantly affect the global hydrologic cycle. Higher temperatures will increase the amount of atmospheric water vapor, which may in turn increase global precipitation; more intense precipitation events are likely to occur over much of the Northern Hemisphere (Houghton et al. 2001; Hennessy et al. 1997). Looking beyond 2100, the changes could be much greater. Even if global $CO_2$ emissions are held constant at present-day values, atmospheric concentrations would continue to increase for hundreds of years before stabilizing. Global temperature and sea level would continue to rise due to the thermal lag associated with the oceans.

How might predictions of global climate change potentially affect the Yucca Mountain region? At a regional scale, predictions of climate change are hampered by many uncertainties. These include the many complexities of the hydrologic cycle, such as the formation and distribution of clouds, as well as processes that occur on spatial scales smaller than general circulation model grid cells (Water Sector Assessment Team 2000). Although many of these regional effects are currently known with less certainty than global predictions, regional assessments do provide some information on climate change at Yucca Mountain. Future climate scenarios for the southwestern United States predict a mean temperature increase of 4°C to 5.5°C over the next one hundred years (National Assessment Synthesis Team 2001; Water Sector Assessment Team 2000). Models predict that higher temperatures will result in overall wetter conditions, evidenced by increased winter precipitation and annual precipitation increases of approximately 30 to 70 percent (Water Sector Assessment Team 2000; Watson et al. 1998). At Yucca Mountain, winter precipitation presently provides the bulk of groundwater recharge water as well as unsaturated zone infiltration (Winograd et al. 1998). Thus, predicted increases in both overall precipitation and specifically winter precipitation may significantly affect how much moisture the waste disposal facility will encounter.

Another way to assess future climate change at Yucca Mountain is to consider the correlation of climate variability in the southwestern United States with larger patterns of climate variability. The largest source of climate variability from one year to the next is the El Niño/Southern Oscillation (ENSO), which is centered in the tropical Pacific but affects weather patterns worldwide. During ENSO warm phases (El Niño events), the eastern Pacific warms as trade winds slacken and the depth of the thermocline increases. El Niño events bring droughts to Southeast Asia and floods to the hyperarid coastal deserts of Peru. El Niño events also bring higher than average levels of winter precipitation to the southwestern United States (e.g., Cayan and Webb 1992). This relationship has allowed for the development of ENSO indexes based on tree ring chronologies from the southwestern United States (Cook 1992).

Over the last decade, there has been growing concern about whether ENSO is being affected by anthropogenic climate change. Trenberth and Hoar (1996, 1997) have suggested that the 1976 "climate shift," which involved a deepening of the thermocline in the eastern Pacific (Guilderson and Schrag 1998), and led to stronger and more frequent El Niño events in the 1980s and 1990s, is related to anthropogenic climate change. Others, however, have disputed this claim (e.g., Rajagopalan et al. 1997; Harrison and Larkin 1997).

Most climate models now used to predict the effects of increasing $CO_2$ over the next century do not have high enough resolution to accurately simulate ENSO variability. But Timmermann and colleagues (1999), using a new model with better resolution, found that an increase in $CO_2$ levels over the next century would not change the frequency and intensity of El Niño events by much. Still, in this study, because the eastern Pacific warmed considerably relative to the western Pacific, the mean state became more like an El Niño event. If correct, this prediction would indicate that the mean precipitation in the southwestern United States could be closer to the anomalous winter precipitation associated with El Niño events. This would add to concerns of wetter conditions over the next century in the Yucca Mountain region.

## An Extended View of Climate Change

The $CO_2$ levels projected by the Intergovernmental Panel on Climate Change for 2100 (Houghton et al. 2001) have not been seen on Earth for millions of years. If one extends the projections forward several centuries from now, the scale of the anthropogenic climate experiment is even more astounding. Direct observation of ancient levels of atmospheric composition are not possible beyond the age of the Vostok ice core (~420,000 years). More indirect methods suggest that the $CO_2$ levels have been similar to modern levels over at least the last twenty million years (Pagani et al. 1999; Pearson and Palmer 2000).

Evidence from fossil seashells, however, indicates that during the early Eocene— fifty million years ago—$CO_2$ levels may have been considerably higher than today.

The boron isotopic composition of calcium carbonate shells of marine animals reflects the pH of the seawater in which the animals lived; the pH is in turn strongly dependent on $CO_2$. Changes in the boron isotopic composition of these shells from fifty-two to thirty-eight million years ago suggest an increase in the pH over this interval, consistent with a decrease in $CO_2$ levels from one thousand to three thousand parts per million, down to near-modern levels (Pearson and Palmer 2000). Thus, the $CO_2$ levels we may see over the next several centuries have not occurred on Earth for fifty million years.

What lessons can we learn from this ancient time of high $CO_2$ that are relevant to our current predicament and Yucca Mountain in particular? The early Eocene was extremely warm. Deep ocean temperatures were 10°C to 12°C above what they are today (Miller et al. 1987). The world was essentially ice free with little or no polar ice and no continental glaciation (Sloan and Barron 1992). Antarctica was covered with pine forests (Case 1988). The Northern Hemisphere's continental interiors and high latitude regions were warm enough to support subtropical vegetation such as palm trees as well as cold-blooded reptiles such as crocodiles (Spicer et al. 1987; Estes and Hutchinson 1980; Marwick 1998). Geologic evidence from the Eocene, such as an abundance of the clay mineral, kaolinite, which commonly forms in warm and humid conditions (Robert and Kennett 1992, 1994), suggests that the climate was wetter than it is today.

In the western United States, oxygen isotope measurements of mammalian teeth from the early Eocene Bighorn Basin are consistent with wetter conditions in this region (Koch et al. 1995). Fossil leaves are another source of information about Eocene climate and rainfall in the western United States. The morphology of leaves is strongly influenced by the available moisture, which in turn can be used to assess climate conditions such as precipitation (Richards 1996). Estimates of precipitation for the western United States during the Eocene from fossil leaves consistently indicate much wetter conditions than currently exist (Wing and Greenwood 1993; Wilf et al. 1998; Wilf 2000). Overall, the climate of the Eocene western United States resembled that of a warm, wet, low-latitude region such as today's western Amazonia (Koch et al. 1995).

When climate modelers attempt to simulate the Eocene climate (or the earlier greenhouse climates of the Cretaceous) using models designed for the modern climate, they have difficulty simulating the warm conditions implied by the fossil and geochemical records. When the $CO_2$ levels in the model are raised even to four times the modern values or higher, the models are incapable of producing temperatures at the poles as warm as what the paleoclimate data suggest (e.g., Bush and Philander 1997). Moreover, the models are unable to keep the winter temperatures warm at high latitudes, particularly in the continental interiors (Sloan and Barron 1992). These problems with the models are troubling as they suggest that certain feedbacks are missing from the current climate models that may be important as $CO_2$ levels increase. Recently, Sloan and Morrill (1998) proposed that optically thick polar stratospheric clouds could produce these two features of Eocene climate if the clouds formed at times of higher $CO_2$. They suggested that the clouds formed due to the increased production of methane from tropical wetlands and the subsequent oxidation of that methane in the stratosphere. Kirk-Davidoff and colleagues (2002) proposed that stratospheric clouds might form through a direct feedback following increased $CO_2$, through a change in stratospheric circulation. If either of these hypotheses is correct, they imply that predictions of future climates may be underestimating the climatic response to high levels of $CO_2$ over the next several centuries. This uncertainty must also be acknowledged in the assessment of future climate conditions at the Yucca Mountain repository.

The extreme climate of the early Eocene, and the inability of most models to adequately simulate those conditions, raises the specter of more extreme changes in the hydrologic conditions at Yucca Mountain over the next several thousand years. These conditions may be far more extreme than the fivefold increase in infiltration rate simulated by Ritcey and Wu (1999). Nonetheless, it is probably not appropriate to use the Eocene as a direct analog for future conditions at Yucca Mountain over the next few centuries. First, there have been several large changes in physical geography that could have contributed to the regional climate in the western United States. Although the configuration of the continents during the Eocene was similar to today, the regional topography was quite different. The Great Basin's arid climate stems from its current location in the rain shadow of the Sierra Nevada; the uplift of these mountains is generally considered to have occurred in the last ten million years, or well after the early Eocene (e.g., Wolfe et al. 1997; Wernicke et al. 1996). Much of the volcanic activity in the southwestern United States has occurred in the last fifty million years. Even the location of the coastline was different in the Eocene, as much of California has been accreted onto the North American plate since that time. Due to the lack of continental ice sheets, Eocene sea level was more than one hundred meters higher than it is today. All these factors could have affected the position of the jet stream and the moisture balance in the southwestern United States, and contributed to wetter conditions.

But most important is the difference between a warm climate in the Eocene that persisted for millions of years, allowing local vegetation, ice sheets, and the deep ocean to reach a quasi-steady state, and the perturbation from a relatively cold climate that we are now experiencing as a result of anthropogenic increases in $CO_2$. The time scale for the deglaciation of the polar regions or the establishment of new ecosystems is difficult to calculate exactly, but it is probably on the order of thousands of years. This is similar to the time scale for the deglaciation of North America and shifts in ecosystems during the termination of the last ice age. The time scale for heating the deep ocean is similar, although the possible reduction of thermohaline circulation adds complications (Manabe and Stouffer 1999).

It is unlikely that we will see the complete demise of ice on Antarctica or the establishment of a rain forest in Nevada over the next few centuries. But the design of the Yucca Mountain repository is supposed to anticipate conditions beyond the next ten thousand years. Over this time scale, the possibility that Earth may experience a drastic climate change to conditions more similar to the Eocene must be considered seriously. There are of course abundant uncertainties in this assessment, most significantly the actions that society may take over the next century to reduce the emissions of $CO_2$ from the combustion of fossil fuels. But the risk of such $CO_2$-induced change under the conditions that exist today is not trivial. The design and construction of the repository must respond to this risk.

### Conclusions and Implications

Scientists cannot predict the future climate or hydrologic conditions at Yucca Mountain with absolute certainty. Nevertheless, the DOE must consider the possibility that the climate over the next several thousand years will be very different from anything experienced by humans, as levels of atmospheric $CO_2$ are likely to be higher than have existed for millions of years.

The most recent time in Earth history when the $CO_2$ levels approached what is anticipated to occur in the next few hundred years was the Eocene. As a result, a consideration of Eocene-like climate scenarios may provide some useful lessons about the full spectrum of possible climate changes resulting from increased $CO_2$. The design of the repository should account for this more complete assessment of risk of dramatically wetter conditions at Yucca Mountain.

## References

Bard, E., Hamelin, B., and Fairbanks, R.G. (1990) U-Th Ages Obtained by Mass Spectrometry in Corals from Barbados: Sea Level during the Past 130,000 Years. *Nature* 346, pp. 456–458.

Benson, L., and Thompson, R.S. (1987) The Physical Record of Lakes in the Great Basin. In *North America and Adjacent Oceans during the Last Deglaciation*, ed. Ruddiman, W.F., and Wright, H.E., Jr. Boulder, CO: Geological Society of America, Geology of North America, K-3, pp. 241–260.

Bodvarsson, G.S., Boyle, W., Patterson, R., and Williams, D. (1999) Overview of Scientific Investigations at Yucca Mountain: The Potential Repository for High-Level Nuclear Waste. *Journal of Contaminant Hydrology* 38, pp. 3–24.

Broecker, W.S., and Hemming S. (2001) Paleoclimate: Climate Swings Come into Focus. *Science* 294, pp. 2308–2309.

Brook, G.A., Burney, D.A., and Cowart, J.B. (1990) Desert Paleoenvironmental Data from Cave Speleothems with Examples from the Chihuahuan, Somali-Chalbi, and Kalahari Deserts. *Palaeogeography, Palaeoclimatology, Palaeoecology* 76, pp. 311–329.

Bush, A.B.G., and Philander, S.G.H. (1997) The Late Cretaceous; Simulation with a Coupled Atmosphere-Ocean General Circulation Model. *Paleoceanography* 12, pp. 495–516.

Case, J.A. (1988) Paleogene Floras from Seymour Island, Antarctic Peninsula. In *Geology and Paleontology of Seymour Island, Antarctic Peninsula*, ed. Feldmann, R.M., and Woodburne, M.O., Geological Society of America Memoir 169, pp. 523–530.

Cayan, D.R., and Webb, R.H. (1992) El Niño/Southern Oscillation and Streamflow in the Western United States. In *El Niño: Historical and Paleoclimatic Aspects of the Southern Oscillation*, ed. Diaz, H.F., and Markgraf, V., Cambridge: Cambridge University Press, pp. 29–68.

Cook, E.R. (1992) Using Tree Rings to Study Past El Niño/Southern Oscillation Influences on Climate. In *El Niño: Historical and Paleoclimatic Aspects of the Southern Oscillation*, ed. Diaz, H.F., and Markgraf, V., Cambridge: Cambridge University Press, pp. 203–214.

Edwards, R.L., Cheng, H., Murrell, M.T., and Goldstein, S.J. (1997) Protactinium-231 Dating of Carbonates by Thermal Ionization Mass Spectrometry: Implications for Quaternary Climate Change. *Science* 276, pp. 782–786.

Estes and Hutchinson, J.H. (1980) Eocene Lower Vertebrates from Ellesmere Island, Canadian Arctic Archipelago. *Palaeogeography, Palaeoclimatology, Palaeoecology* 30, pp. 325–347.

Fairbanks, R.G. (1989) A 17,000-Year Glacio-Eustatic Sea Level Record: Influence of Glacial Melting Rates on the Younger Dryas Event and Deep-Ocean Circulation. *Nature* 342, pp. 637–642.

Flint, A.L., Flint, L.E., Bodvarsson, G.S., Kwicklis, E.M., and Fabryka-Martin, J. (2001a) Evolution of the Conceptual Model of Unsaturated Zone Hydrology at Yucca Mountain, Nevada. *Journal of Hydrology* 247, pp. 1–30.

Flint, A.L., Flint, L.E., Kwicklis, E.M., Bodvarsson, G.S., and Fabryka-Martin, J. (2001b) Hydrology of Yucca Mountain, Nevada. *Reviews of Geophysics* 39, pp. 447–470.

Guilderson, T.P., and Schrag, D.P. (1998) Abrupt Shift in Subsurface Temperatures in the Eastern Tropical Pacific Associated with Recent Changes in El Niño. *Science* 281, pp. 240–243.

Harrison, D.E., and Larkin, N.K. (1997) Darwin Sea Level Pressure, 1876–1996: Evidence for Climate Change? *Geophysical Research Letters* 24, pp. 1779–1782.

Hays, J.D., Imbrie, J., and Shackleton, N.J. (1976) Variations in the Earth's Orbit: Pacemaker of the Ice Ages. *Science* 194, pp. 1121–1132.

Hennessy, K.J., Gregory, J.M., and Mitchell, J.F.B. (1997) Changes in Daily Precipitation under Enhanced Greenhouse Conditions. *Climate Dynamics* 13, pp. 667–680.

Houghton, J.T., Ding, Y., Griggs, D.J., Noguer, M., Van der Linden, P.J., and Xiaosu, D. (2001) *Climate Change 2001: The Scientific Basis, Contribution of Working Group I to the Third Assessment Report of the Intergovernmental Panel on Climate Change.* Port Chester, NY: Cambridge University Press, 944 p.

Kirk-Davidoff, D.B., Schrag, D.P., and Anderson, J.G. (2002) On the Feedback of Stratospheric Clouds on Polar Climate. *Geophysical Research Letters* 29, p. 4.

Koch, P.L., Zachos, J.C., and Dettman, D.L. (1995) Stable Isotope Stratography and Paleoclimatology of the Paleogene Bighorn Basin (Wyoming, USA). *Palaeogeography, Palaeoclimatology, Palaeoecology* 115, pp. 61–89.

Li, J., Lowenstein, T.K., Brown, C.B., Ku, T.-L., and Luo, S. (1996) A 100 ka Record of Water Tables and Paleoclimates from Salt Cores, Death Valley, California. *Palaeogeography, Palaeoclimatology, Palaeoecology* 123, pp. 179–203.

Ludwig, K.R., Simmons, K.R., Szabo, B.J., Winograd, I.J., Landwehr, J.M., Riggs, A.C., and Hoffman, R.J. (1992) Mass-Spectrometric $^{230}$Th-$^{234}$U-$^{238}$U Dating of the Devils Hole Calcite Vein. *Science* 258, pp. 284–287.

Manabe, S., and Stouffer, R.J. (1999) The Role of Thermohaline Circulation in Climate, Tellus Series A. *Dynamic Meteorology and Oceanography* 51, pp. 91–109.

Marwick, P.J. (1998) Fossil Crocodilians as Indicators of Late Cretaceous and Cenozoic Climates: Implications for Using Palaeontological Data in Reconstructing Palaeoclimate. *Palaeogeography, Palaeoclimatology, Palaeoecology* 137, pp. 205–271.

Miller, K.G., Fairbanks, R.G., and Mountain, G.S. (1987) Tertiary Oxygen Isotope Synthesis, Sea-Level History, and Continental Margin Erosion. *Paleoceanography* 2, pp. 741–761.

Musgrove, M., Banner, J.L., Mack, L.E., Combs, D.M., James, E.W., Cheng, H., and Edwards, R.L. (2001) Geochronology of Late Pleistocene to Holocene Speleothems from Central Texas: Implications for Regional Paleoclimate. *Geological Society of America Bulletin* 113, pp. 1532–1543.

National Assessment Synthesis Team (2001) *Climate Change Impacts on the United States: The Potential Consequences of Climate Variability and Change, U.S. Global Change Research Program.* Port Chester, NY: Cambridge University Press, 618 p.

Pagani, M., Freeman, K.H., and Arthur, M.A. (1999) Late Miocene Atmospheric $CO_2$ Concentrations and the Expansion of C4 Grasses. *Science* 285, pp. 876–879.

Pearson, P.N., and Palmer, M.R. (2000) Atmospheric Carbon Dioxide Concentrations over the Past 60 Million Years. *Nature* 406, pp. 695–699.

Petit, J.R., Jouzel, J., Raynaud, D., Barkov, N.I., Barnola, J.-M., Basile, I., Bender, M., Chappellaz, J., Davis, M., Delaygue, G., Delmotte, M., Kotlyakov, V.M., Legrand, M., Lipenkov, V.Y., Lorius, C., Pepin, L., Ritz, C., Saltzman, E., and Stievenard, M. (1999) Climate and Atmospheric History of the Past 420,000 Years from the Vostok Ice Core, Antarctica. *Nature* 399, pp. 429–436.

Quade, J., Mifflin, M.D., Pratt, W.L., McCoy, W., and Burckle, L. (1995) Fossil Spring Deposits in the Southern Great Basin and Their Implications for Changes in Water-Table Levels near Yucca Mountain, Nevada, during Quaternary Time. *Geological Society of America Bulletin* 107, pp. 213–230.

Rajagopalan, B., Lall, U., and Cane, M.A. (1997) Anomalous ENSO Occurrences: An Alternate View. *Journal of Climate* 10, pp. 2351–2357.

Richards, P.W. (1996) *The Tropical Rain Forest*. 2nd ed. Cambridge: Cambridge University Press, 575 p.

Ritcey, A.C., and Wu, Y.S. (1999) Evaluation of the Effect of Future Climate Change on the Distribution and Movement of Moisture in the Unsaturated Zone at Yucca Mountain, NV. *Journal of Contaminant Hydrology* 38, pp. 257–279.

Robert, C., and Kennett, J.P. (1992) Paleocene and Eocene Kaolinite Distribution in the South Atlantic and Southern Ocean: Antarctic Climatic and Paleoceanographic Implications. *Marine Geology* 103, pp. 99–110.

Robert, C., and Kennett, J.P. (1994) Antarctic Subtropical Humid Episode at the Paleocene-Eocene Boundary: Clay Mineral Evidence. *Geology* 24, pp. 347–350.

Sloan, L.C., and Barron, E.J. (1992) A Comparison of Eocene Climate Model Results to Quantified Paleoclimatic Interpretations. *Palaeogeography, Palaeoclimatology, Palaeoecology* 93, pp. 183–202.

Sloan, L.C., and Morrill, C. (1998) Orbital Forcing and Eocene Continental Temperatures. *Palaeogeography, Palaeoclimatology, Palaeoecology* 144, pp. 21–35.

Smith, G.I., and Street-Perrott, F.A. (1983) Pluvial Lakes of the Western United States. In *Late-Quaternary Environments of the United States, Vol. 1: The Late Pleistocene*, ed. Porter, S.C. Minneapolis: University of Minnesota Press, pp. 190–214.

Spaulding, W.G. (1985) *Vegetation and Climates of the Last 45,000 Years in the Vicinity of the Nevada Test Site, South-Central Nevada*. U.S. Geological Society Professional Paper 1329, 83 p.

Spicer, R.A., Wolfe, J.A., and Nichols, D.J. (1987) Alaskan Cretaceous-Tertiary Floras and Arctic Origins. *Paleobiology* 13, pp. 73–83.

Szabo, B.J. (1990) Ages of Travertine Deposits in Eastern Grand Canyon National Park, Arizona. *Quaternary Research* 34, pp. 24–32.

Szabo, B.J., Kolesar, P.T., Riggs, A.C., Winograd, I.J., and Ludwig, K.R. (1994) Paleoclimatic Inferences from a 120,000-Yr Calcite Record of Water-Table Fluctuation in Browns Room of Devils Hole, Nevada. *Quaternary Research* 41, pp. 59–69.

Szymanski, J.S. (1989) *Conceptual Considerations of the Yucca Mountain Groundwater System with Special Emphasis on the Adequacy of This System to Accommodate a High-Level Nuclear Waste Repository.* Department of Energy Internal Report. Yucca Mountain Project Office, Las Vegas. July.

Timmermann, A., Oberhuber, J., Bacher, A., Esch, M., Latif, M., and Roeckner, E. (1999) Increased El Niño Frequency in a Climate Model Forced by Future Greenhouse Warming. *Nature* 398, pp. 694–696.

Trenberth, K.E., and Hoar, T.J. (1996) The 1990–1995 El Niño Southern Oscillation Event: Longest on Record. *Geophysical Research Letters* 23, pp. 57–60.

Trenberth, K.E., and Hoar, T.J. (1997) El Niño and Climate Change, *Geophysical Research Letters* 24, pp. 3057–3060.

U.S. Geological Survey (2000) *Future Climate Analysis.* U.S. Geological Survey, Denver, CO. ANL–NBS–GS–000008–Rev 00. http://www.ymp.gov.

Water Sector Assessment Team (2000) *Water: The Potential Consequences of Climate Variability and Change for the Water Resources of the United States.* Report of the Water Sector Assessment Team of the National Assessment of the Potential Consequences of Climate Variability and Change for the U.S. Global Change Research Program. http://www.usgcrp.gov/usgcrp/ nacc/water/default.htm.

Watson, R.T., Zinyowera, M.C., and Moss, R.H. (1998) *The Regional Impacts of Climate Change: An Assessment of Vulnerability.* Intergovernmental Panel on Climate Change (IPCC) Special Report. Port Chester, NY: Cambridge University Press, 527 p. http://www.grida.no/climate/ipcc/regional/index.htm.

Wernicke, B., Clayton, R., Ducea, M., Jones, C.H., Park, S., Ruppert, S., Saleeby, J., Snow, J.K., Squires, L., Fliedner, M., Jiracek, G., Keller, R., Klemperer, S., Luetgert, J., Malin, P., Miller, K., Mooney, W., Oliver, H., and Phinney, R. (1996) Origin of High Mountains in the Continents: The Southern Sierra Nevada. *Science* 271, pp. 190–193.

Whipple, C.G. (1996) Can Nuclear Waste Be Stored Safely at Yucca Mountain? *Scientific American*, June, pp. 72–79.

Wilf, P. (2000) Late Paleocene–Early Eocene Climate Changes in Southwestern Wyoming: Paleobotanical Analysis. *Geological Society of America Bulletin* 112, pp. 292–307.

Wilf, P., Wing, S.L., Greenwood, D.R., and Greenwood, C.L. (1998) Using Fossil Leaves as Paleoprecipitation Indicators: An Eocene Example. *Geology* 26, pp. 203–206.

Wilson, N.F., Cline, J.S., and Amelin, Y.V. (2003) Origin, Timing, and Temperature of Secondary Calcite-Silica Mineral Formation at Yucca Mountain, Nevada. *Geochimica et Cosmochimica Acta* 67, pp. 1145–1176.

Wing, S.L., and Greenwood, D.R. (1993) Fossils and Fossil Climate: The Case for Equable Continental Interiors in the Eocene. *Philosophical Transactions of the Royal Society of London*, series B, 341, pp. 243–252.

Winograd, I.J., Coplen, T.B., Landwehr, J.M., Riggs, A.C., Ludwig, K.R., Szabo, B.J., Kolesar, P.T., and Revesz, K.M. (1992) Continuous 500,000-Year Climate Record from Vein Calcite in Devils Hole, Nevada. *Science* 258, pp. 255–260.

Winograd, I.J., Riggs, A.C., and Coplen, T.B. (1998) The Relative Contributions of Summer and Cool-Season Precipitation to Groundwater Recharge, Spring Mountains, Nevada, USA. *Hydrogeology Journal* 6, pp. 77–93.

Winograd, I.J., Szabo, B.J., Coplen, T.B., and Riggs, A.C. (1988) A 250,000-Year Climatic Record from Great Basin Vein Calcite: Implications for Milankovitch Theory. *Science* 242, pp. 1275–1280.

Wolfe, J.A., Schorn, H.E., Forest, C.E., and Molnar, P. (1997) Paleobotanical Evidence for High Altitudes in Nevada during the Miocene. *Science* 276, pp. 1672–1675.

# II

## Hydrology

# 10

## Hot Upwelling Water: Did It Really Invade Yucca Mountain?

Nicholas S. F. Wilson and Jean S. Cline

Arguments have raged for more than a decade over whether hot water from deep within the Earth could in the future flood the proposed nuclear waste repository at Yucca Mountain. Rising, heated water that reaches the repository could contribute to the degradation and even destruction of the waste canisters, and could contaminate the groundwater that could then transport radioactive material beyond the perimeter of the site.

As is often the case in geology (and other endeavors too, for that matter), the clues to the future lie in the past. In this instance, knowing the history of fluid movement through the Yucca Mountain site is important in evaluating its long-term safety. If geologic processes caused recent upwelling hot water to flood the site, then those processes might still be active and could cause the flooding of the repository after it has been filled with nuclear waste. If, on the other hand, there is no such evidence of ascending fluids—the conclusion advanced by U.S. government scientists (Paces et al. 1996; Paces et al. 2001; Whelan et al. 2002)—then that is one potential threat to the repository that need not worry us.

### When Did the Upwelling Hot Water Question Arise?

The possibility of upwelling hot fluids was first proposed in 1989 by then Department of Energy (DOE) scientist Jerry Szymanski (1989). Szymanski observed geologic features that he concluded indicated that hot thermal waters from deep in the crust had repeatedly risen along faults and saturated the potential repository site (figure 10.1).

At the request of the DOE, the National Academy of Sciences established a panel to study whether the water table has risen to the proposed repository level in the past and whether this was likely to happen during the next ten thousand years. The

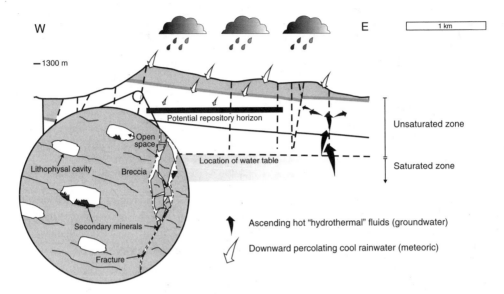

**Figure 10.1**
Simplified cross-section of Yucca Mountain. The host volcanic rocks (enlargement at the left) contain lithophysal (gas) cavities, fractures, and breccias. Secondary minerals are rare and are not present in most of the open spaces in the rocks. On the right, arrows illustrate the two competing models for the formation of secondary minerals: U.S. Geological Survey scientists conclude that the secondary minerals formed from downward percolating rainwater (white arrows). State of Nevada scientists believe that the minerals formed from hot rising fluids (black arrows) along faults (vertical dashed lines) that cut the rock sequence.

panel concluded that the features ascribed to the ascending water were related to the deposition of the volcanic rocks or were classic examples of desert soil recognized throughout the world (National Research Council 1992). The National Research Council (1992) report concluded that there was no evidence that the water table has periodically risen hundreds of meters and flooded the repository rocks.

**Evidence for High-Temperature Fluids**

In 1998, scientists representing the state of Nevada presented new data indicating that fluids of temperatures up to 85°C were present in the repository level at some time in the geologic past (Dublyansky 1998). These scientists had collected thin crusts of minerals (figure 10.2A) from the first two hundred meters of the Exploratory Studies Facility, a twenty-five-foot diameter, five-mile long tunnel driven into

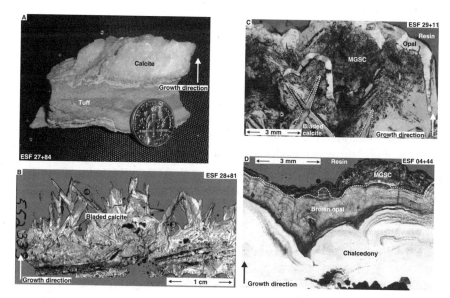

**Figure 10.2**
*A,* sample of the host tuff overgrown by a thin layer of secondary minerals. Most mineral crusts are less than or about 0.5 centimeters in thickness (though a few reach three- to four-centimeters thick), but most cavities, fractures, and breccias contain no mineral crusts. *B,* thin section of mineral crust composed of distinctive long blades of calcite ($CaCO_3$). Note that in this section, no minerals overgrew these crystals. *C,* microscope image of dark calcite with clear opal overgrowing bladed calcite from a lithophysal cavity. The outermost calcite is enriched in magnesium (Mg) (MGC; magnesium growth-zoned calcite), associated with opal, and is important as MGC and opal occur as the youngest minerals across the site. *D,* microscope image from a fracture sample containing a layer of chalcedony ($SiO_2$) adjacent to the host tuff overgrown by brown opal ($SiO_2$) and by MGC on the outermost surface. (Figures 10.2C–D published with the permission of Elsevier and taken from Wilson et al. 2003).

Yucca Mountain, which allowed the host rocks to be better studied. They reported that within these mineral crusts were tiny samples of geologic fluids that had been trapped at elevated temperatures at some time in the past. The scientists interpreted their data as convincing evidence of the recent upwelling of hot fluids (Dublyansky 1998; Dublyansky et al. 2001), although they provided no information on when the fluids were present in the site.

Experts who reviewed the data at the request of the Nuclear Waste Technical Review Board (1998) agreed that the rock record indicated that fluids with elevated temperatures had been present at Yucca Mountain, but without timing information the data could not be interpreted. As a result, the Nuclear Waste Technical Review Board (1998) recommended that the DOE consider funding a study to validate the former presence of hot fluids at Yucca Mountain and, if possible, determine when these fluids were present.

The question of timing is critical. What if the hot fluids were due to "old" rainwater that was heated as it percolated through the hot volcanic rocks when Yucca Mountain was originally formed? Such a scenario would pose no problem for the storage of nuclear waste as the water infiltration occurred millions of years ago and was limited to the event that formed the mountain. But what if the heated fluids were geologically young? If that is the case, then it might indicate that Yucca Mountain is not a suitable site for waste storage.

To answer these questions, the DOE funded an independent study that began in April 1999 at the University of Nevada Las Vegas (our group). To keep the community apprised of progress on the study, the University of Nevada Las Vegas scientists held regular meetings to which scientists representing the state of Nevada, the DOE, and other interested parties were invited. The group also provided samples to the state of Nevada and U.S. Geological Survey scientists so that they could conduct parallel studies on equivalent material. This allowed data collected independently by the different groups to be compared, with the hope of producing a consensus on when hot fluids were present at the site.

### The Question of When

The University of Nevada Las Vegas study was designed to address the following questions:

- Were fluids with elevated temperatures present at Yucca Mountain some time in the geologic past?
- If so, what temperatures were the fluids?
- Were they present throughout the site?
- Most important, when were these fluids present?

First, we collected samples from across the site of all types of mineral crusts that could be identified. We gathered 155 samples from lithophysal cavities (gas cavities formed as the volcanic rock cooled and solidified), fractures, and breccias, which are cemented, broken, angular fragments of rock. Samples of rock and mineral crusts (figure 10.2A) were removed from tunnel walls in the Exploratory Studies

Facility, the cross-drift (a roughly east-west tunnel that cuts through Yucca Mountain about fifteen meters above the planned repository horizon), and the exploratory alcoves that were mined to investigate specific areas of interest. The samples were transported and stored in coolers. This was done to make sure that heat did not destroy the tiny fluid samples in the rocks that provided the temperature information. With this large number of samples, we were able to identify similarities and differences in the mineral crusts across the site as well as determine the growth history of the minerals crusts.

Once the samples arrived in the lab, thin slices of the rocks were prepared and polished for viewing under a microscope (figure 10.2B). The goal in studying each sample was to identify the textural relationships between the minerals and decipher the history of mineral precipitation at each sample site. The crusts formed by the precipitation of minerals, layer by layer, in the open spaces that were present in the volcanic rocks (tuffs). The oldest mineral layers precipitated on the volcanic rock and were followed by younger minerals that formed blanketlike layers. The most recently deposited layer generally forms the outermost mineral surface. In figure 10.2B, the thin section contains long, slender, blades of calcite ($CaCO_3$) that precipitated on the volcanic tuff. In figure 10.2C, a photograph of part of a thin section shows the bladed calcite (highlighted by the dashed white line) and the darker calcite and clear opal ($SiO_2$) that precipitated on the calcite blades. These mineral relationships tell us that the slender blades of calcite are older than the outer calcite and associated opal. A comparison of the sections in figures 10.2B and 10.2C indicates that not all samples experienced the same mineral history, and some samples (e.g., figure 10.2B) recorded only part of the mineral history.

After examining a large number of samples, we were able to identify a consistent sequence of mineral precipitation across the site. Not all layers were present in every sample, but those that were present always appeared in the same order. This result was surprising considering that most of the open spaces in the tunnel walls contained no minerals at all (Whelan et al. 2002). The earliest minerals that were variably present in the samples were calcite with some quartz ($SiO_2$), chalcedony and opal (cryptocrystalline forms of quartz; figure 10.2D), and fluorite ($CaF_2$). These minerals precipitated on the volcanic tuff. The intermediate-age material was composed mainly of bladed calcite, although intermediate-age calcite also exhibited a number of other forms, and occurred with small amounts of opal and chalcedony. The youngest minerals were composed of distinctive dark or clear calcite intergrown with clear opal. This unit is significant, as it always forms the outermost (youngest) surface of the mineral crusts and can be correlated across the site (figures 10.2C and 10.2D; Wilson et al. 2003). These observations provided the strong, fundamental geologic foundation and constrained the relative timing of mineral growth.

## Chemical Variations

To further correlate mineral crusts from sample site to sample site, we investigated their chemical composition. We searched for chemically distinctive mineral layers that could be correlated among individual samples. We used an electron microprobe to produce quantitative chemical analyses of microscopic regions in the mineral crusts. By scanning the electron beam over parts of the thin section, the microprobe produced maps showing the relative abundance of different elements in the area scanned (figure 10.3A).

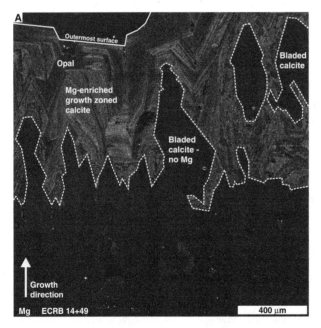

**Figure 10.3**
A, Microprobe map for Mg illustrating the variation in Mg content across a mineral crust. Brighter areas of the map indicate a higher concentration of Mg. Spiky bladed calcite crystals that do not contain Mg (dark) are overgrown (dashed line) at the outer, youngest edge of the section, by Mg-enriched growth-zoned calcite (bright). The presence of MGC can be

One element, magnesium (Mg), was uniquely enriched in two of the calcite layers. Early calcite that precipitated on the wall rock contained patchy enrichments of Mg. The most important occurrence of Mg in calcite was identified in the youngest (that is, outermost) calcite layer. This calcite exhibited growth zones that are enriched by as much as 1 percent Mg by weight and alternate with zones that are Mg free. This calcite was termed magnesium growth-zoned calcite (MGC) and it has been identified in more than 65 percent of the samples collected across the site (Wilson et al. 2003).

This chemical evidence indicated that the youngest event—the precipitation of the MGC and opal layer—was recorded at most of the sample-collection sites and could be correlated between different samples across the site (Wilson et al. 2003). Those sample sites in which MGC was not present (e.g., figure 10.2B) were not accessed by fluids that deposited MGC. The thickness of the MGC layer ranges

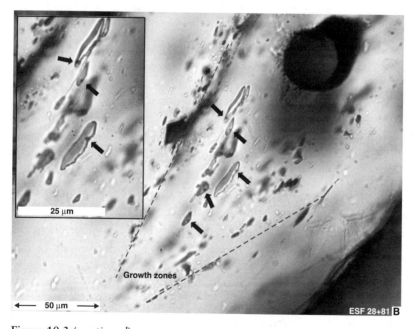

**Figure 10.3** (*continued*)
correlated between samples, indicating that the youngest fluid flow event was recorded across the site. *B*, photomicrograph of typical primary—formed along crystal growth zones—two-phase (liquid + vapor) fluid inclusions in calcite at Yucca Mountain. These inclusions formed at higher than present-day temperatures, and on cooling, small dark vapor bubbles formed (arrows; see detailed inset).

from less than 0.5 millimeter to more than one centimeter in some crusts (figure 10.3A). The presence of Mg in the youngest calcite indicates a change in mineral precipitation at Yucca Mountain; the most recent fluids entering the site contained fluctuating concentrations of Mg (Wilson et al. 2003).

The combined petrography and chemistry studies provided a sound understanding of the growth history of the samples collected at each sample site. Armed with this knowledge, we could now examine the tiny samples of fluid trapped in the rocks to determine the temperatures at which the different mineral layers precipitated.

## Fluid Inclusions

Most crystals precipitate from some type of fluid. As crystals grow, imperfections form in the crystal structure, creating cavities that can trap small amounts of this fluid. These trapped fluids are referred to as *fluid inclusions* and are important because they are actual samples of the geologic fluid that was present when the mineral formed. If we analyze these inclusions we can learn about the temperature, pressure, and chemical environment at the time the mineral formed.

The temperature of the fluid at the time of its capture is indicated by the phases of the fluid present in the inclusions. The presence (at 25°C) of a single-phase liquid indicates that the fluid was trapped originally at ambient or near-ambient temperatures. On the other hand, if the included fluid consists of two phases—a liquid and a vapor (vacuum) bubble—then we can conclude that the fluids were trapped at elevated temperatures. The reasoning is straightforward. Over time, as the crystals containing the fluid inclusions cool, the hot fluid cools and contracts. The cavity in which the fluid inclusion is trapped, however, does not shrink. This results in the formation of a vacuum in the inclusion cavity, which appears as a vapor bubble (figure 10.3B).

We can deduce the approximate temperature at which the fluid was trapped—that is, at which the mineral precipitated—by reversing the natural cooling of the system. We heat the inclusions in the laboratory while observing them under a microscope. As the temperature rises, the fluid expands and the vapor bubble shrinks. The so-called homogenization temperature, at which the bubble disappears altogether to leave a liquid-only inclusion, is equivalent to the temperature at which the mineral originally precipitated.

Since the presence of hot fluids correlates to two-phase fluid inclusions, we need to know the distribution of two-phase fluid inclusions within the mineral crusts and across the site. Specifically, we need to relate the presence of one- and two-phase inclusions to the growth history determined for each sample. This is called fluid inclusion petrography and is critical in determining the relative ages of the two-phase inclusions.

## Fluid Inclusion Study

Our study began with months of microscope work locating the two-phase and liquid-only fluid inclusions in the thin sections, determining which mineral layers they were present in and noting their locations on images of polished sections (figure 10.4). These observations indicated the distribution and relative timing of the two-phase and liquid-only inclusions. Of the 155 samples studied, 78 contained two-phase fluid inclusions. About half of these 78 samples came from lithophysal cavities, with the remainder from fractures or breccia occurrences. Most of these two-phase inclusions were in calcite, and are evidence of elevated temperature fluids. The remaining samples contained only liquid-only inclusions.

Determining the location of the two-phase fluid inclusions in various mineral layers in the mineral crusts provided a critical piece of information. Two-phase fluid inclusions were, for the most part, present only in the earliest calcite; the exception was a single sample in which two-phase inclusions were observed in early-intermediate calcite (Wilson et al. 2003). This means that the record of fluids with elevated temperatures is confined to early and early-intermediate calcite; elevated temperatures were not recorded in late-intermediate calcite or the youngest MGC. All of the samples taken from the younger minerals contained liquid-only inclusions, which indicate the presence of low-temperature (<35°C) fluids (Wilson et al. 2003). These observations showed that fluids with elevated temperatures were present only during the earliest and early-intermediate history of the site (Wilson et al. 2003).

Fluid inclusion temperatures obtained in this study were similar to those presented by earlier studies (Dublyansky 1998; Dublyansky et al. 2001). For the fluid inclusions in most samples collected from across the Yucca Mountain site, the homogenization temperature ranged between 39°C and 59°C (figure 10.5). One small area in the northeastern part of the site recorded homogenization temperatures as high as 83°C; the central part of the site recorded lower temperatures, generally below 50°C (Wilson et al. 2003). Our fluid inclusion study demonstrated that the

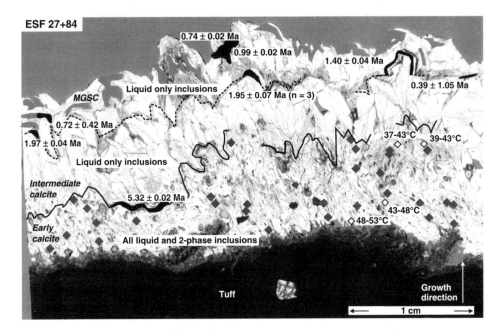

**Figure 10.4**

This section illustrates the process involved in determining the age of elevated temperature fluid inclusions. First, the important mineral relationships are documented (i.e., darker basal calcite, clear middle calcite, and dark and clear outer calcite) and the growth history of the section is constructed. This is integrated with microprobe data (the dashed line approximately represents the boundary between MGC and intermediate calcite) and further constrains the formation of the mineral crust. Fluid inclusion petrography located two-phase inclusions (black diamonds) that provided temperature information (white diamonds indicate measured inclusions). Other minerals in the crust were then dated; in this sample, opal (black = dated; gray = not dated) constrains the ages of the two-phase inclusions. This figure shows that elevated temperature fluids were present more than 5.32 million years ago and were trapped in two-phase inclusions in the darker calcite below the first opal layer (solid line). The fluid inclusion data also show that fluids decreased in temperature as the minerals continued to grow (a cooling trend). (Figure 10.4 is taken from Wilson et al. 2003, which tabulates forty-one U-Pb dates presented graphically for eighteen samples and is published with the permission of Elsevier.)

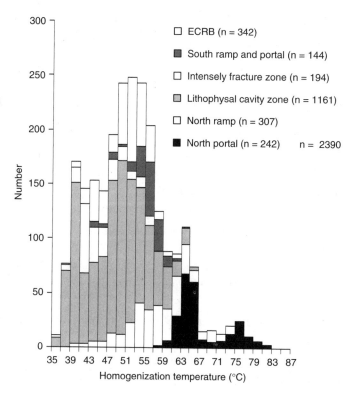

**Figure 10.5**
Summarized fluid inclusion data (Wilson et al. 2003) for the Yucca Mountain samples. Most inclusions homogenized from 39°C to 59°C. Samples from the north portal and ramp have higher temperatures than inclusions from the rest of the site. Conversely, samples from the intensely fractured zone recorded the coolest temperatures, usually <50°C. Samples from the lithophysal cavity zone and the ECRB cross-drift have bimodal distributions with modes of 47°C to 55°C in older calcite and 39°C to 41°C in younger calcite, and indicate a cooling trend with time (figure 10.4). Locations of these areas are given by Wilson et al. 2003.

homogenization temperatures declined continuously over time (Wilson et al. 2003). In many samples, the two-phase fluid inclusions in older mineral layers homogenized at temperatures greater than 50°C; the fluid inclusions in younger mineral layers in these samples homogenized at lower temperatures (figure 10.4). Significantly, the fluid inclusions provide no evidence for multiple incursions of hot fluids.

## How Can We Determine When the Elevated Temperature Minerals Were Deposited?

Research by the U.S. Geological Survey indicated that opal and chalcedony were enriched (typically around 150 to 200 parts per million) in uranium (U), and that these minerals could be dated using the uranium-lead (Pb) method (Paces et al. 1996). U-Pb dating is one of the most robust methods of dating geologic events and has helped scientist determine the age of the earth. By measuring the ratio of U and Pb in a mineral, and by knowing the constant rate at which uranium atoms radioactively decay to form lead atoms, the mineral's age can be calculated (Faure 1986).

The strategy for dating the Yucca Mountain minerals and determining the timing of fluid flow was straightforward. The first goal was to determine the age of the outermost MGC layer, which contained liquid-only inclusions. In all cases, two-phase inclusions were older than this calcite and opal layer. Therefore, by dating the opal that formed with the MGC, a minimum age for the two-phase inclusions could be determined (Wilson et al. 2003). A second goal was to obtain more precise ages for the fluids from samples in which the inclusions were embedded amid layers of opal and chalcedony (Wilson et al. 2003). By dating these layers and noting their spatial relation to the inclusions, it was possible to impose boundaries on the ages of the fluid inclusions and the fluids of various temperatures.

Samples of opal were taken from the base, the middle and the top of the MGC layer, and the outermost surface of the sample. Ages determined for these opals tell us that MGC began to precipitate between 2.90 ± 0.06 and 1.95 ± 0.07 million years ago, and continued to precipitate until at least a few hundred thousand years ago (figure 10.4; Wilson et al. 2003). Since we know from the lack of two-phase inclusions that MGC formed from fluids at temperatures below 35°C, we can conclude that fluids with elevated temperatures were last present more than about two to three million years ago (Wilson et al. 2003).

Ages determined from other opals and chalcedony in older mineral layers in the crust provide additional constraints for the ages of the two-phase fluid inclusions.

These mineral ages show that the incursion of hot fluids occurred more than 5.3 million years ago (figure 10.4; Wilson et al. 2003). These hotter fluids may have been associated with the cooling of the volcanic tuff sequence (Wilson et al. 2003; Marshall 2000). Ever since, the only fluid that percolated into Yucca Mountain appears to have been descending rainwater at temperatures below 35°C (Wilson et al. 2003).

Our results corroborate earlier studies (e.g., Paces et al. 1996; Whelan et al. 2002) that also concluded that there was no evidence for the recent movement of hot water through the site (Wilson et al. 2003).

## Conclusions

As geologists, we usually look to the present to understand the past. In the case of Yucca Mountain, we are asked to look at the past and predict the future—certainly a more difficult task. Millions of dollars have been spent over the last fifteen years assessing the question of whether hot fluids rose up from deep within the Earth to the repository level. Such a geologic event, if it happened after high-level wastes were emplaced in Yucca Mountain, could seriously compromise the safety of the repository.

The question of upwelling fluids is one of the most heavily studied and best-understood aspects of the Yucca Mountain site. Independent research by three groups (the University of Nevada Las Vegas, the U.S. Geological Survey, and the state of Nevada) has produced consistent data that answer the questions addressed by the University of Nevada Las Vegas study.

Based on all available data, there is no evidence to support the hypothesis of recent upwelling hydrothermal fluids. Results of our study solidly reinforce these conclusions by precisely determining the timing of past fluid flow; hot fluids were present more than five million years ago, and there is no record of hot fluids in younger rocks. We therefore conclude that there is no geologic evidence supporting the prediction that hydrothermal fluids will flood the site during the next ten to hundreds of thousands of years—the minimal regulatory life of the repository—or even during the next million years.

## References

Dublyansky, Y. (1998) *Fluid Inclusion Studies of Samples from the Exploratory Study Facility, Yucca Mountain, Nevada.* Institute for Energy and Environmental Research Report, Washington. http://www.ieer.org/reports/yucca/index.html.

Dublyansky, Y., Ford, D., and Reutski, V. (2001) Traces of Epigenetic Hydrothermal Activity at Yucca Mountain, Nevada: Preliminary Data on the Fluid Inclusion and Stable Isotope Evidence. *Chemical Geology* 173, pp. 125–149.

Faure, G. (1986) *Principles of Isotope Geology*. New York: John Wiley and Sons, 589 p.

Marshall, B.D. (2000) Isotope Geochemistry of Calcite Coatings and the Thermal History of the Unsaturated Zone at Yucca Mountain, Nevada. *Geological Society of America Abstracts with Programs* T111, p. A–259.

National Research Council (1992) *Ground Water at Yucca Mountain: How High Can It Rise?* Washington, DC: National Academy Press, 242 p.

Nuclear Waste Technical Review Board (1998) Review on Hydrothermal Activity. http://www.nwtrb.gov/reports/review1. pdf, 15 p. http://www.nwtrb.gov/reports/review2.pdf, 46 p.

Paces, J.B., Neymark, L.A., Marshall, B.D., Whelan, J.F., and Peterman, Z.E. (1996) *Ages and Origins of Subsurface Secondary Minerals in the Exploratory Studies Facility.* U.S. Geological Survey, Yucca Mountain Project Branch. Milestone Report 3GQH450M.

Paces, J.B., Neymark, L.A., Marshall, B.D., Whelan, J.F., and Peterman, Z.E. (2001) *Ages and Origins of Calcite and Opal in the Exploratory Studies Facility Tunnel, Yucca Mountain, Nevada.* U.S. Geological Survey. Water-Resources Investigations Report 01–4049.

Szymanski, J.S. (1989) *Conceptual Considerations of the Yucca Mountain Groundwater System with Special Emphasis on the Adequacy of This System to Accommodate the High-Level Nuclear Waste Repository.* Department of Energy Internal Report, Yucca Mountain Project Office, Las Vegas.

Whelan, J.F., Paces, J.B., and Peterman, Z.E. (2002) Physical and Stable-Isotope Evidence for Formation of Secondary Calcite and Silica in the Unsaturated Zone, Yucca Mountain, Nevada. *Applied Geochemistry* 17, pp. 735–750.

Wilson, N.S.F., Cline, J.S., and Amelin, Y. (2003) Origin, Timing, and Temperature of Secondary Calcite-Silica Minerals at Yucca Mountain. NV. *Geochimica et Cosmochimica Acta* 67, pp. 1145–1176.

# 11

# Water and Radionuclide Transport in the Unsaturated Zone

June Fabryka-Martin, Alan Flint, Arend Meijer, and Gilles Bussod

From a human perspective, the desert environment around Yucca Mountain seems harsh and uninviting. Summer temperatures often exceed 100°F (38°C). The total annual rainfall averages only nineteen centimeters, and the amount of water that manages to escape the thirsty desert vegetation and soil and to infiltrate into the underlying bedrock is typically only a few millimeters—about the thickness of a cracker. Indeed, the original appeal of placing a geologic repository above the water table as proposed over thirty years ago was based largely on the perceptions that subsurface water in such a dry environment moves at rates too slow to measure and is completely buffered from the vagaries of the capricious desert weather at the surface.

The study of unsaturated-zone flow and transport in a desert environment is a young science because hydrologists have traditionally focused on finding usable sources of water. The current interest in quantifying vanishingly small flow and transport rates is a relatively new direction, limited mostly to evaluating the water-quality impacts of past, present, and future waste disposal practices. Nevertheless, growing field and laboratory evidence indicates that it would be a mistake to assume that water and dissolved substances (solutes) are immobile in such an environment. Even though this movement cannot usually be directly seen or measured, we can infer it from clues left behind or picked up by the water along its path.

This chapter reviews some of the dramatic changes that have occurred in our technical understanding of water and solute movement in unsaturated fractured rocks, and how these scientific advances have influenced repository design, primarily in the United States. In addition, we point out some of the remaining uncertainties associated with transport issues, and offer some critical perspectives and possible solutions to these problems.

## Conceptual Model of the Unsaturated Zone at Yucca Mountain

Groundwater at Yucca Mountain lies at a depth of about five hundred to a thousand meters. The zone of variable saturation between the ground surface and the water table is known as the unsaturated zone. Alluvial cover (made of unconsolidated sediments) varies in depth from none over most of the ridgetops and sideslopes, to more than ten meters in some canyon bottoms. The bedrock consists of individual volcanic flows separated by thin layers of bedded tuff as well as eroded and redeposited material. The welded units are highly fractured but have very low matrix permeabilities—that is, in the absence of fractures, the rock is essentially impenetrable. In contrast, the nonwelded and bedded tuff units are only sparsely fractured but have high matrix permeabilities. The unsaturated zone as represented in most site-scale models encompasses four hydrologic tuff units. From top to bottom these include the welded Tiva Canyon tuff, the nonwelded Paintbrush tuff, the welded Topopah Spring tuff, and the nonwelded Calico Hills.

A reliable and defensible prediction of the performance of a repository at Yucca Mountain requires an understanding of how water moves from the surface to the repository level and then down to the water table. The most critical part of any prediction is the conceptual model on which it is based. Such a model must address how water movement is affected by the sequence of multiple layers of rock with sharp or gradual contrasts in porosity, conductivity, fracture length and interconnectivity, and degree of saturation (Flint 2002). Contrasting rock properties impede the downward flow and transport because excursions in infiltration rates are damped by subsurface geologic layers that absorb these water pulses, and then redistribute the flow more evenly in time and, to a lesser degree, space (Wolfsberg et al. 1998; Ritcey and Wu 1999; Pruess et al. 1999; Flint et al. 2001).

The current conceptual model contains the following key features (Flint et al. 2001):

• Water migrates rapidly downward (within a few years) through fractures in the welded Tiva Canyon tuff to the top of the nonwelded Paintbrush tuff; this is shown by moisture monitoring data (Flint and Flint 1995), and by the presence in these rock units of chlorine-36 produced by aboveground nuclear weapons tests and carried underground by rainwater (Fabryka-Martin et al. 1998; CRWMS M&O 2000b).

• The rate of water movement slows dramatically once it enters the nonwelded Paintbrush tuff because water in the fractures diffuses into the matrix. As a result, most water takes five hundred to twenty thousand years to pass through this unit (e.g., Flint et al. 2001; Wolfsberg et al. 2000). An exception to this general behavior may occur near fault zones, in which a small amount of fracture flow may be sustained throughout the entire nonwelded Paintbrush tuff thickness, with travel times of only a few years (Wolfsberg et al. 2000; Zhou et al. 2003).

• At the bottom of the nonwelded Paintbrush tuff and the top of the welded Topopah Spring tuff, flow transitions back into fractures, allowing for faster travel times from the base of the nonwelded Paintbrush tuff through the level of the potential repository horizon.

• Isolated perched water bodies below the welded Topopah Spring tuff at Yucca Mountain indicate the presence of barriers to vertical flow. In the northern part of the site, minerals in the nonwelded Calico Hills tuff have largely been altered to zeolites, which greatly decrease the permeability of this unit and can lead to perching. In the southern part, the nonwelded Calico Hills tuff is largely unaltered and expected to slow the downward movement of water in the same way as the nonwelded Paintbrush Tuff slows water entering from the fractured welded Tiva Canyon tuff. The boundary between the unaltered and zeolitized parts of the nonwelded Calico Hills tuff could also cause extensive lateral flow and the formation of perched water bodies. In addition to acting as a barrier to lateral flow, faults may also provide fast paths to the water table as a result of the increased fracturing commonly associated with them.

## Water Flux through the Repository Horizon

### Technical Issues

If radioactive waste were to escape from a geologic repository and reach the accessible environment, infiltrating water from the surface would be the most probable cause. First, surface infiltration is the primary mechanism for the entry of rainwater and snowmelt into the repository. Because of its extremely low salt content, this water will aggressively dissolve or alter minerals along its flow path and attack the integrity of waste containers and waste forms (Meijer 2002). Second, this water carries dissolved agents that keep radionuclides in solution by forming either soluble complexes or colloids. Finally, water moving along fractures and fault zones can be an efficient mechanism for the transport of radionuclides away from the repository. The major technical problems in understanding unsaturated zone flow involve evaluating how much water could come into direct contact with the waste and what proportion of this water could subsequently reach the water table by moving through fractures and faults.

Arguably one of the most significant recent advances in our understanding of flow through unsaturated fractured rocks concerns the role of fractures in conducting water. When a repository above the water table was first proposed for the Yucca Mountain site in the late 1970s, the prevailing belief was that water flow in the unsaturated zone at this location was largely limited to that moving through the rock matrix and that fracture flow was insignificant, if not physically impossible (e.g., Wang and Narasimhan 1985; Peters and Klavetter 1988). This contention was the basis for initial travel time estimates of several hundred thousand years for water traveling from the repository horizon to the water table beneath the mountain. Theoretical calculations as well as observations supported the expectation that water flowing in a fracture under unsaturated conditions would be quickly assimilated into the adjacent matrix by capillary forces, in the same manner as water is pulled up into a thin straw when the straw touches a water surface.

## Evidence

This view of nearly stagnant water beneath Yucca Mountain changed radically as different scientists independently obtained evidence that could be explained only by admitting the presence of a significant amount of fracture flow in the unsaturated zone. The first clue came from measured infiltration rates that appeared to exceed the ability of the rock matrix to transmit water from the surface to the repository horizon (Flint and Flint 1995). The results of the infiltration measurements were reinforced by a variety of methods that Yucca Mountain Project scientists applied to quantify what proportion of precipitation infiltrates into the ground, moving below the root zone of local vegetation (Flint et al. 2002). At Yucca Mountain, the net infiltration values average five millimeters per year, with the highest values approaching twenty near Yucca Crest (Flint et al. 2002). These data sets have been used to constrain several site-scale models that take into account the highly variable temporal and spatial infiltration inputs (Robinson et al. 1997; Bandurraga and Bodvarsson 1999).

Other evidence for fracture flow at Yucca Mountain is the presence of dilute water observed in deep water bodies perched above the local water table (Yang et al. 1996; CRWMS M&O 2000b). The extremely low salt content of dilute water indicates that the rain did not undergo significant evaporation at the surface before it infiltrated into the bedrock. More convincing perhaps is the presence of bomb-pulse tritium and radiocarbon in deep boreholes (Yang et al. 1996). Isolated fast pathways extend from the surface to at least the repository horizon as inferred from the presence of elevated levels of bomb-pulse chlorine-36 in subsurface tunnels at the proposed repository at Yucca Mountain (figure 11.1) (Fabryka-Martin et al. 1997; unpublished data). This chlorine-36 signal originated from the aboveground testing of nuclear weapons primarily in the 1950s and 1960s (Phillips 2000). In the tunnel, the elevated chlorine-36 signals are commonly associated with fracture systems, including those adjacent to faults (Campbell et al. 2003). It appears likely from this association that fault zones may in some cases act as vertical conduits for the rapid liquid-water flow resulting from large infiltration events (Fabryka-Martin et al. 1998). At Yucca Mountain, major faults cut the nonwelded Paintbrush tuff, thereby potentially reducing the damping effect of this unit, although only where the faults occur. (Note that the potential repository horizon is located principally away from that portion of Yucca Mountain containing major fault zones.)

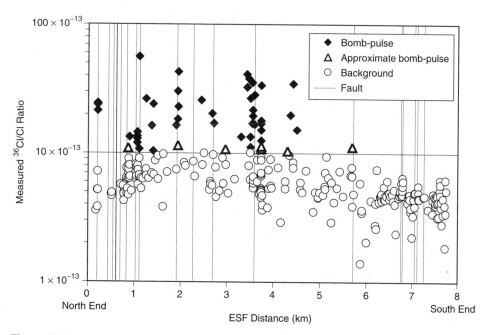

**Figure 11.1**
Distribution of chlorine-36/chlorine ratios measured in pore-water salts from the Exploratory Studies Facility.

Yucca Mountain is not unique with regard to fast flow through unsaturated fractures. For example, evidence of the deep transport of bomb-pulse chlorine-36 is also found in tunnels beneath Rainier Mesa in the northern part of the Nevada Test Site (Norris et al. 1990) and isolated pools in deep sections of Lechuguilla Cave at Carlsbad Caverns National Park in New Mexico (Turin 2001). Other geochemical tracers indicate fracture transport in the unsaturated fractured tuff at the semiarid Apache Leap Research Site near Superior, Arizona (Davidson et al. 1998), and unsaturated fractured chalk in the Negev Desert, Israel (Nativ et al. 1995).

## Uncertainties

Hydrologists and modelers for the Yucca Mountain Project generally agree that fracture flow and fracture-matrix interactions are pervasive in the rock layers that would host the repository (Flint et al. 2001; Bodvarsson et al. 1999). Still open to debate, however, are the following three questions:

· What proportion of the total flux is carried in the active fracture networks?

· Is the bulk of the water carried through only a few fractures or are active fracture paths ubiquitous?

· Are the locations of flowing fractures predictable?

Answers to these questions determine the optimum engineering design for the repository.

Twenty years ago, fracture flow was portrayed as a curtain of water moving between two parallel permeable plates. This concept implied an ample opportunity for physical and chemical interaction between water in fractures and water in the matrix. Under such conditions, matrix diffusion is a highly effective mechanism for slowing the transport rates of solutes that enter flowing fractures because water in the matrix is comparatively stagnant. In nature, however, many fracture surfaces are commonly coated with silicate or calcitic mineral deposits, and numerous laboratory studies over the last ten years have shown that in arid environments, fracture flow may be better described as a chaotic fingering or channeling process (Doe 2001; Berkowitz 2002). This process is akin to what one observes on a window surface during a heavy rainstorm, where water drops coalesce and run down the surface in channels or rivulets instead of covering the entire glass surface.

The area over which fracture water is in contact with the matrix is far less with channeled flow than with sheet flow, thereby further reducing the ability of the fracture water to diffuse into the matrix. To complicate matters further, it is not known whether the coatings commonly observed on fracture surfaces enhance or retard the exchange of water and solutes between the fracture and the matrix. Do these coatings essentially "grease the skids" for fast flow and transport by plugging matrix pores? Or is their presence an indicator that mineral precipitation is still active in this setting and can slow the movement of some radionuclides through coprecipitation or sorption?

Another area of uncertainty concerns the use of spatial and temporal averages of hydrologic parameters that may not capture the low-probability, extreme behaviors that could dominate water movement into and away from a geologic repository. Based on a few decades of rainfall monitoring as well as local studies (Russell et al. 1987), water flow and transport in the unsaturated zone at Yucca Mountain are presently dominated by rainfall events that happen once in ten or more years. Still uncertain, though, are those rare threshold storms that may occur once in a thousand years, in which fracture flow would be prevalent. This expectation is supported by the unusually high activity ratio of uranium-234 to uranium-238 in the perched water and groundwater beneath Yucca Mountain (Paces et al. 2002). Uranium-234 is an intermediate decay product resulting from the emission of an alpha particle by the parent uranium-238. Processes related to the effects of this mechanism of radioactive decay cause uranium-234 to be more soluble than uranium-238. Chief among these are the ejection of thorium-234 (which rapidly decays to uranium-234) from the crystal surface by recoil (analogous to the recoil of a gun when a bullet is fired from it) and radiation damage of the crystal lattice in which the uranium-234 atom resides, making it more accessible to water. Paces and his coauthors suggest that large uranium activity ratios develop in some Yucca Mountain waters because infrequent flow events allow larger amounts of recoil uranium-234 to accumulate on fracture surfaces. Thus, the largest ratios of uranium-234 to uranium-238 are expected to occur in zones of low-volume, low-frequency flow over long path lengths in the unsaturated zone. Should this be the case, modeling the risk for "steady-state" conditions characterized by "annual average" boundary conditions could be unrealistic when dealing with an arid-zone subsurface environment on a time scale of ten thousand years or greater.

What are the implications for repository safety of the deep transport of bomb-pulse nuclides? Some scientists (e.g., CRWMS M&O 2000a; Paces et al. 2002) argue that because bomb-pulse chlorine-36 is found at only a few locations at depth, and because no significant correlation has been found between high matrix saturation and elevated chlorine-36 signals, these discrete fast paths may not be associated with large drainage areas. From this perspective, even though fracture zones may provide fast paths for infiltrating water, they do not necessarily involve a significant percentage of the total infiltration over the repository footprint. If this is the case, then the overall flow pattern in the unsaturated zone, as characterized by current site-scale models, may be adequate. This interpretation is supported by the observation that bomb-pulse chlorine-36 was detected in none of the perched water bodies underlying the potential repository horizon, and that post-bomb tritium was found (in low levels) only in the smallest perched water body (CRWMS M&O 2000a). Given tritium's short half-life (12.3 years) and the expected dilution of bomb-pulse chlorine-36 and tritium with depth, it should be noted that the absence of detectable levels of these indicators does not necessarily indicate the absence of fast flow paths either during the present-day dry climate or past wetter ones.

Currently, neither the location nor the amount of water that could seep into waste emplacement drifts are known with certainty. Only a limited amount of relevant information is available on fracture properties, and our understanding of the nature of flow partitioning between the fractures and the matrix remains speculative. In the Department of Energy's performance assessment models, this uncertainty is represented by means of probability distributions and conservative approximations.

## Retardation Processes in Unsaturated Fractured Rock

### Technical Issues

Understanding how water moves is necessary but insufficient for predicting the transport of dissolved substances. During saturated flow in a porous medium, the transport rates for reactive solutes are slower than that of the water. This differential arises from a number of processes, including the diffusion of solutes into dead-end pores, sorption onto mineral surfaces and colloids, ion exchange, mineral precipitation, and the incorporation into an existing mineral by the replacement of another species. Conversely, the rate of transport of other solutes can be enhanced by processes that include ion exclusion, molecular filtration, and sorption onto colloids. Note that sorption onto colloids can either retard or enhance the transport of solutes depending on whether the transport occurs principally through the matrix—in which colloids can be trapped by narrow bottlenecks connecting pores—or fractures, in which the colloidal transport of solutes is enhanced (see chapter 12, this volume). Transport under unsaturated conditions adds complexity due to the presence of gas phases. None of these phenomena are fully quantified for the unsaturated zone at the Yucca Mountain site. The proper quantification of these phenomena would require inordinately large numbers of laboratory and field tests. Hence, in transport calculations performed by the Yucca Mountain Project, assumptions or bounding approximations are used to quantify these phenomena (CRWMS M&O 2000c). The key technical question is whether these assumptions and approximations are adequately supported by the available data.

### Evidence

Yucca Mountain Project scientists determined sorption coefficients and other relevant transport properties for the radioelements of concern—especially neptunium, plutonium, uranium, technetium, iodine, and selenium—in three ways: by direct measurements in the laboratory and the field under saturated conditions, through a review of the data reported in the literature, and through expert elicitation (CRWMS M&O 2000c). Because of the need for sorption data early in the Yucca Mountain Project, and because of the difficulty in applying theoretical models to complex rock-water systems, the Yucca Mountain planners decided early on to adopt an empirical approach in which sorption parameters are measured in laboratory batch experiments. In this widely used technique, a crushed rock sample is bathed in groundwater from the site and then spiked with a known amount of the radioelement of interest. After an appropriate period of time, the solution is separated from the rock sample and radioactivity is measured in the solution. Scientists then calculate a sorption coefficient based on the loss of radioactivity from the initial solution. Yucca Mountain Project scientists also conducted a limited number of transport experiments using rock columns to verify the batch sorption coefficient data and further evaluate the kinetics of the sorption reactions involved (CRWMS M&O 2000c).

### Uncertainties

The properties summarized in CRWMS M&O (2000c) are the best estimates based on the available data. These values are considered to be conservative and should thus provide conservative estimates for assessing repository performance. Yet there are several key limitations and uncertainties associated with these transport parameters.

Within the time frame of the Yucca Mountain Project, it is impossible to test the many variations and permutations of all the rock types and water compositions relevant to the proposed repository. To address this issue, Meijer (1992) reevaluated the direction of the project's laboratory sorption studies in 1990 and refocused them on those radioelements that were thought to be most readily transported in groundwater. This list includes radioelements that only sorb weakly (neptunium, uranium, and selenium) or not at all (iodine, technetium, and carbon) in most tuffaceous rock-water systems under oxidizing conditions. Meijer (1992) judged that the remaining radioelements that could be released from the proposed repository had sufficiently high sorption coefficients that they would not reach the accessible environment. This approach to addressing the need for sorption coefficient data appears reasonable, although a larger range of rock-water systems needs to be tested to confirm this.

The DOE assumes sorption parameters measured using a single radioelement to be applicable to the case where more than one radioelement is present (CRWMS M&O 2000c). For transport in the far field, away from the repository's immediate zone of influence, the rationale for accepting this assumption is that solutes released from the repository would be transported at different rates (due to different sorption characteristics) such that the groundwater in the far field would not contain multiple radionuclides at significant concentrations. Nevertheless, this technical argument may not apply near the repository, where relatively high concentrations of various radioelements may coexist in the water contained in the rock matrix.

The use of a sorption coefficient term in the transport calculations assumes that the sorption reaction attains equilibrium conditions instantaneously and is reversible. The reversibility constraint is not particularly critical to the repository safety analysis because the concern is more about the time and dose from those radionuclides that arrive the earliest, rather than the time of arrival of the peak dose. The establishment of equilibrium conditions on many types of mineral surfaces is a slow process relative to most sorption reactions, however; in complex rock-water systems such as those at the Yucca Mountain site, nonequilibrium conditions are likely the norm rather than the exception. Yet the impact this condition could have on the sorption behavior of a given radioelement is not obvious.

The assumption of instantaneous sorption reactions could present a problem; it may lead to underestimates of travel times because noninstantaneous reactions result in nonconservative estimates of travel times. Kinetic data on sorption reactions for important radioelements are required to address this issue. The existing kinetic information (Rundberg 1987, table 12), suggests crucial radioelements such as plutonium display slow sorption kinetics. A proper evaluation of the potential impact of slow sorption kinetics would require additional studies.

Due to a dearth of experimental data on the sorption behavior of radioelements under unsaturated conditions, sorption experiments conducted in the laboratory under saturated conditions are assumed to reflect the same reactions that would occur under unsaturated conditions. This assumption needs to be tested.

Radionuclide sorption parameters measured in laboratory experiments are assumed not to be affected significantly by microbial activity. This assumption primarily applies to sorption data obtained for elements that can have different oxidation-reduction states under the environmental conditions expected at Yucca Mountain. This assumption is probably valid for laboratory tests, given the short times that are typical of sorption experiments (days to a few weeks) and the expectation that the presence of significant microbial activity would be marked by turbid conditions in the solutions. In the potential repository horizon, though, this assumption may not be correct because active microbial populations are known to alter local environmental conditions (e.g., water chemistry) and, in some cases, to synthesize solutes (Hersman 1997). Such changes could have particularly significant implications for plutonium because microbial activity has the potential to reduce plutonium from the +6 oxidation state, which sorbs poorly, to the +4 oxidation state, which sorbs strongly.

Studies of natural analogs to transport in unsaturated volcanic tuffs can yield useful data, assuming the conditions during transport are truly analogous to conditions at Yucca Mountain after waste emplacement. Of the many potential studies that could be cited, most suffer from a lack of adequate site characterization or from a flaw in the analogy. Studies of the uranium deposit at Peña Blanca, Mexico, however, suggest that matrix diffusion was quite limited during the transport of uranium along fractures radiating from the deposit (Pearcy et al. 1995). This conclusion implies that the diffusion of solutes from fractures into the rock matrix may not be as effective as originally expected for retarding the rate of radionuclide transport through the unsaturated rocks within Yucca Mountain.

The geologic disposal concept is based on the premise that multiple natural and engineered barriers to flow and transport exist between the repository and the environment, including those provided by the host rock. Over the past decade, the DOE's design for a repository at Yucca Mountain has shifted toward an increasingly robust engineered system in order to offset some of the difficulties inherent in assessing the performance of a complex natural system. The current DOE emphasis on enhanced assurance through engineering does not, however, detract from the fact that the geosphere ultimately remains the key component of any geologic disposal system. This point is underscored by the lack of experimental data that can support the integrity and survival of metallic man-made components much beyond a thousand years. For example, the only evidence relating to canister corrosion generally involves the study of thousand-year-old bronze and copper artifacts (see Murphy et al. 1998; Hughson et al. 2000). The applicability of such observations to the repository evaluation is limited by the fact that canisters are not typically composed of pure copper; even if they were, this evidence does not apply to the integrity of the welds and seals involved in engineered systems.

## Modeling Flow and Transport through the Unsaturated Zone

Technical confidence in the long-term radiological safety of a geologic repository requires the ability to model transport processes realistically enough to capture the effects of complex physical phenomena such as thermally induced stresses, canister failure, changes in infiltration rates, changes in water chemistry, temperature, gas transport, and bacterial activity. The discussion in this chapter focuses in particular on the difficulties inherent in modeling transport through variably fractured rocks in a semiarid environment under natural conditions, which are assumed to represent the most likely postclosure conditions after the repository has cooled back down to ambient temperatures. Largely as a result of the Yucca Mountain site assessment, fracture modeling has been elevated from a knotty problem of mostly academic interest to an issue with national, if not international, attention.

An early and contentious debate among Yucca Mountain hydrologists centered on the optimal method for representing fracture networks in flow and transport calculations. Scientists gradually came to recognize that fracture flow through unsaturated rock differs greatly from that through saturated rocks (Berkowitz 2002). The inability to incorporate each individual fracture and its complex geometry and surface characteristics in the site-scale transport model inevitably requires the use of averaged properties in imaginary representative rock types; the current model represents Yucca Mountain geology with thirty-five rock layers. The current site-scale model developed by Yucca Mountain scientists allows the fractured rock mass or block to respond differently to different levels of infiltration (Liu et al. 1998; Zhou et al. 2003). At low infiltration rates (e.g., < one millimeter per year), transport occurs solely in the rock matrix, where long travel times allow for radionuclide sorption onto mineral surfaces. At higher infiltration rates, transport shifts to the fractures, where retardation processes may not be as effective as in the matrix. This dual behavior of a single hydrologic unit requires that values for several model input parameters—such as solubility, sorption, infiltration, and fracture permeability—be variable in time and space. Not only does this requirement add a considerable degree of complexity to the model itself but it also moves far beyond our ability to collect data that represent the variability on the same scale as the model. Consequently, many inputs to the Yucca Mountain site model are themselves outputs of other models intended to simulate the natural yet unobservable variability of these characteristics or processes.

Evaluating the reliability of the model's predictions of radionuclide movement through Yucca Mountain is a daunting challenge. Model predictions have been compared against measured results from laboratory experiments, small- and large-scale field tests, and information from other sites (e.g., natural analogs). None of these approaches, however, can completely capture the spatially variable behavior of the system exposed to varying conditions through geologic time. Models can never be completely validated because no site is ever completely understood in all of its complexity. For example, the Busted Butte Unsaturated Zone Transport Test (figure 11.2) was designed specifically to test assumptions about flow and transport through the Calico Hills tuff, one of the principal barriers to unsaturated zone flow from the repository to the water table (Bussod et al. 1999). In this test, tracers were injected into the rock through horizontal boreholes in an underground rock laboratory and then monitored for over three years using geophysical and geochemical techniques. Model predictions made prior to the field experiments compared successfully to results observed when the block was excavated at the end of the experiment (Robinson and Bussod 2000; Tseng et al. 2003). Nevertheless, although this result enhances confidence in the matrix flow and transport model for this particular block of rock, it may not adequately represent the behavior of this identical horizon near a fault zone where fracture flow may dominate during periods of higher infiltration. Similarly, the long time required for measurable transport to occur in the unsaturated rocks at Yucca Mountain under natural conditions prevents a convincing test of the model's reliability for predicting long-term transport in heterogeneous media such as variably fractured rock. Indeed, the longest transport test at Yucca Mountain (i.e., at Busted Butte) lasted only three years and involved transport distances of only five meters from the injection source.

The strength of the case for the long-term safety of a geologic repository relies on the extent to which we are confident that the models adequately represent the future evolution of the system through time. Assuming that the important physical

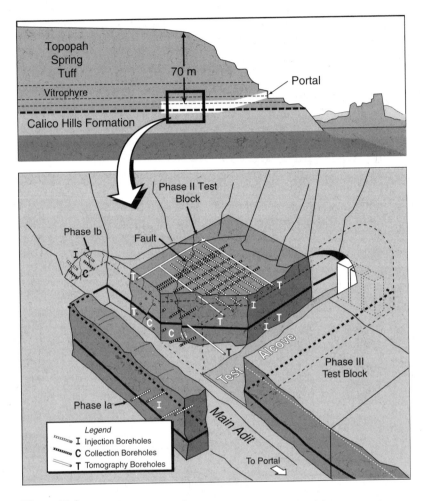

**Figure 11.2**
The Busted Butte Unsaturated Zone Transport Test facility. *A*, geologic cross-section. *B*, schematic of the facility, showing boreholes for the different experimental phases. The solid line marks the top of the Calico Hills formation, while the small dashed line is the contact between two tuff subunits.

processes have all been identified and correctly formulated, the reliability of these models can only be as good as the assumptions and databases on which they rely. Some continuing level of uncertainty associated with these assumptions and data must be accepted as inevitable. Consequently, a defensible demonstration that stable and favorable conditions will be maintained over the regulatory timescale of thousands to tens of thousands of years can only be addressed in terms of a probability function, which will always show an extremely small but finite possibility of a failure. Therefore, the question is not whether there is a chance that some part of the system will fail; that should be taken as a given. Rather, the question is whether we have yet reached a point at which we have enough confidence in the model's prediction to move forward toward building and eventually licensing the repository.

## Conclusion

Doubt is not a pleasant state, but certainty is absurd.
—Voltaire, letter to Frederick the Great, 1767

Although the challenge of the safe disposal of radioactive waste was not to arise for another two centuries, Voltaire's pithy statement applies well to our state of knowledge about geologic repositories in general and the Yucca Mountain site in particular. Our ability to predict the rate of transport of radionuclides through the unsaturated zone at Yucca Mountain is constrained by inevitable scientific uncertainties about several key phenomena. As of this writing (2002), we cannot confidently state that the unsaturated zone at Yucca Mountain will be an effective barrier for radioactive waste migration from the repository to the accessible environment for the requisite amount of time.

Although bomb-pulse radionuclides indicate rapid percolation of some fraction of the water infiltrating to the repository horizon, it is unclear how much of the total flux moves along these fast paths, and how these fast paths are distributed above, in, and below the repository horizon under various climate scenarios. As a result, the potential impact of these rapid paths on radionuclide transport has not been adequately bounded.

For many of the radionuclides present in radioactive waste, there is sufficient evidence that the transport rate through the unsaturated zone would be sufficiently slow that an adequate safety case can be made using our current state of knowledge. Only limited evidence is available for the retardation of other radionuclides under unsaturated conditions, however. For these radionuclides, the retardation mechanisms have not been quantified to the extent needed for the DOE to take credit for any retardation of their transport in the unsaturated zone. These require further investigation using a wider range of water chemistries, rock types, and the degree of saturation than has been used thus far.

Any safety argument relies heavily on confidence in the predictive models and the science that supports them. During the coming years, predictions based on conceptual and numerical models must continue to be tested against site characterization data, field tests, natural analog studies, and laboratory transport tests. Furthermore, to increase our confidence that the natural and engineered barriers will function as needed through time, it is essential that traceable and transparent documentation be peer reviewed and widely disseminated.

# References

Bandurraga, T.M., and Bodvarsson, G.S. (1999) Calibrating Hydrogeologic Parameters for the 3-D Site-Scale Unsaturated Zone Model of Yucca Mountain, Nevada. *Journal of Contaminant Hydrology* 38, pp. 25–46.

Berkowitz, B. (2002) Characterizing Flow and Transport in Fractured Geologic Media: A Review. *Advances in Water Resources* 25, pp. 861–884.

Bodvarsson, G.S., Boyle, W., Patterson, R., and Williams, D. (1999) Overview of Scientific Investigations at Yucca Mountain: The Potential Repository for High-Level Nuclear Waste. *Journal of Contaminant Hydrology* 38, pp. 3–24.

Bussod, G.Y, Turin, H.J., and Lowry, W.E. (1999). *Busted Butte Unsaturated Zone Transport Test.* Yucca Mountain Site Characterization Project Milestone SPU85M4. Los Alamos National Laboratory. LA–13670–SR. November.

Campbell, K., Wolfsberg, A., Fabryka-Martin, J., and Sweetkind, D. (2003) Chlorine-36 Data at Yucca Mountain: Statistical Tests of Conceptual Models for Unsaturated-Zone Flow. *Journal of Contaminant Hydrology* 62–63, pp. 43–61.

CRWMS M&O (Civilian Radioactive Waste Management System, Management and Operating Contractor) (2000a) *Conceptual and Numerical Models for UZ Flow and Transport.* Office of Civilian Radioactive Waste Management, U.S. Department of Energy. MDL–NBS–HS–000005. March.

CRWMS M&O (Civilian Radioactive Waste Management System, Management and Operating Contractor) (2000b) *Analysis of Geochemical Data for the Unsaturated Zone.* Office of Civilian Radioactive Waste Management, U.S. Department of Energy. ANL–NBS– HS–000017. April.

CRWMS M&O (Civilian Radioactive Waste Management System, Management and Operating Contractor) (2000c) *Unsaturated Zone and Saturated Zone Transport Properties.* Office of Civilian Radioactive Waste Management, U.S. Department of Energy. ANL–NBS– HS–000019. June.

Davidson, G.R., Bassett, R.L., Hardin, E.L., and Thompson, D.L. (1998) Geochemical Evidence of Preferential Flow of Water through Fractures in Unsaturated Tuff, Apache Leap, Arizona. *Applied Geochemistry* 13, pp. 185–195.

Doe, T.W. (2001) What Do Drops Do? Surface Wetting and Network Geometry Effects on Vadose-Zone Fracture Flow. In *Conceptual Models of Flow and Transport in the Fractured Vadose Zone*. (Panel on Conceptual Models of Flow and Transport in the Fractured Vadose Zone, U.S. National Committee for Rock Mechanics, National Research Council (U.S.) Board on Earth Sciences and Resources), Washington, DC: National Academy Press, pp. 243–270.

Fabryka-Martin, J., Wolfsberg, A.V., Dixon, P.R., Levy, S., Musgrave, J., and Turin, H.J. (1997) *Summary Report of Chlorine-36 Studies: Sampling, Analysis, and Simulation of Chlorine-36 in the Exploratory Studies Facility*. Yucca Mountain Site Characterization Project Milestone 3783M. Los Alamos National Laboratory. LA–13352–MS. December.

Fabryka-Martin, J.T., Turin, H.J., Wolfsberg, A.V., Brenner, D., Dixon, P.R., and Musgrave, J.A. (1998) *Summary Report of Chlorine-36 Studies as of August 1996*. Yucca Mountain Site Characterization Project Milestone 3782M. Los Alamos National Laboratory. LA–13458–MS. May.

Flint, A.L., Flint, L.E., Bodvarsson, G.S., Kwicklis, E.M., and Fabryka-Martin, J. (2001) Evolution of the Conceptual Model of Unsaturated Zone Hydrology at Yucca Mountain, Nevada. *Journal of Hydrology* 247, pp. 1–30.

Flint, A.L., Flint, L.E., Kwicklis, E.M., Fabryka-Martin, J., and Bodvarsson, G.S. (2002) Estimating Recharge at Yucca Mountain, Nevada: Comparison of Methods. *Hydrogeology Journal* 10, pp. 180–203.

Flint, L.E. (2002) Characterization of Unsaturated Zone Hydrologic Properties and Their Influence on Lateral Diversion in a Volcanic Tuff at Yucca Mountain, Nevada. PhD diss., Oregon State University, 161 p.

Flint, L.E., and Flint, A.L. (1995) *Shallow Infiltration Processes at Yucca Mountain, Nevada: Neutron Logging Data, 1984–93*. U.S. Geological Survey. Water-Resources Investigations Report 95–4035.

Hersman, L.E. (1997) Subsurface Microbiology; Effects on the Transport of Radioactive Waste in the Vadose Zone. In *Microbiology of the Terrestrial Deep Subsurface*, ed. Amy, P.S., and Haldeman, D.L. Boca Raton, FL: CRC Press, pp. 299–323.

Hughson, D.L., Browning, L., Murphy, W.M., and Green, R.T. (2000) Archeological Site at Akrotiri, Greece, as a Natural Analog for Radionuclide Transport: Implications for Validity of Performance Assessments. In *Scientific Basis for Nuclear Waste Management XXIII*, ed. Smith, R.W., and Shoesmith, D.W. Warrendale, PA: Materials Research Society, pp. 557–561.

Liu, H.H., Doughty, C., and Bodvarsson, G.S. (1998) An Active Fracture Model for Unsaturated Flow and Transport in Fractured Rocks. *Water Resources Research* 34, pp. 2644–2646.

Meijer, A. (1992) A Strategy for the Derivation and Use of Sorption Coefficients in Performance Assessment Calculations for the Yucca Mountain Site. In *Proceedings of the DOE/*

*Yucca Mountain Site Characterization Project Radionuclide Adsorption Workshop.* Los Alamos, NM: Los Alamos National Laboratory. LA–12325–C, pp. 9–40.

Meijer, A. (2002) Conceptual Model of the Controls on Natural Water Chemistry at Yucca Mountain, Nevada. *Applied Geochemistry* 17, pp. 793–805.

Murphy, W.M., Pearcy, E.C., Green, R.T., Prikryl, J.D., Mohanty, S., Leslie, B.W., and Nedungadi, A. (1998) A Test of Long-Term, Predictive, Geochemical Transport Modeling at the Akrotiri Archaeological Site. *Journal of Contaminant Hydrology* 29, pp. 245–279.

Nativ, R., Adar, E., Dahan, O., and Geyh, M. (1995) Water Recharge and Solute Transport through the Vadose Zone of Fractured Chalk under Desert Conditions. *Water Resources Research* 31, pp. 253–261.

Norris, A.E., Bentley, H.W., Cheng, S., Kubik, P.W., Sharma, P., and Gove, H.E. (1990) $^{36}$Cl Studies of Water Movement Deep within Unsaturated Tuffs. *Nuclear Instruments and Methods in Physics Research* B52, pp. 455–460.

Paces, J.B., Ludwig, K.R., Peterman, Z.E., and Neymark, L.A. (2002) $^{234}$U/$^{238}$U Evidence for Local Recharge and Patterns of Groundwater Flow in the Vicinity of Yucca Mountain, Nevada, USA. *Applied Geochemistry* 17, pp. 751–779.

Pearcy, E.C., Prikryl, J.D., and Leslie, B.W. (1995) Uranium Transport through Fractured Silicic Tuff and Relative Retention in Areas with Distinct Fracture Characteristics. *Applied Geochemistry* 10, pp. 685–704.

Peters, R.R., and Klavetter, E.A. (1988) A Continuum Model for Water Movement in an Unsaturated Fractured Rock Mass. *Water Resources Research* 24, pp. 416–430.

Phillips, F.M. (2000) Chlorine-36. In *Environmental Tracers in Subsurface Hydrology*, ed. Cook, P.G., and Herczeg, A.L. Norwell, MA: Kluwer Academic Publishers, pp. 299–348.

Pruess, K., Faybishenko, B., and Bodvarsson, G.S. (1999) Alternative Concepts and Approaches for Modeling Flow and Transport in Thick Unsaturated Zones of Fractured Rocks. *Journal of Contaminant Hydrology* 38, pp. 281–322.

Ritcey, A.C., and Wu, Y.S. (1999) Evaluation of the Effect of Future Climate Change on the Distribution and Movement of Moisture in the Unsaturated Zone at Yucca Mountain, NV. *Journal of Contaminant Hydrology* 38, pp. 257–279.

Robinson, B.A., and Bussod, G.Y. (2000) Radionuclide Transport in the Unsaturated Zone at Yucca Mountain: Numerical Model and Preliminary Field Observations. In *Dynamics of Fluids in Fractured Rocks, Geophysical Monograph 122*, ed. Fabyshenko, B. Washington, DC: American Geophysical Union, pp. 323–336.

Robinson, B.A., Wolfsberg, A.V., Viswanathan, H.S., Bussod, G.Y., Gable, C.W., and Meijer, A. (1997) *The Site-Scale Unsaturated Zone Transport Model of Yucca Mountain.* Los Alamos National Laboratory Yucca Mountain Project Milestone. SP25BM3. August 29.

Rundberg, R.S. (1987) *Assessment Report on the Kinetics of Radionuclide Sorption on Yucca Mountain Tuff.* Los Alamos National Laboratory. LA–11026–MS. July.

Russell, C.E., Hess, J.W., and Tyler, S.W. (1987) Hydrogeologic Investigations of Flow in Fractured Tuffs, Rainier Mesa, Nevada Test Site. In *Proceedings Flow and Transport in Unsaturated Fractured Rock*, ed. Evans, D.D., and Nicholson, T.J. Washington, DC: American Geophysical Union, Geophysical Monograph 42, pp. 43–50.

Tseng, P.H., Soll, W.E., Gable, C.W., Turin, H.J., and Bussod, G.Y. (2003) Modeling Unsaturated Flow and Transport Processes at the Busted Butte Field Test Site, Nevada. *Journal of Contaminant Hydrology* 62–3 (Special Issue), pp. 303–318.

Turin, H.J. (2001) Isotope Profiles from Three Deep Western U.S. Caves: Insights into Arid-Region Vadose-Zone Processes. *GSA Abstracts with Programs* 33, p. A110.

Wang, J.S.Y., and Narasimhan, T.N. (1985) Hydrologic Mechanisms Governing Fluid Flow in a Partially Saturated, Fractured, Porous Medium. *Water Resources Research* 21, pp. 1861–1874.

Wolfsberg, A., Campbell, K., and Fabryka-Martin, J. (2000) Use of Chlorine-36 Data to Evaluate Fracture Flow and Transport Models at Yucca Mountain, Nevada. In *Dynamics of Fluids in Fractured Rocks: Geophysical Monograph 122*, ed. Fabyshenko, B. Washington, DC: American Geophysical Union, pp. 349–362.

Wolfsberg, A.V., Robinson, B.A., Roemer, G.J.C., and Fabryka-Martin, J.T. (1998) Modeling Flow and Transport Pathways to the Potential Repository Horizon at Yucca Mountain. In *Proceedings, 1998 8th Annual International High-Level Radioactive Waste Management Conference*. La Grange Park, IL: American Nuclear Society, pp. 81–84.

Yang, I.C., Rattray, G.W., and Yu, P. (1996) *Interpretation of Chemical and Isotopic Data from Boreholes in the Unsaturated Zone at Yucca Mountain, Nevada*. U.S. Geological Survey. Water-Resources Investigations Report 96–4058.

Zhou, Q., Liu, H.-H., Bodvarsson, G.S., and Oldenburg, C.M. (2003) Flow and Transport in Unsaturated Fractured Rock: Effects of Multiscale Heterogeneity of Hydrogeologic Properties. *Journal of Contaminant Hydrology* 60, pp. 1–30.

# 12

## Colloidal Transport of Radionuclides

Annie B. Kersting

No matter how effective the final engineered barrier system at Yucca Mountain will be in isolating more than seventy thousand metric tons of high-level nuclear waste, it will eventually fail. When this failure occurs, canisters containing the waste will corrode as water comes into contact with the waste packages. The long-lived radio-nuclides remaining at the Yucca Mountain repository will then begin to migrate through the natural environment. Some radionuclides persist in the environment for an extremely long time—so long, in fact, that they will be present well after the engineered barrier system has been breached. The half-life of plutonium-239, for example, is 24,100 years. That means that even if a barrier system lasts ten thousand years, more than one-half of the originally deposited plutonium-239 will still exist. The spent fuel deposited at the proposed Yucca Mountain repository will contain more than six hundred metric tons of plutonium.

For the Yucca Mountain repository to rely, in part, on the geologic media to provide a long-term barrier to radionuclide migration, it is critical that the fundamental processes that control their transport be well understood. Moreover, this understanding must be incorporated into predictive models in order to assess the safety of the Yucca Mountain repository. A number of crucial questions must be answered. Will the radionuclides stored in the Yucca Mountain repository be transported to an environment accessible by humans, and if so, by what means and how fast? Will the natural geologic environment contain or at least retard the migra-tion of the waste until most of the radioactivity has decayed away? Will the waste migrate from its original emplacement in the unsaturated fractured volcanic rocks to the saturated zone below, making it available for transport in groundwater?

Conceptual models have so far proven to be ineffective for predicting contaminant transport in complex environments. This lack of a robust modeling capability remains a major scientific challenge (National Research Council 2001). In the past, it was thought that radionuclides that do not readily dissolve in groundwater (low solubility) and have a high-sorption (attachment) capacity for minerals would adhere to the rock matrix and not migrate in groundwater. As a result, most models used for predicting transport were based on laboratory studies of the solubility of these elements in water. These solubility limits were considered an upper bound on how much contaminant could be released in groundwater. Low-solubility radionuclides, such as plutonium and americium, would therefore not be expected to move from their original source. Under the geochemical conditions expected at the Yucca Mountain repository, the solubility of plutonium at moderate pH was experimentally determined to be extremely low—about $10^{-8}$ to $10^{-9}$ molar (one molar is one mole of plutonium-239 atoms per liter of solution) (Efurd et al. 1998). Yet some field investigations have observed low-solubility radionuclides that have evidently been mobilized and transported over surprisingly long distances (e.g., Kersting et al. 1999). This means that transport models predicting that low-solubility radionuclides are immobile in the subsurface fail to capture the fundamental processes operating in a complex field setting.

Plutonium is a significant concern for two reasons. First, it is highly toxic. Second, with plutonium's long half-life, it will, over time, constitute a larger and larger fraction of the radioactivity produced by the discarded waste. The waste inventory (total radioactive content of ~$10^{10}$ curies) will be initially dominated by fission products, most of which have short half-lives. Due to radioactive decay, however, after about ten thousand years, most of the radioactivity generated by the waste will come from technetium-99, iodine-129, and three actinides: neptunium-237, plutonium-239, and plutonium-242 (Eckhardt et al. 2000). In ten thousand years, the plutonium-239 will represent approximately 90 percent of the total waste inventory. After one million years, neptunium-237 will dominate the waste inventory (Eckhardt et al. 2000).

The groundwater chemistry, flow paths, rock properties, and behavior of the radionuclides under the conditions expected at the Yucca Mountain repository all play an important role in determining how and if radionuclides will move, once the engineered barrier system fails. Although a growing number of field studies document the movement of low levels of low-solubility radionuclides in the subsurface (e.g., Kersting et al. 1999; Santschi et al. 2002), a consensus as to the mechanisms responsible for this mobility has been slow to emerge. In part, the natural complexity of geologic field environments has made definitive conclusions regarding transport difficult. Transport models based on laboratory experiments have generally only considered the partitioning of radionuclides between the dissolved phase (mobile) and the solid mineral or rock phase (immobile). This simplistic approach has lead to a lack of agreement between field observations, laboratory experiments, and modeling studies.

One explanation that has recently gained support is that low-solubility radionuclides, such as plutonium, can be transported in groundwater either by "hitching a ride" on naturally occurring colloids (sometimes called pseudocolloids) or are themselves colloidal-size polymers (intrinsic colloids) (figure 12.1). Colloids are small particulates (between one and one thousand nanometers in diameter) composed of inorganic minerals and organic material, and due to their small size have the ability to remain suspended in water and transported (Stumm 1992). Although the idea that colloids may facilitate the transport of contaminants in groundwater is not new (Ramsey 1988; McCarthy and Zachara 1989; Kim 1991; Bates et al. 1992), compelling field evidence has previously been lacking. In addition, new techniques have recently allowed for more detailed investigations, and such studies are providing better insights into the physicochemical nature of colloid and particulate transport. This chapter describes the current understanding of colloids and their ability to facilitate the transport of low-solubility radionuclides, and its relevance to the Yucca Mountain repository.

**What Are Colloids?**

Colloids are ubiquitous, naturally occurring small particles found in groundwater, and are composed of inorganic minerals or organic species (figure 12.2). Inorganic colloids consist of common minerals such as clays, zeolites, silica, and Fe-oxides, and are usually reflective of the host rock or fracture lining minerals (McCarthy and Degueldre 1993). Particles larger than colloids generally settle out of a fluid due to gravitational forces, while those smaller than one nanometer behave as dissolved species. Due to their large surface area per unit mass, colloids can sorb significant quantities of contaminants. Contaminants that strongly sorb to rocks can also sorb to suspended colloids and these can ultimately be transported away from their original location.

Groundwater colloids originate from the mechanical weathering of rocks, plants, and soil (McCarthy and Degueldre 1993). Organic colloids are common in surface and shallow waters, where their concentrations approach four hundred milligrams per liter (Buckau et al. 2000). Well-developed surface soils have the

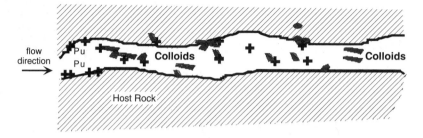

**Figure 12.1**
Cartoon drawing of colloids moving in groundwater through a fractured rock. The cartoon represents several possible scenarios for plutonium (Pu). In this cartoon, Pu, shown as a cross (+) is drawn attached to the fracture walls (immobile); sorbed to different minerals (pseudo-colloid, shown by different shapes) (mobile); sorbed to minerals that are attached to the fracture walls (immobile); and by itself as a polymer or intrinsic colloid (mobile).

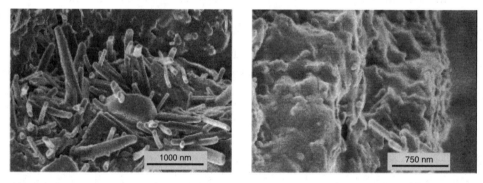

**Figure 12.2**
Scanning electron micrographs of naturally occurring colloids filtered from groundwater. The inorganic colloids with a rhombohedral shape are the zeolite minerals: clinoptilolite and mordenite. The platy minerals are clays.

highest concentration of organic material, usually containing dissolved organic carbon concentrations greater than twenty milligrams per liter, whereas subsurface soils contain less, ~five milligrams per liter. There is recent recognition that a continuum can exist between organic and inorganic colloids. In environments where the organic content of the groundwater is high, inorganic mineral colloids are found to have organic surfaces (Degueldre et al. 2000). In fact, laboratory experiments and field studies show that the presence of organic substances enhances the ability of inorganic colloids to transport radionuclides (Kim 1991). The organic content of deep groundwater compared to surface water is generally low, and these systems are dominated by inorganic colloids (Smith and Degueldre 1993). For example, J-13 water—a deep groundwater representative of the conditions at the Yucca Mountain repository—has a low dissolved organic content between 0.15 and 0.5 milligrams per liter (Minai et al. 1992).

In addition to sorbing to inorganic and organic colloids, plutonium and other actinide ions can also form colloid-size polymers that are sometimes called intrinsic colloids. Intrinsic colloids can form when the concentration of the actinide ions in solution exceeds the solubility product for the formation of a solid phase. The solubility product of an actinide ion depends on the oxidation state of the ion and the composition of the groundwater. For example, the intrinsic colloid of plutonium (IV) can be easily produced and remain stable in near-neutral solutions (Clark 2000).

To evaluate the potential for colloids to transport radionuclides at a given location, such as the Yucca Mountain repository, several criteria must be met: (1) colloids must exist (pseudocolloids or intrinsic colloids); (2) radionuclides must adhere to the colloids (pseudocolloids); and (3) colloids must be transported (Ryan and Elimelech 1996). In addition, the colloids must be stable and exist in quantities sufficiently high to transport detectable levels of contaminants. Studies have shown that pseudocolloids are found in all natural waters, but they have a huge range in concentration, with measured values ranging from 0.0002 to 200 milligrams per liter (e.g., McCarthy and Degueldre 1993). With criterion one above shown to be true, the following sections focus on laboratory and field examples where criteria two and three are investigated.

## Laboratory Studies

If plutonium has a high sorption affinity for pseudocolloids (minerals and organics) and is not quickly desorbed from the colloids, it can be transported in water.

Extensive literature exists on the sorption of radionuclides to bulk rock and individual minerals (both colloid size and larger). In laboratory experiments, sorption of plutonium, neptunium, and uranium on different minerals and tuff rock core samples was investigated (Triay et al. 1996; Duff et al. 1999). Under oxidizing conditions and a pH between 7 and 8.5, plutonium sorbed to a variety of common minerals (zeolites, iron oxide, clay, silica, albite, and quartz). (See chapter 13, this volume, for more on the ability of radionuclides to sorb to specific minerals.) Experiments showed that once plutonium sorbed onto iron oxides, it desorbed slowly; thus sorption to iron oxides could be a fairly permanent process (Lu et al. 1999). The sorption of plutonium (IV) on the colloid-size clinoptilolite—a common zeolite—was also found to be strong with the desorption rate low (Kersting et al. 2003b). Low-solubility radionuclides, such as plutonium, can sorb strongly to a variety of common minerals that comprise both the host rock and form the pseudocolloid component of the groundwater. As shown in figure 12.1, some of the pseudocolloids will either remain immobile by attaching to the wall of the fracture and/or exist suspended in water available for transport as mobile colloid species.

Laboratory experiments were conducted to determine if colloids doped with radionuclides could be transported in natural fractured rock on the scale of centimeters (Reimus et al. 2003). Plutonium was sorbed on mineral colloids (silica, zeolites, and clays) and then injected into several fractured volcanic cores approximately fifteen centimeters long. Although plutonium colloids were transported through the columns, most of the injected material was retained at the top of the fractures; less than 20 percent of the plutonium was recovered at the other end of the column. On a laboratory scale, plutonium sorbed to colloids is capable of migrating through fractures, although at a significantly attenuated concentration. In other laboratory experiments, radionuclides were injected into columns of crushed volcanic tuff from the proposed Yucca Mountain repository in order to simulate alluvium conditions where groundwater flows in the absence of fractures (Thompson 1989). Solutions containing plutonium, americium, and neptunium polymer (intrinsic colloids) were passed through crushed volcanic tuff in short-term experiments. These studies showed that colloidal transport of plutonium (IV) and americium (III) through crushed tuff does occur, but the overwhelming majority of the colloids (>90 percent) remains on the tuff and is not transported.

## Field Studies

The laboratory studies discussed above show that the colloidal transport of plutonium can occur, although the concentrations of radionuclides are significantly attenuated during the transport. Does the mechanism of colloid-facilitated transport occur in nature, in an environment much more complicated than in a laboratory? Pioneering field studies examining the association of colloids and the transport of radionuclides were carried out at the Nevada Test Site (Buddemeier and Hunt 1988), where more than eight hundred underground nuclear tests (one-third were below the water table) were conducted from 1951 to 1992 (U.S. Department of Energy 1994). This site, adjacent to the proposed Yucca Mountain repository, contains approximately $10^8$ curies of residual radioactive material deposited below the water table (Smith 1995; Bowen et al. 2001). In one study, filtered groundwater was collected from a well that was screened in fractured volcanic tuff three-hundred-meters downgradient of an underground nuclear test (Buddemeier and Hunt 1988). Approximately 98 percent of the low-solubility radionuclides (cobalt, cesium, and europium) detected in the groundwater were associated with the colloidal fraction. (The groundwater was not analyzed for plutonium.) These findings strongly support the hypothesis that colloids play a role in the transport of low-solubility radionuclides in a saturated, fracture-flow aquifer.

In a more recent study at the Nevada Test Site, low levels of plutonium, americium, cesium, cobalt, and europium, as well as high levels of tritium, were detected in groundwater collected from two different aquifers downgradient of a nuclear test (Kersting et al. 1999; Thompson 1999). Groundwater from one well came from the Topopah Spring tuff rock unit, in which the Yucca Mountain repository is situated. Water from the second well came from the Calico Hills unit, which is three hundred meters deeper than the Topopah Spring tuff. Groundwater was collected from both aquifers and filtered. Greater than 98 percent of the plutonium was associated with the colloidal fraction. The colloids consisted of clays, zeolites, and silica with a mean size range of 100 to 150 nanometers. These minerals are consistent with the host rock composition. The residual plutonium remaining after a nuclear test *can* have a unique isotopic signature (the ratio of plutonium-240 to plutonium-239). In both groundwater samples, the plutonium isotope ratio was the same and exactly matched only one underground test—a test detonated twenty-eight years earlier, 1.3-kilometers upgradient from the sampling wells. This work clearly established that plutonium was associated with the colloidal fraction of the groundwater and can be transported significant distances in fractured rock. It is not known if the plutonium was transported as a pseudocolloid or intrinsic colloid.

In a field study conducted at the Rocky Flats Environmental Technology Site (then called the Rocky Flats Plant), groundwater from a shallow well was collected, filtered, and analyzed for radionuclides (Harnish et al. 1994). The Rocky Flats Plant was established in 1951 to manufacture triggers for use in nuclear weapons and purify plutonium recovered from "retired" weapons. Harnish and colleagues (1994) detected low levels of plutonium in the shallow groundwater, the overwhelming majority of which was associated with the particulate and colloidal fractions of the groundwater. Organic material was suspected, but not analyzed. This study documented that the low levels of plutonium detected in shallow groundwater is in the colloid and not the dissolved fraction.

Another study at the Rocky Flats Environmental Technology Site showed that most of the plutonium and americium transported in surface waters is associated with the particulate and colloidal fractions of groundwater, and not the dissolved fraction. Between 1998 and 2000, water samples were collected from storm runoff and pond discharge sites (Santschi et al. 2002). The water collected contained low levels of plutonium (0.01 to 0.21 picocuries per liter) and americium (up to 0.08 picocuries per liter). Most of these radionuclides were detected in the particulate and colloidal fractions, with less than 8 percent detected in the dissolved fraction of the groundwater. Isoelectric experiments showed that the colloidal plutonium was in the plus-four valence state and was associated with the organic component of the colloids.

Groundwater containing radionuclides was collected from a vadose zone—that is, an unsaturated, fractured rock environment—at the Nevada Test Site in order to evaluate whether colloids are associated with radionuclides in a partially saturated groundwater system (Kersting et al. 2003a). Groundwater was collected dripping from the surface of fractures exposed in tunnel bores near the location of previous underground nuclear tests conducted in volcanic tuff in the Tunnel formation. The Tunnel formation lies stratigraphically below the Calico Hills formation (Laczniak et al. 1996). Greater than 90 percent of the plutonium was associated with the colloidal fraction of the groundwater. Colloids ranged in size from 120 to less than 80 nanometers, and consisted primarily of clays, zeolites, and silica, similar to the minerals that comprise the host rock. In a partially saturated fracture-flow, vadose zone system, low-solubility radionuclides are associated with the colloidal fraction and not the dissolved fraction of the groundwater.

While the field studies discussed above demonstrate that low levels of plutonium can migrate under a variety of hydrologic environments, other field studies suggest that colloids and their sorbed contaminants are not always so easily transported. Nonsorbing polystyrene microspheres, which are expected to mimic the transport of natural colloids, were injected in an unsaturated fractured rock environment at Busted Butte, eight kilometers southeast of the proposed Yucca Mountain repository. The fluid containing the microspheres had a high ionic strength, significantly different from the local water composition. During the first six months after injection, no microspheres were detected at the collection boreholes (Eckhardt 2000). In a second colloid transport experiment, microspheres were injected into two saturated aquifers (C–wells) that underlie the proposed Yucca Mountain repository. The microspheres were transported and reached the collection well slightly ahead of the dissolved solutes. Nevertheless, only a small fraction of the original mass of microspheres was transported (Eckhardt 2000). The experimental results suggested that there were multiple transport pathways and that colloidal microspheres move slightly faster than the average rate of groundwater. These two field experiments highlight the importance of understanding the local hydrologic and geochemical factors that ultimately influence transport.

A naturally occurring high-grade uranium ore in Oklo, Gabon, reached criticality approximately two billion years ago, producing several natural fission reactors as a result of highly reducing environmental conditions and the presence of abundant organic material (Nagy et al. 1991). This unique geologic situation produced several natural fission reactors by the thermalized neutron-induced fission of uranium-235, as well as neutron capture species such as plutonium-239 from uranium-238. The fissiogenic daughters and the uranium as well as plutonium are retained in the uraninite and secondary (uranium, zirconium) silicate minerals, and have not migrated, even after such an extremely long time (Jensen and Ewing 2001).

The field and laboratory studies discussed above demonstrate that colloids exist, radionuclides can sorb to pseudocolloids, and colloids have been transported in nature. What isn't known from field studies is whether the plutonium is transported as an intrinsic colloid or pseudocolloid. Due to low concentrations of contaminants, spectroscopic methods have not been able to determine with which colloids the plutonium is associated. Collectively, these field studies highlight the importance of the specific local geologic, geochemical, and hydrologic conditions as well as the emplacement conditions of actinides in determining if and when colloids will facilitate the transport of low-solubility radionuclides. These studies show that the colloid-facilitated transport of low-solubility radionuclides is the dominant transport mechanism under ambient groundwater conditions. Yet in some cases, actinides are not transported. Therefore, it is equally crucial to understand when the geologic/hydrologic and source term conditions are not conducive to the transport of low-solubility radionuclides. It is necessary to evaluate the local environmental conditions before determining if colloids will or will not transport contaminants.

## Implications for the Yucca Mountain Repository: Developing a Conceptual Model for Colloidal Transport

The colloid-facilitated transport of low-solubility radionuclides, such as plutonium, can occur in nature, but will this process occur under the conditions expected at the proposed Yucca Mountain repository, both in the near field and farther away (far field) from the emplaced waste? Knowledge of the thermal regime, water composition, and water flux in both the near and far field is necessary for determining if and how much colloid-facilitated transport of low-solubility radionuclides can occur. Will water be available for colloid transport away from the near-field environment, without which the process of colloid-facilitated transport becomes unimportant? What effect will the near-field thermal environment have on stabilizing/destabilizing the colloid population? What is the life cycle of the colloids generated in the near-field environment—from generation and detachment of the colloids in contact with the waste package in the near field to transport and interaction with the surrounding geologic environment in the far field? At what rate will colloids form? If colloids are formed and detached slowly, the total available mass of colloids may be too low to transport a significant dose of radioactivity. How much radioactivity can be transported? Will the colloids be stable and remain stable if transported away from the waste package? Ultimately, how much, how fast, and how far can colloids transport radioactivity? These important questions need answers before colloid transport models can adequately determine whether or not colloids will transport enough radioactivity into the far field to jeopardize human health.

## Colloids in the Near Field

The chemical environment of the near field after the waste has been emplaced will be quite different from the current conditions. The nuclear waste will generate a significant amount of heat, which will alter the current hydrologic environment. The rate at which colloids can be transported depends on the amount of water available. The stability of colloids is a function of both the heat budget and the water composition. Currently, near-field thermal models predict an initial thermal pulse (the first thousand years) of greater than 95°C (Nuclear Waste Technical Review Board 2003). The thermal heat is thought to help in keeping the environment near the stored waste dry by driving water away. If the repository remains dry, the transport of radionuclides in the near field is minimized. The thermal pulse of 95°C or greater is also thought to increase the rate of corrosion of the waste canisters, resulting in corrosion and release of radionuclides during the first thousand years, and lessening the effectiveness of the engineered barrier system (Nuclear Waste Technical Review Board 2003). This effect would increase the amount of time the radioactive waste is available for transport.

Colloids formed from the degradation of waste materials represent the material most likely to sorb radionuclides and be available for transport (Ramsay 1988; Bates et al. 1992). If colloids generated from the degradation of the waste package are produced and detached slowly or are unstable in the groundwater, then the mass of colloids could be too low to carry a significant concentration of radioactivity. Chemical changes in the ionic strength, the pH, the groundwater composition, the thermal history, and the status of the steady-state hydrogeochemical system have been found to greatly influence colloid generation and stability (e.g., Ryan and Gschwend 1994; Roy and Dzombak 1996; Degueldre et al. 2000). Clearly, the final engineered barrier system chosen and the resulting hydrologic conditions in the near field over the lifetime of the repository will influence the generation rate of colloids, the water flux available for colloidal transport, the stability of the colloid population, and the length of time the radionuclides are available for transport.

Few studies have focused on colloid generation from the degradation of the spent fuel, waste package, or shielding materials associated with the radionuclides in the repository. Buck and Bates (1999) showed that colloid particles were produced from the corrosion of the borosilicate glass, which is being used to solidify high-level nuclear waste. The colloids consisted of clays with small amounts of calcite, dolomite, and transition metal oxides. Clays are stable in most groundwater environments and have been identified in groundwater representative of the far-field environment at the proposed Yucca Mountain repository (Viani and Martin 1998). The colloids generated from the interaction with the waste glass would most likely be stable in the far-field environment; yet it is uncertain how stable clay colloids are under the geochemical, hydrologic, and thermal conditions expected in the near field.

Most actinides will be contained not in borosilicate glass but in the spent fuel (see chapter 21, this volume). Currently, scientists cannot estimate the concentration or composition of the colloids generated from the spent fuel waste package. In addition, scientists cannot predict the concentration of plutonium that can attach to colloids, which are then available for transport. More data on colloid composition, colloid generation and detachment rate, and colloid stability along with the expected water flux and thermal history in the near-field environment would greatly improve our understanding of colloid-facilitated transport at the repository.

## Colloids in the Far Field

Our understanding of colloid behavior under conditions expected in the far field at the proposed Yucca Mountain repository is based on studies conducted by the Yucca Mountain Project as well as other analog studies. Several studies analyzed the concentration and composition of groundwater colloids from wells at the Nevada Test Site and one well (J-13) near the Yucca Mountain repository (Viani and Martin 1998; Brachmann and Kersting 2000; Kung 2000). Studies showed that colloids consisted of clays, zeolites, and cristobalite. Although the relative abundances of the mineral colloids varied, the composition is consistent with the mineralogy of the host rock or fracture-lining minerals (Brachmann and Kersting 2000). The colloid concentration varied significantly, between 0.03 and 3.12 milligrams per liter, with colloids in Yucca Mountain well water (J-13) reported to have the lowest concentration (Viani and Martin 1998). Measured concentrations should be considered a maximum as pump rates tend to overestimate the natural concentration of colloids (Buddemeier 1986; Puls et al. 1992; Degueldre et al. 1996). More information is known regarding colloid composition and concentration expected in the far-field environment than in the near-field one, yet information on colloid stability and colloid loads is still inadequate to predict the maximum amount of radioactivity and the rate of transport.

## Summary

Work presented in this chapter shows that under ambient groundwater conditions, colloids are the dominant transport process for low-solubility radionuclides such as plutonium. Yet it is important to evaluate the local environmental conditions before determining if colloids will or will not transport contaminants. Over the lifetime of the proposed Yucca Mountain repository, conditions are expected to change dramatically in the near-field environment. Knowledge of the thermal regime, water composition, and water flux in both the near and far field is critical for determining if and how much colloid-facilitated transport of low-solubility radionuclides can occur. Although the Yucca Mountain Project has worked to understand the evolving physical and chemical conditions expected at the proposed Yucca Mountain repository during the first ten thousand years, it is still unclear what the thermal and hydrological environment will be during the first thermal pulse (hot and dry or cooler and wet), making it difficult to determine to what extent the near-field environment will facilitate or hinder the colloidal transport of plutonium.

To date, there have been no studies to systematically examine the life cycle of colloids—from the generation of colloids at the waste package through transport to the near field and finally the far field along a continuous geochemical and hydrologic reaction pathway. Are colloids stable under conditions expected along this pathway? Will the colloid population be dominated by intrinsic or pseudocolloids? Will the changing thermal, chemical, and hydrologic environment along the pathway from stored waste to a far-field environment help to minimize colloid transport or just the opposite? Specific emphasis should be on determining the conditions that would minimize colloid transport in both the near- and far-field environments.

The failure to determine the conditions under which colloids facilitate transport at the proposed Yucca Mountain repository could lead to a significant underestimation of the migration of low-solubility radionuclides such as plutonium. Yet many fundamental scientific questions still remain before models are adequate to reliably predict actinide migration. How far, how much, and how fast are long-lived radionuclides such as plutonium likely to be transported given our best understanding of the evolving conditions over the lifetime of the Yucca Mountain repository? Ultimately, can colloids transport a high enough dose of radioactivity to jeopardize human health? Without such knowledge, large and unsatisfactory uncertainties will accompany any attempt to predict the concentration of plutonium that may migrate into the groundwater. These questions ultimately can—and must—be answered.

## References

Bates, J.K., Bradley, J.P., Teetsov, A., Bradley, C.R., and Bucholtz Ten Brink, M. (1992) Colloid Formation during Waste Form Corrosion: Implications for Nuclear Waste Disposal. *Science* 256, pp. 649–651.

Bowen, S.M., Finnegan, D.L., Thompson, J.L., Miller, C.M., Baca, P.L., Olivas, L.F., Geoffrion, C.G., Smith, D.K., Goishi, W., Esser, B.K., Meadows, J.W., Namboodiri, N., and Wild, J.F. (2001) *Nevada Test Site Radionuclide Inventory, 1951–1992*. Los Alamos National Laboratory. LA–13859–MS.

Brachmann, A., and Kersting, A.B. (2000) Characterization of Groundwater Colloids from the ER20–5 Well Cluster and Cheshire Underground Nuclear Test. In *Hydrologic Resources Management Program and Underground Test Area FY1999*, ed. Smith, D.K., and Eaton, G.F., Lawrence Livermore National Laboratory. UCRL–ID–139226, pp. 13–33.

Buck, E.C., and Bates, J.K. (1999) Microanalysis of Colloids and Suspended Particles from Nuclear Waste Glass Alteration. *Applied Geochemistry* 14, pp. 635–653.

Buckau, G., Artinger, R., Fritz, P., Geyer, S., Kim, J.I., and Wolf, M. (2000) Origin and Mobility of Humic Colloids in the Gorleben Aquifer System. *Applied Geochemistry* 15, pp. 183–191.

Buddemeier, R.W. (1986) Sampling the Subsurface Environment for Colloidal Materials: Issues, Problems, and Techniques. In *Transport of Contaminants in the Subsurface: The Role of Organic and Inorganic Colloidal Particles*, ed. McCarthy, J.F. Department of Energy. U. S. DOE/ER–0331.

Buddemeier, R.W., and Hunt, J.R. (1988) Transport of Colloidal Contaminants in Groundwater: Radionuclide Migration at the Nevada Test Site. *Applied Geochemistry* 3, pp. 535–548.

Clark, D. (2000) The Chemical Complexities of Plutonium. In *Los Alamos Science: Challenges in Plutonium Science 2*, ed. Cooper, N.G. Los Alamos, NM: Los Alamos National Laboratory. LA–UR 00–4100, pp. 364–391.

Degueldre, C., Pfeiffer, H.R., Alexander, W., Wernli, B., and Bruetsch, R. (1996) Colloid Properties in Granitic Groundwater Systems I: Sampling and Characterization. *Applied Geochemistry* 11, pp. 677–695.

Degueldre, C., Triay, I., Kim, J.I., Vilks, P., Laaksoharju, M., and Miekeley, N. (2000) Groundwater Colloid Properties: A Global Approach. *Applied Geochemistry* 15, pp. 1043–1051.

Duff, M.C., Hunter, D.B., Triay, I., Bertsch, P.M., Reed, D.T., Sutton, S.R., Shea-McCarthy, G., Kitten, J., Eng, P., Chipera, S.J., and Vaniman, D.T. (1999) Mineral Associations and Average Oxidation States of Sorbed Pu on Tuff. *Environmental Science and Technology* 33, pp. 2163–2169.

Eckhardt, R.C. (2000) Yucca Mountain: Looking Ten Thousand Years into the Future. In *Los Alamos Science: Challenges in Plutonium Science 2*, ed. Cooper, N.G. Los Alamos, NM: Los Alamos National Laboratory. LA–UR 00–4100, pp. 464–496.

Efurd, D.W., Runde, W., Banar, J.C., Janecky, D.R., Daszuba, J.P., Palmer, P.D., Roensch, F.R., and Tait, C.D. (1998) Neptunium and Plutonium Solubilities in a Yucca Mountain Groundwater. *Environmental Science and Technology* 32, pp. 3893–3900.

Harnish, R.A., McKnight, D.M., and Ranville, J.F. (1994) *Particulate, Colloidal, and Dissolved-Phase Associations of Plutonium and Americium in a Water Sample from Well 1587 at the Rocky Flats Plant, Colorado.* U.S. Geological Survey Water-Resources Investigations. Report 93–4175, 27 p.

Jensen, K.A., and Ewing, R.C. (2001) The Okelobondo Natural Fission Reactor, Southeast Gabon: Geology, Mineralogy, and Retardation of Nuclear-Reaction Products. *GSA Bulletin* 113, pp. 32–62.

Kersting, A.B., Efurd, D.W., Finnegan, D.L., Rokop, D.J., Smith, D.K., and Thompson, J.L. (1999) Migration of Plutonium in Groundwater at the Nevada Test Site. *Nature* 397, pp. 56–59.

Kersting, A.B., Smith, D.K., and Wang, L. (2003a) Characterization of Colloids in Water from a Vadose Zone: Rainier Mesa, Nevada Test Site. In *Colloid-Facilitated Transport of Low-Solubility* Radionuclides: *A Field, Experimental, and Modeling Investigation*, ed. Kersting, A.B., and Reimus, P.W. Livermore, CA: Lawrence Livermore National Laboratory. UCRL–ID–149688, pp. 33–43.

Kersting, A.B., Zhao, P., Zavarin, M., Sylwester, E.R., Allen, P.G., Wang, L., Nelson, E.J., and Williams, R.W. (2003b) Sorption of Pu (IV) on Mineral Colloids. In *Colloid-Facilitated Transport of Low-Solubility Radionuclides*: *A Field, Experimental, and Modeling Investigation*, ed. Kersting A.B., and Reimus, P.W. Livermore, CA: Lawrence Livermore National Laboratory. UCRL–ID–149688, pp. 44–66.

Kim, J.I. (1991) Actinide Colloid Generation in Groundwater. *Radiochimica Acta* 52/53, pp. 71–81.

Kung, K.S. (2000) *Colloid Characterization and Quantification in Groundwater Samples.* Los Alamos National Laboratory. LA–13727–MS.

Laczniak, R.J., Cole, J.C., Sawyer, D.A., and Trudeau, D.A. (1996) *Summary of Hydrologic Controls on Groundwater Flow at the Nevada Test Site, Nye County, Nevada.* U.S. Geological Survey Water-Resources Investigations. Report 96–4109.

Lu, N., Cotter, C.R., Kitten, H.D., Bentley, J., and Triay, I.R. (1999) Reversibility of Sorption of Plutonium-239 onto Hematite and Goethite Colloids. *Radiochimica Acta* 83, pp. 167–173.

McCarthy, J.F., and Degueldre, C. (1993) Sampling and Characterization of Colloids and Particules in Groundwater for Studying Their Role in Contaminant Transport. In *Environmental Particles 2*, ed. Buffle, J., and van Leeuwen, H.P. Boca Raton, FL: Lewis Publisher, pp. 315–427.

McCarthy, J.F., and Zachara, J.M. (1989) Subsurface Transport of Contaminants. *Environmental Science and Technology* 23, pp. 496–502.

Minai, Y., Choppin, G.R., and Sisson, D.H. (1992) Humic Material in Well Water from the Nevada Test Site. *Radiochimica Acta* 56, pp. 195–199.

Nagy, B., Gauthier-Lafaye, F., Holliger, P., Davis, D.W., Mossman, D.J., Leventhal, J.S., Rigali, M.J., and Parnell, J. (1991) Organic Matter and Containment of Uranium and Fissiogenic Isotopes at the Oklo Natural Reactors. *Nature* 354, pp. 472–475.

National Research Council (2001) *Deposition of High-Level Waste and Spent Nuclear Fuel: The Continuing Societal and Technical Challenges.* Washington, DC: National Academy Press, 198 p.

Nuclear Waste Technical Review Board (2003) Letter from the Board to Dr. Chu, October 21. http://www.nwtrb.gov/.

Puls, R.W., Clark, D.A., Bledsoe, B., and Powell, R.M. (1992) Metals in Groundwater: Sampling Artifacts and Reproducibility. *Hazardous Waste and Hazardous Materials* 9, pp. 149–162.

Ramsay, J.D.F. (1988) The Role of Colloids in the Release of Radionuclides from Nuclear Waste. *Radiochimica Acta* 44/45, pp. 165–170.

Reimus, P.W., Ware, S.D., Lu, N., Kung, K.S., Abdel-Fattah, A., Anghel, I., Neu, M.P., and Reilly, S.D. (2003) Colloid-Facilitated Plutonium Fracture Transport Experiments. In *Colloid-Facilitated Transport of Low-Solubility Radionuclides: A Field, Experimental, and Modeling Investigation*, ed. Kersting, A.B., and Reimus, P.W. Livermore, CA: Lawrence Livermore National Laboratory. UCRL–ID–149688, pp. 93–128.

Roy, S.B., and Dzombak, D.A. (1996) Colloid Release and Transport Processes in Natural and Model Porous Media. *Colloids and Surfaces A: Physicochemical Engineering Aspects* 107, pp. 245–262.

Ryan, J.N., and Elimelech, M. (1996) Colloid Mobilization and Transport in Groundwater. *Colloids and Surfaces A: Physicochemical and Engineering Aspects* 107, pp. 1–56.

Ryan, J.N., and Gschwend, P.M. (1994) Effects of Ionic Strength and Flow Rate on Colloid Release: Relating Kinetics to Intersurface Potential Energy. *Journal of Colloid and Interface Science* 164, pp. 21–34.

Santschi, P.H., Roberts, K.A., and Guo, L.D. (2002). Organic Nature of Colloidal Actinides Transported in Surface Water Environments. *Environmental Science and Technology* 36, pp. 3711–3719.

Smith, D.K. (1995) *Challenges in Defining a Radiologic and Hydrologic Source Term for Underground Nuclear Test Centers, Nevada Test Site, Nye County, Nevada.* Livermore, CA: Lawrence Livermore National Laboratory. UCRL–JC–120389.

Smith, P.A., and Degueldre, C. (1993) Colloid-Facilitated Transport of Radionuclides through Fractured Media. *Journal of Contaminant Hydrology* 13, pp. 143–166.

Stumm, W. (1992) *Chemistry of the Solid-Water Interface: Processes at the Mineral-Water and Particle-Water Interface in Natural Systems.* New York: Wiley, 428 p.

Thompson, J.L. (1989) Actinide Behavior on Crushed Rock Columns. *Journal of Radioanalytical and Nuclear Chemistry Articles* 130, pp. 353–364.

Thompson, J.L. (1999) *Laboratory and Field Studies Related to Radionuclide Migration at the Nevada Test Site: Oct. 1, 1997–Sept. 30, 1998.* Los Alamos, NM: Los Alamos National Laboratory. LA–13576–PR.

Triay, I.R., Cotter, C.R., Kraus, S.M., Huddleston, M.H., Chipera, S.J., and Bish, D.L. (1996) *Radionuclide Sorption in Yucca Mountain Tuffs with J–13 Well Water: Neptunium, Uranium, and Plutonium.* Los Alamos, NM: Los Alamos National Laboratory. LA–12956–M.

U.S. Department of Energy (1994) *United States Nuclear Tests, July 1945 through September 1992.* U.S. Department of Energy/Nevada Field Office. DOE/NV–209 Rev. 14.

Viani, B.E., and Martin, S.I. (1998) *Groundwater Colloid Characterization MOLO3.* Livermore, CA: Lawrence Livermore National Laboratory. UCRL–ID–132087.

# 13

## The Importance of Mineralogy at Yucca Mountain

David L. Bish, J. William Carey, Steve J. Chipera, and David T. Vaniman

Designers of nuclear waste repositories assume that over the millennia, the waste containers will eventually corrode and release radionuclides into the local environment. The geologic repository was conceived to take advantage of the ability of natural systems (e.g., rocks) to isolate radionuclides from human contact for long times. The question facing those trying to characterize any proposed location for a repository then becomes when and how will radioactivity travel to the biosphere. Isolation can be achieved by either chemical reactions that bind the radionuclides to the surrounding rock or low rock permeability that traps radionuclides near the waste source. Minerals play a role in both of these processes, either directly through reactions or indirectly through the relationship between mineral shape and rock permeability.

Zeolites are unique minerals that occur at Yucca Mountain. They have unusual crystal structures containing large channels and pores, giving them the ability to reversibly take up water and many chemical species. Because of their ability to sorb positively charged radionuclides, zeolites have the potential to slow the release of radionuclides. As early as the 1970s, the presence of zeolites in volcanic tuffs attracted scientists' attention to Yucca Mountain as a possible host for radioactive waste (e.g., Johnstone and Wolfsberg 1980).

Further studies of Yucca Mountain over the past two decades, however, have shown that the importance of mineralogy is much broader than radionuclide adsorption by zeolites. For example, minerals present in trace amounts, such as manganese oxides, can be particularly significant either because they are concentrated in fractures that are possible flow paths or because they interact strongly with some radionuclides. Mineralogy plays a fundamental role in determining the permeability, mechanical properties, and thermal and chemical stability of the rocks at Yucca Mountain. In addition, studies of the area's present-day mineralogy (Broxton et al.

1987; Bish and Aronson 1993; Vaniman et al. 2001) have been crucial in unraveling the past thermal and geochemical response of rocks at Yucca Mountain, and they have provided key insights into predicting the mountain's future behavior.

## Yucca Mountain: Background

The rocks at Yucca Mountain are known as volcanic tuffs, formed from volcanic ash erupted onto Earth's surface. If the ash deposits were very hot when deposited, they created hard and dense tuffs that have been welded together. These rocks are known as welded tuffs, and they commonly contain abundant feldspars and silica minerals. Ash deposits that were cooler when deposited created porous and soft tuffs that were not welded together, often containing original volcanic glass. Where nonwelded glassy rocks occur near or beneath the water table, they have been altered by groundwater to zeolites via a process known as zeolitization. Large-scale zeolitization has occurred at elevations up to about a hundred meters above the present water table.

The hydrogeologic setting of Yucca Mountain is unique in that a repository could potentially be sited more than 300 meters below the surface and yet still be more than 150 meters above the water table within welded tuff. Welded tuff units contain ~60 percent feldspar, ~30 percent silica minerals (quartz, cristobalite, and tridymite), and minor or trace minerals such as smectite (a swelling clay), calcite, and manganese oxides. This repository horizon tuff unit is underlain across Yucca Mountain by tuffs containing common zeolite minerals—predominantly clinoptilolite, with smaller amounts of mordenite. Zeolitic rocks occur everywhere across the mountain between the potential repository horizon and the water table. Several hundred meters below the water table, the tuffs contain analcime, a more stable zeolite that does not affect the migration of radionuclides.

The zeolitized rocks usually have low permeabilities. As a result, computer models of radionuclide movement at Yucca Mountain have assumed that water will not flow through such rocks. This may not be a valid assumption, however. Recent geochemical studies suggest that such units have in fact transmitted water throughout much of the mountain's geologic history (Vaniman et al. 2001). Rocks above the water table that are only partially zeolitized contain unaltered volcanic glass and up to ~50 percent clinoptilolite, and they retain much of their originally high permeability. Such rocks can be of great importance in radionuclide retardation due to the combination of high permeability and the presence of zeolites. Figure 13.1

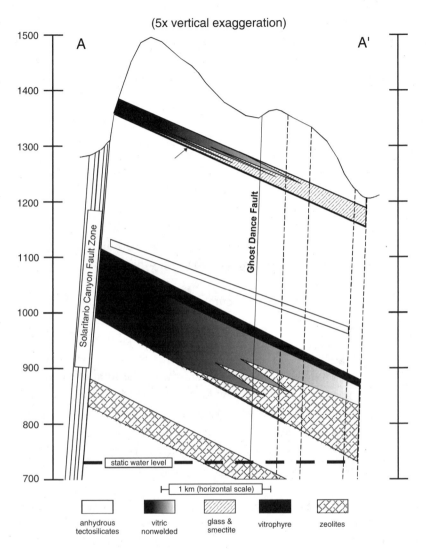

**Figure 13.1**
East-west cross-section along Antler Ridge at Yucca Mountain. Solid lines represent mapped faults and fault zones; inferred faults are dashed. The repository horizon is indicated by the enclosed rectangular area above the vitrophyre.

shows the gradational nature of the glass-zeolite transition above the water table in an east-west cross-section. This illustration also indicates the approximate depth of the repository horizon and shows that the first occurrence of zeolites ranges from 100 to 250 meters below this horizon.

## Mineralogical Barriers to Radionuclide Migration

An important reaction involving radionuclides is cation (positively charged elements) exchange. In this process, minerals exchange an atom in their structure (e.g., sodium) with a radionuclide in the fluid (e.g., cesium). Although it is a common perception that highly sorbing minerals such as zeolites and clays *stop* the movement of radionuclides in groundwater via cation exchange, these minerals in fact only *retard* or slow radionuclide migration. This is because most sorption reactions on minerals are reversible—that is, the element sorbed may be released back into solution.

The natural zeolites present at Yucca Mountain such as clinoptilolite and mordenite gained widespread attention in radioactive waste applications after the pioneering research of Ames in the late 1950s. Ames (1960) showed that clinoptilolite was effective in removing cesium and strontium from wastewater; these two elements account for between two-thirds and three-fourths of the total radioactivity in high-level waste in the respository's first few hundred years. In addition to possessing high cation-exchange capacities (often twice that of typical clay minerals), clinoptilolite and mordenite have high *selectivities*: these zeolites can remove cesium, barium, and strontium from solutions even when these elements are present in small amounts together with large amounts of other competing elements such as sodium or calcium.

The relative importance of different mineral-radionuclide interactions depends on various properties of the particular radionuclides that are present in the waste: half-life, chemical form (e.g., gas versus aqueous), toxicity, solubility, and the form in which the radionuclide exists in water. For example, the predominant isotopes of both cesium (cesium-137) and strontium (strontium-90) have short half-lives ($\leq$ thirty years), and it is reasonable to assume that they will have decayed to insignificant quantities long before the engineered portion of a repository (e.g., the waste canister) is breached. Thus, in spite of clinoptilolite's and mordenite's large cation-exchange capacities and selectivity for cesium and strontium, they are only of minor importance in a geologic repository (although at least one isotope of cesium [cesium-135] has a sufficiently long half-life [2.3 million years] to be of long-term concern). Still, the presence of these zeolites is key in providing assurance against significant early movement of cesium and strontium should the engineered barriers fail much sooner than expected.

Positively charged (*cationic*) radionuclides such as cesium and strontium are of less concern than *anionic* (negatively charged) radionuclides such as technetium oxide ($TcO_3^-$), neptunium oxycarbonate ($NpO_2CO_3^-$), and iodide ($I^-$). Few, if any, minerals at Yucca Mountain or in other natural environments interact strongly with anionic species. For this reason, some scientists (e.g., Bish 1980) have proposed the use in the engineered system of layered double hydroxide materials containing interlayer anions to sorb such anionic species.

Another constituent of waste would be long-lived heavy elements that form large, complex aqueous species such as uranium dioxide ($UO_2^{2+}$), neptunium dioxide ($NpO_2^+$), plutonium dioxide ($PuO_2^+$), and neptunium oxycarbonate ($NpO_2CO_3^-$). Recent work (Vaniman et al. 1996; Denniston et al. 1997; Duff et al. 1999) has shown that minor and trace minerals such as smectite and manganese oxides are much more likely than zeolites to dominate the retardation of heavy elements such as plutonium, americium, and neptunium. For example, batch sorption studies on a Yucca Mountain zeolitic tuff originally suggested that clinoptilolite had a small preference for $Pu^{+5}$. Detailed microscopic evaluation of Yucca Mountain tuffs, however, has shown that zeolites are not important sorbing minerals for plutonium. Indeed, microautoradiography studies have allowed identification, on a microscopic scale, of the individual minerals onto which a given radionuclide has sorbed (become attached) rather than relying on determination of the bulk behavior of a particular sample. Vaniman and others (1996) showed that plutonium sorbed to minerals that occur in low concentrations, such as to altered (partially reacted) iron and manganese silicates and partially crystalline oxides in devitrified tuffs (tuffs whose original glassy makeup has reacted to form crystalline minerals, mainly silica minerals and feldspars). Plutonium also sorbed to fracture-lining smectites in zeolitic tuffs (tuffs whose original glassy makeup has been altered by groundwater to form zeolites). Plutonium sorption appears to be dominated by coexisting trace minerals such as smectite and iron or manganese oxides, particularly the mineral rancieite $[(Ca,Mn^{+2})Mn_4^{+4}O_9 \cdot 3H_2O]$. Given the common occurrence of manganese-oxide minerals in fractures at Yucca Mountain, it is likely that these minerals will be important in limiting the concentration of plutonium in groundwater. A related study (Duff et al. 1999) showed that sorbed plutonium is primarily associated with trace rancieite and smectite, but not with iron oxides. These data demonstrate that sorption experiments on natural materials, and even on synthesized "pure" materials, must consider the possibility that minerals that occur in only trace amounts of a rock may dominate a rock's sorption capacity for a particular radionuclide.

It is of course impractical to conduct ten-thousand-year experiments to judge how well Yucca Mountain will perform. The only way to assess the repository is through numerical models of water flow and radionuclide transport through the rocks. Application of laboratory data on hydrologic and mineral properties in such models requires determination of whether both the data and the simulations represent processes that may occur in the real system. For example, although we know that zeolites strongly sorb strontium and cesium in the laboratory, can we be certain that they perform effectively in real rocks? In addition, the mere presence of sorbing minerals is not a guarantee that sorption will take place. Radionuclide-bearing fluids must encounter the minerals. Fluids may flow through the bulk of the rock (matrix flow) or through fractures. As noted above, the bulk-rock mineralogy of Yucca Mountain rocks is often distinct from the mineralogy in fractures. Thus, any model of retardation must consider the different mineralogies in the matrix and fractures at Yucca Mountain, and the fact that migration of many radionuclides may be controlled by minerals that occur only in trace amounts. Models based on bulk-rock data that show little sorption of plutonium, for instance, will predict very different behavior than would models based on sorption data from the manganese-oxide minerals found in existing fractures. Performance assessment models of Yucca Mountain greatly simplify the mineralogy of the mountain and neglect the detailed differences in bulk-rock or fracture mineralogy.

Knowledge of Yucca Mountain's mineralogical history underscores the complexity of simulating the natural system. Trace-element studies of zeolites from Yucca Mountain confirm that these minerals have acted as efficient barriers to the migration of large, positively charged elements (such as strontium or cesium) (Vaniman et al. 2001). Figure 13.2 shows that even low concentrations of clinoptilolite (<5 percent) have retarded the downward migration of natural strontium in the past, as illustrated by the large strontium concentrations at the tops of the zeolitic rock unit. Calculations of strontium accumulation suggest that zeolitic units have been transmissive and have concentrated strontium from percolating groundwaters over much of Yucca Mountain's history. Not all stratigraphic horizons at Yucca Mountain show a simple history; figure 13.3 illustrates that the strontium distribution is

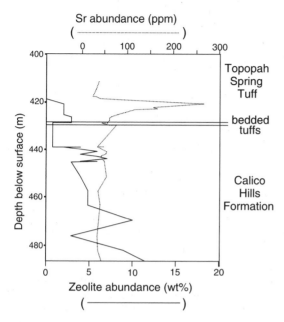

**Figure 13.2**
Comparison of total zeolite abundance with total strontium (Sr) concentration in a portion of drill hole USW SD-12 at Yucca Mountain. Note that a large spike in Sr concentration occurs near the top of the zeolitic horizon.

more complex in other drill holes, due partly to complex groundwater flow and partly to the original compositional heterogeneity of the rocks. Strontium distribution patterns such as these place limits on the lateral movement of water. Plutonium does not occur naturally, but indications of how it might behave in the repository can be inferred from investigations of the distribution of cerium, which is one of plutonium's best natural analogs. Studies show that manganese-oxide minerals in Yucca Mountain have efficiently removed cerium from downward-percolating groundwater, creating abnormally low cerium concentrations in calcite (Vaniman and Chipera 1996; Denniston et al. 1997). These observations of present-day trace element distributions demonstrate conclusively that the rocks and minerals at Yucca Mountain have played a major role in controlling which elements have migrated down through the mountain and into the water table. These data also contradict some models that suggest or assume that downward-percolating fluids will not flow through zeolitic tuffs (Bodvarsson et al. 1999).

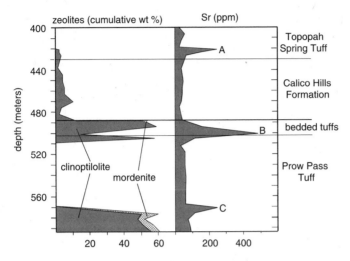

**Figure 13.3**
Comparison of clinoptilolite plus mordenite abundance with total strontium (Sr) concentration in a portion of drill hole USW SD-12. Sr enrichments at A (see detail in figure 13.2) and C are caused by element exchange with zeolites, but the high Sr at B (low zeolite abundance) is caused by an original accumulation of Sr-rich feldspar in the bedded tuffs, determined by analyzing the Sr content of individual feldspar grains.

## Repository-Induced Changes to Yucca Mountain Minerals

The stored radioactive waste will generate a large amount of heat for a long period of time. Depending on the distribution of waste canisters, this heat could modify the minerals and rocks at Yucca Mountain. This concern is amplified by the observation that many of the minerals (zeolites, smectite, forms of silica, and volcanic glass) are probably metastable—that is, they are not in the most stable form, and a rise in temperature or a change in the chemical environment could transform them into different minerals. Indeed, the concern that heat from the disposed waste might induce thermal alterations in the minerals at Yucca Mountain has forced the repository's designers to put a lower limit on the site's areal waste loading (the amount of waste emplaced per unit area) than would have been appropriate if mineral alteration were not possible (Nuclear Waste Technical Review Board 2001). Lower waste loading, of course, increases the total area of the repository.

Figure 13.4 shows a cross-section of the heated region surrounding the repository horizon (modified from Buscheck and Nitao 1993), together with the geology, for

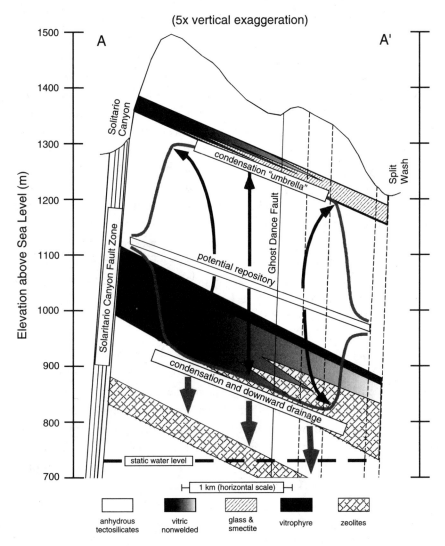

**Figure 13.4**
Cross-section as in figure 13.1, but including the maximum dry-out zone for a waste loading of 114 kilowatt per acre for thirty-year-old spent nuclear fuel at a time two thousand years after waste emplacement (Buscheck and Nitao 1993). The dry-out zone is surrounded by a condensation "umbrella" at the top, and a condensation and downward drainage zone at the bottom. For this particular waste-loading scenario, much higher than currently being considered, a large volume of vitric and zeolitized tuff is at or above the boiling point of water.

what would now be considered a very high areal waste loading (114 kilowatts per acre). Spent nuclear fuel and high-level waste, depending on their age and thus radionuclide inventory, emit heat (measured in kilowatts) generated from the decay of radionuclides. The details of the temperature profiles depend on the age of the spent fuel, the areal waste loading, and the local geology and hydrology, but this figure illustrates that a radioactive waste repository has the potential to heat a large volume of rock. Today, the areal waste loadings considered by the Yucca Mountain Project are much lower than those considered ten years ago. To reduce the extent of thermal, hydrologic, chemical, and mechanical effects, the Nuclear Waste Technical Review Board (2001) continues to push for a cooler repository. Other measures being considered to keep the temperatures of the surrounding rocks below the boiling point of water include aging the waste (to allow the heat-producing radionuclides to decay), increasing the separation between waste canisters, and using forced ventilation of the repository. Although reducing temperature by lowering the areal waste loading is an important step in reducing mineral reactions, in some cases, as will be clear from the following discussion, simply changing the relative humidity in the rocks is sufficient to cause major changes in rock properties.

Some simple examples illustrate why the rocks at Yucca Mountain are so sensitive to elevated temperatures. Calculations considering both the water content of clinoptilolite at 100 percent relative humidity and the measured porosity of zeolitic tuff at Yucca Mountain show that a clinoptilolite-rich tuff contains about as much water in its zeolite minerals as in the rock pores at 92 percent saturation (the approximate measured saturation value in the unsaturated zone of the zeolitic tuff). This "zeolitic" water represents a tremendous reservoir of water. Heating a rock containing 80 percent clinoptilolite to 200°C will release an amount of water from the clinoptilolite structure that is equivalent to ~27 percent of the available porosity, accompanied by a small yet significant volume decrease in the rocks themselves. On cooling, partially dehydrated zeolites will rehydrate, expanding and readily taking up any free water initially available in the rock pores. As noted above, recent repository designs will produce temperature rises much smaller than projected only a few years ago. Given the importance of water as a transport medium in any repository environment, however, it is clear that clinoptilolite and mordenite are significant as both *sources* and *sinks* of water. Furthermore, as will be shown below, volume changes and rock stresses accompany dehydration/rehydration, and zeolites can be crucial sources and sinks of thermal energy during dehydration/rehydration.

Models of the long-term thermohydrologic behavior of Yucca Mountain should consider these sources and sinks of water and rock-property changes, but present models do not. Contrary to expectations, elevated temperatures do not appear to cause significant changes in the sorptive properties of clinoptilolite as long as new minerals are not formed. The longer-term reaction of clinoptilolite and mordenite to other minerals such as analcime, though, will greatly affect the overall sorptive properties of the zeolitic tuffs.

The early work of Smyth (1982) focused on the possibility of repository-induced changes in mineralogy. Large amounts of clinoptilolite at the mountain, for example, might be transformed to analcime, having far less sorptive capacity. This change would produce additional water, reduce mineral volume, and result in rock stress. Subsequent research on the thermal reactions of zeolites and smectites at Yucca Mountain focused on the evolution of water from clinoptilolite and mordenite on heating, and the volume changes and stresses that accompany dehydration and hydration reactions (Bish 1984). Figure 13.5 shows the change in volume of a unit cell (a precisely defined small unit of material) as a function of temperature for several different clinoptilolite samples, both in a vacuum and a 100 percent relative humidity atmosphere. This figure emphasizes several important details concerning the heating and cooling of clinoptilolite (accompanied by dehydration and rehydration). First, we see that compositional differences (potassium versus sodium versus calcium) cause large variations in both the magnitude of the heat-induced volume decrease and the temperature at which the shrinkage takes place. Sodium-exchanged clinoptilolite experiences up to an 8.4 percent decrease in volume on heating to 100°C, but a natural, mixed-element clinoptilolite from Yucca Mountain (USW G-4 450.5) loses only 2.4 percent after heating to 300°C. Whereas potassium-exchanged clinoptilolite experiences most of its volume decrease on evacuation, calcium-exchanged clinoptilolite is more temperature sensitive and decreases in volume up to at least 300°C. As expected, the presence of water vapor greatly inhibits the volume decrease, raising the temperature required for a given volume decrease or reducing the magnitude of the volume decrease at a given temperature.

The 100 percent relative humidity experiments more accurately represent the conditions in Yucca Mountain rocks. Under these conditions, the volume decreases for clinoptilolite are rapidly reversible, with samples returning to their original volumes at room temperature after the experiments. Given the low waste loadings in the recently proposed repository design and the presence of partially saturated zeolitic rocks at Yucca Mountain, it appears that the existence of a repository may

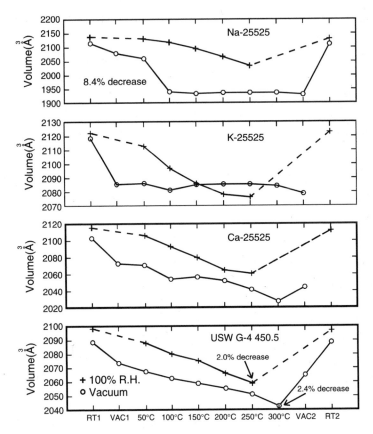

**Figure 13.5**
Unit-cell volume for three element-exchanged clinoptilolites and a natural sample from Yucca Mountain, drill hole USW G–4, 450.5 meters depth. Solid curves with open circles represent data collected as follows: sample RT1 was examined under room conditions (23°C, ~10 percent relative humidity); VAC1, under vacuum (~2.67 Pascal or 20 milltor at 23°C); 50°C through 300°C, under vacuum at the specified temperature; VAC2, returned to 23°C under vacuum; and RT2, 23°C, ~10 percent relative humidity. Dashed/solid curves with plus symbols (+) are similar, but these measurements were done at a relative humidity of 100 percent (determined at 23°C) at the specified temperatures and vacuum measurements were therefore not performed.

not cause large volume decreases. Nevertheless, related experiments on large rock samples show that even small volume changes can greatly affect large-scale rock properties, even without a change in temperature. For example, hydration experiments on vacuum-dried (unheated) bulk samples of zeolitic (clinoptilolite) tuff and welded, devitrified (nonzeolitic) tuff (Kranz et al. 1989) show the importance of the volume effects associated with zeolite hydration (figure 13.6). Hydration of the zeolitic rock generated stresses at least ten times greater than those generated in the nonzeolitic devitrified tuff. Kranz and colleagues (1989) concluded that swelling stresses on the order of ten times the tensile strength of typical zeolitic rocks (five to seven MPa; Blacic et al. 1982) may be possible; because hydration is reversible, dehydration will give rise to comparable tensile stresses. Any concern with the integrity of mined openings as well as issues such as development of differential stresses and opening new fractures dictates greater consideration to the sensitivity of zeolites and clays at Yucca Mountain to dehydration and hydration.

Other data show that zeolites and clays are not only sensitive to temperature but also to relative humidity, and there is no threshold temperature (for example, the

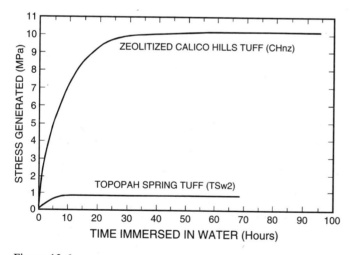

**Figure 13.6**
Axial stress generated in axially confined, unjacketed cylindrical samples of zeolitic (Calico Hills) and devitrified (Topopah Spring) tuff from Yucca Mountain after vacuum drying and immersion in water. Note that the stress value for the devitrified tuff reached a plateau after ~10 hours, whereas the value for the zeolitic sample was still increasing after thirty hours, likely a reflection of the low permeability of the zeolitic sample. Also note that the samples were not heated before immersion.

boiling point of water) above which the zeolites and clays will be affected. In illustrating this point, Carey and Bish (1996; figure 13.7) showed that, for example, the amount of water in sodium-clinoptilolite at 98°C and high relative humidity can be duplicated through appropriate combinations of lower humidities at any lower temperature (or higher humidities at higher temperatures). Thus, as illustrated by the data in figure 13.5 and in Carey and Bish (1996), maintaining the zeolitic rocks at an elevated relative humidity would minimize heat-induced shrinkage. Unfortunately, in a repository environment, the only effective way to ensure high humidity is to reduce the waste loading, which in turn lowers the maximum temperature reached by zeolitic rocks.

In addition to their effect on rock volume, dehydration and hydration reactions of zeolites and clays also have important impacts on the total heat content of the natural system. Because most water molecules in a zeolite or clay crystal structure are held more tightly than in liquid water, the energy required to heat and dehydrate

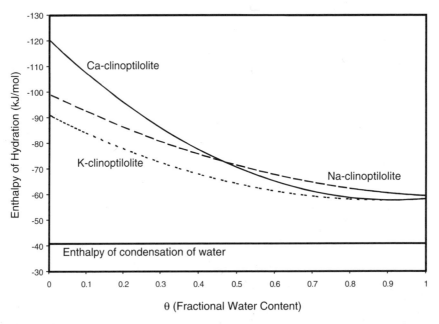

**Figure 13.7**
Calculated partial molar enthalpies of hydration for Na-, K-, and Ca-clinoptilolites at 25°C as a function of fractional water content (Carey and Bish 1996). The enthalpy of vaporization of liquid water is shown on the plot as a horizontal line at ~ −41 kJ/mol.

a zeolite or clay containing an amount of water is greater than that required to vaporize the same amount of liquid water in rock pores (figure 13.7). For example, the energy needed to remove a molecule of water from a half-dehydrated clinoptilolite is approximately twice that required to vaporize a free water molecule in a rock pore. As a consequence of these effects, the presence of water-bearing minerals around a repository will keep rock temperatures lower than they would be in the absence of such minerals.

A simple calculation can be made of the energy necessary to heat a volume of rock to a given temperature with and without zeolite dehydration. For this calculation, two rock units were compared; in one case clinoptilolite was allowed to dehydrate and in the other no minerals dehydrated. Eliminating dehydration from the calculation allows the simulation of nonzeolitic rocks; the energy applied is used to heat the rock only, rather than also to heat the water contained in zeolites or other water-bearing minerals. The result: substantially more heat energy is required to reach a given temperature in rocks containing dehydrating clinoptilolite. The lowest difference occurred for potassium-clinoptilolite, which still required 71 percent more energy to reach 200°C than rocks lacking clinoptilolite. Sodium-clinoptilolite required from 81 to 93 percent more energy than if dehydration were ignored. This effect would be more pronounced for Ca-clinoptilolite because more energy is required to dehydrate Ca-clinoptilolite than Na- or K-clinoptilolite (figure 13.7).

This model shows that rocks containing a substantial amount of zeolites or clays at Yucca Mountain will act as a heat buffer, resulting in lower rock temperatures than would occur in the absence of mineral dehydration. These results also emphasize the importance of considering the complete mineralogical system in any thermal-chemical-hydrologic model of Yucca Mountain. These heat effects are sufficiently well quantified that they can be included in project models to give an accurate picture of the amount of buffering. Although·a three-dimensional mineralogical model of Yucca Mountain is available, this information and the associated thermal effects of hydrous minerals are not included in project models. This omission could become a serious issue because the thermal reactions of minerals impose key constraints on the amount of radioactive waste that can be emplaced in Yucca Mountain per unit area. Dehydration, water production, and the associated volume reduction and rock stress generated place restrictions on waste loading, whereas the energy requirements associated with dehydration partially offset these restrictions by buffering temperature increases.

Longer-term reactions in which one mineral is transformed into a new mineral, such as the conversion of clinoptilolite to analcime, may also be crucial if temperatures are sufficiently high for long periods. These reactions have a threefold importance. First, they cause large volume reductions and produce water. Second, they change the natural collection of minerals present. And finally, they are likely to be irreversible. Observed mineral sequences suggest that clinoptilolite reacts to form analcime at temperatures as low as 90°C to 100°C (Smyth 1982); geothermometry studies indicate that this reaction has indeed occurred in Yucca Mountain rocks at around 100°C (Bish and Aronson 1993). Based on this thermometry, mordenite appears to transform to analcime at 100°C to 130°C (Bish and Aronson 1993). More recent work indicates that the exact temperature of zeolite transformation depends on how much silica is dissolved in the groundwater and the ratio of aluminum to silicon in the zeolite (Chipera and Bish 1997). Therefore, the reaction of clinoptilolite or mordenite to analcime at Yucca Mountain can be caused by elevated temperatures or lower amounts of dissolved silica in the groundwater.

The clinoptilolite-to-analcime reaction would have profound effects on the mineralogical and hydrologic properties of the rock. The reaction produces large amounts of silica, which may either precipitate or migrate away as aqueous silica. The reaction also produces large amounts of free water as the highly sorptive clinoptilolite is replaced by the poorly sorbing analcime. More important, the reaction causes a volume decrease that is most likely manifested as increased rock porosity, permeability, and fracturing. The volume would decrease by 21.5 percent if the silica precipitated as quartz and by 16.9 percent if it precipitated in the form of cristobalite.

In addition to promoting the reactions above, heat from the repository is likely to enhance fluid flow in the region surrounding the repository. These fluids could dissolve and precipitate minerals, most likely one of several varieties of silica minerals. The proposed repository horizon contains abundant cristobalite. Enhanced fluid flow around the repository could result in the dissolution of cristobalite and the precipitation of quartz (or another form of silica) and may significantly change the permeability of the rocks.

As a final example of the importance of mineralogy at Yucca Mountain, the mineralogy in the tunnels at Yucca Mountain affects the structural integrity of the repository. A common design of the repository calls for an extended period in which the waste packages can be retrieved, requiring that the tunnels stay open. At temperatures between 200°C and 300°C, cristobalite and tridymite, both forms of silica that occur in repository-horizon rocks, undergo changes in their crystal structures that result in volume changes. Cristobalite, for example, swells by ~5 percent at ~230°C (Peacor 1973). These mineral changes could cause fractures in the tunnel walls and subsequent rock fall. Therefore, sufficiently high repository temperatures may, by altering the minerals, compromise the ability to retrieve waste packages. Although feldspars are major constituents of the repository-horizon rocks, they are less sensitive than silica minerals are to temperature changes over the range expected and should not affect repository behavior or rock strength.

## Summary

Mineralogy is critical in understanding the effects of high-level radioactive waste at Yucca Mountain. The examples described here show that the importance of minerals extends beyond simple cation-exchange interactions with zeolites and clays to phenomena that can affect the entire thermal-chemical-hydrologic system. In particular, possible thermal reactions of zeolites and clays ultimately provide crucial constraints on the maximum thermal (waste) loading of a repository at Yucca Mountain. Although zeolites occur below the potential repository horizon, they are not critical for the long-term retardation of cesium-137 and strontium-90, as commonly assumed, due to the short half-lives of those radionuclides. The ability of the rocks to interact with radionuclides is probably determined much more by a variety of trace minerals, including manganese-oxides and smectites. The emplacement of radioactive waste will heat large volumes of rock, and hydrous minerals, such as zeolites and smectites, will be both sources and sinks of water and thermal energy during heating and cooling in the unsaturated zone. Also, changes in volume that accompany the dehydration/hydration reactions of these minerals will cause fractures to open and close and will affect rock strength. The longer-term reactions of zeolites—for example, the transformation of clinoptilolite to analcime—can significantly reduce the volume of the minerals, which can in turn have two effects: an increase of fracturing or pore space, and the production of water and silica. For a complete evaluation and prediction of long-term repository performance, the full mineralogical inventory, including major, minor, and trace minerals, must be considered.

# References

Ames, L.L., Jr. (1960) Cation Sieve Properties of Clinoptilolite. *American Mineralogist* 45, pp. 689–700.

Bish, D.L. (1980) Anion Exchange in the Pyroaurite Group Mineral Takovite: Applications to Other Hydroxide Minerals. *Bulletin Mineralogie* 103, pp. 170–175.

Bish, D.L. (1984) Effects of Composition on the Thermal Expansion/Contraction of Clinoptilolite. *Clays and Clay Minerals* 32, pp. 444–452.

Bish, D.L., and Aronson, J.L. (1993) Paleogeothermal and Paleohydrologic Conditions in Silicic Tuff from Yucca Mountain, Nevada. *Clays and Clay Minerals* 41, pp. 148–161.

Blacic, J., Carter, J., Halleck, P., Johnson, P., Shankland, T., Anderson, R., Spicochi, K., and Heller, A. (1982) *Effects of Long-Term Exposure of Tuffs to High-Level Nuclear Waste-Repository Conditions*. Los Alamos National Laboratory. LA–9174–PR. February.

Bodvarsson, G.S., Boyle, W., Patterson, R., and Williams, D. (1999) Overview of Scientific Investigations at Yucca Mountain: The Potential Repository for High-Level Nuclear Waste. *Journal of Contaminant Hydrology* 38, pp. 3–24.

Broxton, D.E., Bish, D.L., and Warren, R.G. (1987) Distribution and Chemistry of Diagenetic Minerals at Yucca Mountain, Nye County, Nevada. *Clays and Clay Minerals* 35, pp. 89–110.

Buscheck, T.A., and Nitao, J.J. (1993) The Analysis of Repository-Heat-Driven Hydrothermal Flow at Yucca Mountain. In *Proceedings of the 4th Annual International Conference on High-Level Radioactive Waste Management*. New York: American Society of Civil Engineers and La Grange Park, IL: American Nuclear Society, pp. 847–867.

Carey, J.W., and Bish, D.L. (1996) Equilibrium in the Clinoptilolite-H₂O System. *American Mineralogist* 81, pp. 952–962.

Chipera, S.J., and Bish, D.L. (1997) Equilibrium Modeling of Clinoptilolite-Analcime Equilibria at Yucca Mountain, Nevada. *Clays and Clay Minerals* 45, pp. 226–239.

Denniston, R.F., Shearer, C.K., Layne, G.D., and Vaniman, D.T. (1997) SIMS Analysis of Minor and Trace Element Distributions in Fracture Calcite from Yucca Mountain, Nevada, USA. *Geochimica et Cosmochimica Acta* 61, pp. 1803–1818.

Duff, M.C., Hunter, D.B., Triay, I.R., Bertsch, P.M., Reed, D.T., Sutton, S.R., Shea-McCarthy, G., Kitten, J., Eng, P., Chipera S.J., and Vaniman, D.T. (1999) Mineral Associations and Average Oxidation States of Sorbed Pu on Tuff. *Environmental Science and Technology* 33, pp. 2163–2169.

Johnstone, J.K., and Wolfsberg, K., eds. (1980) *Evaluation of Tuff as a Medium for a Nuclear Waste Repository: Interim Status Report on the Properties of Tuff*. Sandia National Laboratory. SAND80–1464. July.

Kranz, R.L., Bish, D.L., and Blacic, J.D. (1989) Hydration and Dehydration of Zeolitic Tuff from Yucca Mountain, Nevada. *Geophysical Research Letters* 16, pp. 1113–1116.

Nuclear Waste Technical Review Board (2001) Report to the U.S. Congress and the Secretary of Energy, January to December 2000. http://www.nwtrb.gov.

Peacor, D.R. (1973) High-Temperature Single-Crystal Study of the Cristobalite Inversion. *Zeitschrift für Kristallographie* 138, pp. 274–298.

Smyth, J.R. (1982) Zeolite Stability Constraints on Radioactive Waste Isolation in Zeolite-Bearing Volcanic Rocks. *Journal of Geology* 90, pp. 195–201.

Vaniman, D.T., and Chipera, S. (1996) Paleotransport of Lanthanides and Strontium Recorded in Calcite Compositions from Tuffs at Yucca Mountain, Nevada, USA. *Geochimica et Cosmochimica Acta* 60, pp. 4417–4433.

Vaniman, D.T., Chipera, S., Bish, D., Carey, J.W., and Levy, S. (2001) Quantification of Unsaturated-Zone Alteration and Cation Exchange in Zeolitized Tuffs at Yucca Mountain, Nevada. *Geochimica et Cosmochimica Acta* 65, pp. 3409–3433.

Vaniman, D.T., Furlano, A., Chipera, S., Thompson, J., and Triay, I. (1996) Microautoradiography in Studies of Pu(V) Sorption by Trace and Fracture Minerals in Tuff. *Materials Research Society Symposium Proceedings* 412, pp. 639–646.

# 14

# Contaminant Transport in the Saturated Zone at Yucca Mountain

Lynn W. Gelhar

The movement of contaminated groundwater away from the Yucca Mountain nuclear waste repository site is expected to be the most likely way that people can be exposed to escaping radioactivity. Department of Energy (DOE) researchers have collected data to evaluate the flow of groundwater in and around Yucca Mountain, and to describe how fast radioactive contaminants will be transported away from the site. The researchers have generated computer models based on these data to predict the rate of transport of contaminants away from the site as well as the contamination levels to be expected as groundwater is withdrawn through wells. To evaluate the potential impact on water users around Yucca Mountain, we must quantify the degree of reduction in contaminant concentration due to physical dilution, chemical effects, and radioactive decay.

## The Aqueous Issues

Rainwater infiltrating down through Yucca Mountain reaches the water table some three hundred meters below the repository level and moves horizontally away from the mountain. Because this water moves down through the mountain slowly at rates of one to ten millimeters per day, any contamination it carries will stay near the top of the saturated zone, just below the water table, as it moves away from the site. Most of the water moving through the saturated zone at Yucca Mountain originates in the higher elevation areas to the north where the rainfall and net infiltration are higher.

The aquifer through which contaminants from Yucca Mountain would migrate is a complex of faulted and extensively fractured volcanic tuffs. This volcanic aquifer extends some ten to fifteen kilometers south of the site, after which the groundwater moves into alluvial sediments (see chapter 7, this volume). Groundwater generally moves from areas of high to low head at a rate proportional to the permeability of the aquifer and the head gradient. Volcanic tuff has a high permeability, largely due to extensive fracturing of the rock. Because of the high permeability, high rates of water movement can occur through the rock. The effective quantification of the flow and contaminant transport properties of such a complex fractured and faulted system at a scale of tens of kilometer, pertinent to Yucca Mountain, is an unresolved research problem. A recent National Academy of Sciences study of this issue (National Research Council 1996) contains hundreds of pages of discussion, but does not identify a proven approach to be followed.

This chapter focuses on important limitations of the available data. In addition, it describes efforts to identify uncertainties in the prediction of exposure to radioactivity. The assessment here is based primarily on the review of the pertinent portions of the DOE's (2001a) *Yucca Mountain Science and Engineering Report*. This report, along with the supporting technical documents (DOE 2000, 2001b, 2001c), forms the scientific and technical basis for the DOE's recommendation that the Yucca Mountain site is suitable for radioactive waste disposal. The assumptions and interpretations used in the report are frequently portrayed as being conservative— that is, erring on the side of safety. This chapter will explore several assumptions or interpretations that may not be conservative and explain how these could affect the long-term safety of Yucca Mountain. In this context, the discussion inherently emphasizes potential weaknesses of the existing information as it is used to predict long-term safety.

The U.S. Geological Survey has had primary responsibility for collecting field data to characterize groundwater conditions at Yucca Mountain; much of that work is summarized in Fridrich and colleagues (1994) as well as Luckey and colleagues (1996). More recently, Nye County (2002) has developed a program of well drilling to obtain groundwater data in the area near U.S. Route 95 south of Yucca Mountain. This chapter draws from some of this information. I am familiar with the hydrogeologic conditions at Yucca Mountain—both through work as an outside expert evaluating uncertainties in the saturated zone and through consulting work on the adjacent Nevada Test Site. I have not, however, reviewed all of the many thousands of pages of technical documents pertaining to groundwater at Yucca Mountain.

### The Data Drought

Most of the information about groundwater flow away from the Yucca Mountain site comes from a few dozen observation wells drilled primarily in the area south and east of the site (see figure 14.1). Further data come from the existing wells in the area of groundwater withdrawal for irrigation in the Amargosa Valley some twenty-five kilometers south of the site. The water-level measurements in these wells generally indicate that groundwater initially flows away from the site in a south-easterly direction and then shifts to a more southerly direction as its distance from the site increases (see figure 14.1). The water-level contours shown in that figure are reliable only in areas around Yucca Mountain and to the south in the Amargosa Valley, where water-level data are available. There are many faults in and around Yucca Mountain that may influence the water level and flow, but the well network is not dense enough to sort out the effects of individual faults.

The situation is complicated by the fact that faults can act hydrologically as barriers to or conduits for flow. There is also ambiguity in the area just north of the Yucca Mountain site, where the contour map shows an abrupt change in the water level—a three-hundred-meter drop over a distance of three kilometers. On this map, the high water levels (greater than one thousand meters) at the two northernmost wells have been interpreted to reflect the top of the fully saturated groundwater system. This interpretation is at odds, however, with the findings of a group of outside experts assembled to assess the saturated zone data (Geomatrix Consultants, Inc. 1997). These experts judged that the high water levels more likely reflected perched water—that is, locally saturated zones occurring in the unsaturated zone above the water table.

Perched water has been observed in at least six of the unsaturated zone boreholes at Yucca Mountain. It occurs primarily at the base of the Topopah Spring tuff, the unit in which the repository is to be located (see chapter 7, figure 7.1, this volume). Perched water, which typically occurs above rocks with low permeability that restrict downward flow, seems to be more common in the northern portion of Yucca Mountain, but is also observed in two boreholes in the southern part of the site (SD-7 and SD-12). Unsaturated zone flow modeling simulates the occurrence of perched water, but these predictions may not be reliable because they do not represent the perched conditions observed in the southern part of the site (see DOE 2001a, figure 4.119). Periods of wetter climate would lead to more extensive perched water, but the potential adverse effects of such a change have not been adequately addressed by the DOE. Wetter conditions at or near the repository level that could lead to a higher rate of release of radioactivity are a concern.

Gauging the amount of water flowing through the aquifer requires knowledge of the aquifer's permeability, among other factors. Since the groundwater in the volcanic aquifer flows mainly through fractures, permeabilities derived from laboratory

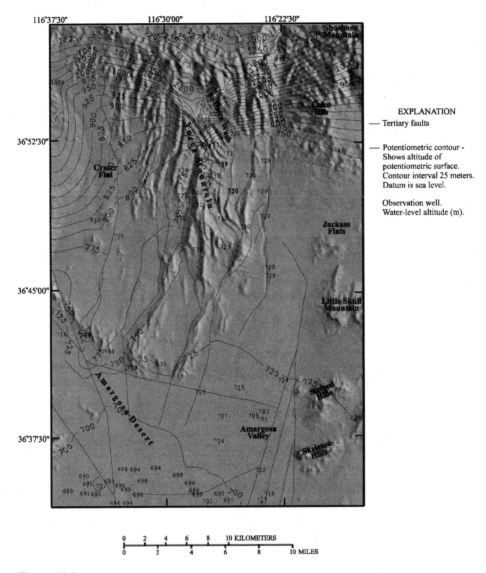

**Figure 14.1**
The estimated contours of the water level in the aquifer underlying Yucca Mountain based on the water levels measured in the observation wells; the locations of the wells are identified by the dots below the numerical values, which are the water levels measured at that location. From DOE 2001a, figure 4.141.

tests on rock cores or in situ single-well tests do not give a good indication of permeability in the field. Though it is geologically plausible that the more extensively fractured welded tuff units would show greater bulk permeability, this notion is not supported by the results of hydrologic borehole testing. Permeable zones with high flow occur frequently in both welded and nonwelded tuffs (DOE 2001d); apparently, hydrologically significant fractures occur throughout the tuff.

The only location in the volcanic aquifer where a multiwell aquifer test has been done to determine field-scale permeability is at the C-wells complex, about three kilometers east of the southern end of the proposed disposal area. Hydraulic testing of the aquifer at the C-wells complex (Geldon et al. 1998) definitively shows that the large-scale permeability of the volcanic aquifer in this area, along the main pathway for potential contaminant migration away from the site, is very high. In fact, the permeability of the aquifer at C-wells is many orders of magnitude larger than that of tuff cores from these rocks and typically two orders of magnitude larger than values derived from single-well in situ tests in the same area. Pumping from the aquifer at the C-wells measurably lowers the water level in observation wells as much as three kilometers away, providing evidence that the area of high permeability southeast of the site is extensive. The low-ambient hydraulic gradient observed in this area further supports this conclusion.

Even when the permeability and the hydraulic gradient are known, it is difficult to assess how fast water will actually move in the aquifer. This is because it is difficult to determine, in a fractured volcanic aquifer of this kind, how much bulk rock the water will flow through. The high porosities found from laboratory studies of tuff cores are not relevant because water moves predominantly through a complex, interconnected network of fractures and fault zones. Tracer tests have been done at the C-wells complex in an attempt to determine the field-scale effective porosity and other transport properties of the volcanic aquifer. But because of the thirty-meter spacing between wells, the results of those tests are largely irrelevant to performance assessment in which transport distances of tens of kilometer are of interest. Furthermore, the tested portion of the aquifer apparently includes anomalously extensive faults and a brecciated zone. As a result, the effective porosities estimated from the test are unreasonably large for a fractured system.

In view of these complications, the DOE uses the C-wells tracer tests mainly in a qualitative sense to explore the role of matrix diffusion effects—that is, the diffusion of contaminants from the fractures into the immobile water in the porous tuff matrix. Some DOE scientists (2000) contend that the tracer tests show some possible matrix diffusion effect, but the results are not relevant at the performance assessment scale because of the small scale of the C-wells tests and the locally anomalous geologic conditions. In reality, there is no pertinent data on the key field-scale physical transport properties of the volcanic aquifer: effective fracture porosity, dispersivities, and matrix diffusion coefficient.

In addition to permeability and porosity, the transport of radionuclides in an aquifer will be affected by the ability of the surrounding rock to retard or sorb radionuclides from the water. DOE scientists have extensively explored the data from chemical and isotopic analyses of water samples from the observation wells around Yucca Mountain as a method of inferring groundwater's flow pathways, mixing behavior, and travel times. Their interpretations are not likely to be of significance, however, primarily because of the sparse and inadequate nature of the network of wells available for sampling. The sampled wells have depths and open intervals where water may enter the well that vary from two to thirteen hundred meters; sampling from wells with such widely ranging depths obliterates vertical variation in chemical composition, creates artificial mixing, and confounds interpretations. The DOE's interpretations unrealistically assume the homogeneity of chemical composition in the vertical direction and look only at two-dimensional horizontal variations. Any meaningful attempt to use geochemical and isotopic data to quantify travel times and mixing at a scale pertinent to potential plumes from Yucca Mountain would require a much more extensive sampling network that carefully considers the vertical positioning of sampling points as well as horizontal coverage.

Data from laboratory measurements of sorption and diffusion in tuff samples are not likely to be directly applicable to field conditions. Batch tests on crushed tuff may be representative of the sorption of radionuclides by the matrix, but the contaminated water moves primarily in fractures in the volcanic aquifer (Luckey et al. 1996). The potential diffusional exchange of dissolved radionuclides between the fractures and the tuff matrix needs to be characterized to indicate how much sorption will retard migration. The DOE has made some laboratory measurements of diffusion into water-saturated, machined tuff slabs; natural fracture surfaces, however, have apparently not been tested. This represents an important gap in the data, as mineral deposits on fracture walls can greatly reduce the diffusive transfer of radionuclides from the fracture to the matrix. It is also critical to know if there is a relationship between the bulk permeability and bulk retardation effects of diffusion/sorption because such effects can strongly influence the longitudinal dispersion of the dissolved contaminants that may be adsorbed on mineral surfaces. Essentially, the sorption/diffusion data have not been collected systematically enough to provide a solid understanding of the relationship between those properties and bulk permeability.

Overall, the data available to quantitatively characterize the transport of contaminants from a repository at Yucca Mountain via flowing groundwater are very limited. From groundwater-level measurements, it is clear that the direction of expected contaminant migration from the proposed repository site is to the southeast and south. Aquifer testing data from the volcanic aquifer southeast of the site show that there is an extensive region of high bulk permeability there, but data are lacking on key field-scale transport properties. Without this information, it is difficult to predict with any confidence the movement of contaminants for the fractured volcanic aquifer.

Why are we faced with this data drought? The answer rests partly in the history of the Yucca Mountain Project. Much of the groundwater data was collected in the 1980s as part of early reconnaissance efforts prior to the 1987 designation by the U.S. Congress of Yucca Mountain as the only high-level radioactive waste site to be investigated. After the site had been designated, more a detailed characterization of the saturated zone was not pursued. One reason for this inaction is that the modeling studies then available indicated that the rate of infiltration of water through the unsaturated zone was extremely low, indicating that little contamination would reach the groundwater. In addition, part of the strategy of confirming the suitability of the site was the construction of a billion dollar underground experimental facility (the Exploratory Studies Facility), which could be expanded to become the actual disposal facility. Program managers appeared not to consider further characterization of the saturated zone to be a high priority.

By the mid-1990s, results from several types of investigation converged to produce irrefutable evidence that water was infiltrating through the unsaturated zone at Yucca Mountain much more rapidly than originally thought (see, for example, Flint et al. 2001). Although this raised concerns about the adequacy of the groundwater data, little has been done to develop the essential field experiments needed to determine the field-scale transport properties of the volcanic aquifer. Some DOE-associated scientists have recognized the need for such experiments; in 1997, for instance, large-scale (hundreds or thousands of meters) tracer tests in the volcanic aquifer were being proposed (I served at the time on a panel of outside experts dealing with saturated zone issues at Yucca Mountain).

For unexplained reasons, however, these crucial experiments in the volcanic aquifer were not undertaken. Rather, an extensive program of aquifer characterization was initiated, in cooperation with Nye County, for the alluvial aquifer at the southern end of the groundwater flow system. This recent focus of the field investigations of the saturated zone exclusively on the alluvial aquifer is technically inexplicable. Admittedly, there were some gaps in the spatial distribution of data for the alluvium, but a much more modest effort could easily correct these. Many large-scale field experiments have already been conducted in similar alluvial materials at other sites, and these studies show that effective porosities and dispersivities fall in a relatively narrow range. No such information from analogous sites exists in the case of the volcanic aquifer, where key transport parameters such as effective porosity can vary over many orders of magnitude.

## The Modeling Deluge

The DOE has applied a hierarchy of computational models to predict the long-term consequences of contaminant leaks from the proposed disposal facility at Yucca Mountain and to quantify the uncertainty in these predictions. These models combine the well-established fundamental physical principles governing the movement of water and contaminants through the earth with input properties specific to the Yucca Mountain site. The following discussion outlines the key features of these models, emphasizing the limitations and uncertainties that should be considered when judging the significance of the modeling results.

Two steps in constructing an adequate groundwater model pose particular difficulty. One is the selection of an appropriate conceptual model—that is, choosing which processes and features to be represented by the model. The second involves determining the numerical values of input parameters that represent site-specific conditions (National Research Council 2001).

The DOE used a three-dimensional regional groundwater flow model (D'Agnese et al. 1997) to provide boundary conditions for the embedded site-scale model used in detailed contaminant transport calculations for the site. This model encompasses the closed Death Valley subsurface hydrologic basin surrounding Yucca Mountain (see figure 14.2). The regional model has the advantage that it uses the natural hydrologic boundaries (that is, assuming no flow). Because of the sparseness of the input data, however, it only crudely quantifies the pattern of flow from the moist upland areas to the discharge areas near the Amargosa River and Death Valley.

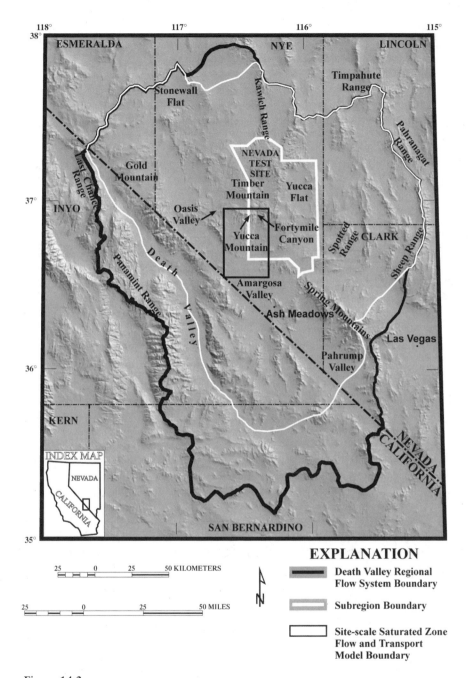

Figure 14.2
Map showing the outline to the Death Valley regional flow system; the white line encompasses the area included in the regional flow model; the black rectangle outlines the area of the site-scale groundwater flow and transport model. From DOE 2001a, figure 4.138.

Water-level data are sparse and unevenly distributed, and because of the complex faulting and extensive volcanic activity in the basin, the subsurface geologic conditions are poorly understood in many areas of this large (twenty-thousand-square-kilometers) region. The permeabilities of the basin's different rock types vary greatly and are largely unknown; as a result, predictions of the flow pathways and head (water-level) distribution through the basin carry large uncertainties. Such a regional flow model should, for example, bring a better understanding of the potential changes in regional flow patterns and the water level as a result of potential climate change (D'Agnese et al. 1999).

The DOE uses a three-dimensional groundwater flow and transport model (the site-scale model), encompassing a rectangular (thirty by forty-five kilometers) region around the site to predict the transport of contaminants southward away from the site. The groundwater flow model, which is restricted to steady flow, treats flow through square blocks 500 meters on a side and 10 to 550 meters thick. Permeabilities in these blocks represent the aggregated effect of interconnected fractures, faults, and other permeable zones in the volcanic aquifer. The limited nature of the available data leads to many problems with the setup and calibration of the site-scale flow model. A few of the most serious flaws are outlined below.

No actual data exist on head (water levels) or water flow at the boundaries of the site-scale model. Therefore, the water levels predicted by the regional model were used to set head-boundary conditions—which are assumed, without justification, to be vertically constant. Because of the crude nature of the regional model, these specified heads are highly uncertain. To compound this problem, the modelers have introduced, in the calibration process, several extremely low permeability features spanning the entire northern boundary area of the model. In so doing, they cut off the inflow from the extensive water-rich upland north of the modeled area. This low permeability region was introduced without any justification from permeability data and with only the uncertain boundary heads predicted by the regional model. In fact, there are no data of any type within eight kilometers of the northern boundary of the model. That leaves a gap in knowledge about the crucial area north of Yucca Mountain. The modelers essentially introduced across the entire northern portion a large underground dam. This dam prevents water in the model from entering the extensive high-permeability zone that extends to the southeast, where contaminants could be carried rapidly away from the site.

There are a number of other problems with the calibration of the site-scale flow model. One is the dubious application of a complex, computer-based parameter scheme to estimate the permeability values for different zones in the model. In this process, the boundary heads are assumed to be known and the permeabilities of the different zones are adjusted until the model produces an acceptable degree of agreement with the observed water levels at wells. It is well-known, however, that when heads rather than flows are specified on the boundaries, only the relative distribution of permeability can be determined. The measured permeabilities are not used directly in this calibration process but were compared graphically with values found from calibration. This comparison is misleading because it includes a large number of the unreliable single-well permeability tests that are not applicable at the 0.5-kilometer scale of the blocks used in the model. Because there are many more of the single-well tests than the more reliable multiwell tests, the comparison, in effect, places more emphasis on these less reliable single-well tests. Notably, the calibrated model uses a permeability value for one of the formations (Tram tuff) that is two orders of magnitude smaller than that determined from the hydraulic testing at the C-wells. This is a crucial discrepancy because it causes the model to predict a water velocity two orders of magnitude lower than the actual velocity in this formation.

The above-described calibration approach used for the site-scale flow model is fundamentally flawed because it does not make direct use of the only reliable information on the permeability—the C-wells tests. The C-wells results are extremely important because the tests are located directly along the pathway that would be taken by contaminants escaping from the site and encompass a pertinent scale of several kilometers. The site-scale model should have been designed to treat the unsteady flow conditions of these tests, and should have used a grid resolution adequate to represent the pumping and observation wells of the tests.

Even more speculative is the contaminant transport component of the site-scale model. There are no data on which to calibrate the model or on the key field-scale transport parameters. The transport model uses a random walk particle-tracking numerical technique to represent dispersive mixing, but no available documents adequately describe the details of how this scheme is implemented. Concern about the modeling technique is appropriate in view of the suspicious kludge that has been introduced reducing the longitudinal dispersivity of contaminants by a factor of ten (DOE 2001b, pp. 12–27). The DOE uses an analytic scheme to represent matrix diffusion effects, but again, key details are not available. To cite yet another flaw, the model adopts a standard conceptualization of parallel, constant aperture fractures that is of doubtful applicability in the field. It is well-known that real fracture flow occurs predominantly through narrow, sinuous channels at velocities that are higher than would be calculated assuming a constant aperture fracture. These channelization effects may greatly reduce the diffusion of dissolved contaminants from the fracture into the porous tuff matrix. Murphy (1995) interprets the undersaturation with respect to calcite of Yucca Mountain groundwater as a clear indication of such channelization.

The three-dimensional model could not adequately predict the transport of radionuclides associated with radioactive decay chains—a phenomenon that has therefore been represented with a separate one-dimensional transport model. Though the details of this one-dimensional approximation are unclear, it does not appear that decay chains play a major role in the performance assessment.

It is difficult to quantify the uncertainty in the predictions from the site-scale flow and transport model. Because of the lack of pertinent data, the assignment of the probability distributions for the input parameters is largely arbitrary. Lacking real data, the DOE adopted a formalized process of speculation using subjective probabilities derived from beliefs expressed by a small group of recognized experts. This process, which the DOE calls "expert elicitation," is scientifically questionable because there is no relevant information from comparable volcanic aquifer systems on which to base meaningful judgments. In reality, these expert speculations have become a substitute for data collection.

The treatment of uncertainty is further limited by the lack of accounting for correlations in the variability of different input parameters. Because of this lapse, important synergistic effects may be missed. Moreover, the uncertainty analysis is conceptually incomplete in that it treats only the uncertainty of the spatially uniform input parameters. In reality, however, the heterogeneity of permeability leads to spatial variations in flow (e.g., Kapoor and Gelhar 1994; Fiori and Dagan 2000) as well as in the interaction between flow and sorption (e.g., Talbott and Gelhar 1994; Miralles-Wilhelm and Gelhar 1996). For all these reasons, the uncertainty analysis is quantitatively meaningless. It is likely that the degree of uncertainty in the predictions of radionuclide concentration is being seriously underestimated.

In summary, the model-based predictions of the likely transport of contamination leaking from the site by flowing groundwater are quantitatively dubious. This low level of confidence derives from the lack of appropriate field data needed to determine the large number of unknown parameters introduced in these models. Disappointingly, the models do not even effectively use the reliable hydraulic data derived from the C-wells testing. Essentially, the modeling approach has not been designed to be consistent with the nature of the data available for its calibration. The models' excessive complexity obscures the tenuous nature of their inputs.

## The Dilution Solution

The dilution or concentration attenuation that occurs during groundwater flow largely determines the potential harm to the people who use that water. The site-scale transport model incorporates the dilution of radionuclide concentration due to physical mixing (dispersion), chemical exchange (sorption), and radioactive decay. In view of the many problems with the model as discussed above, only the last effect is known with reasonable certainty. Even for decay, significant uncertainty remains when predicting concentration at some downstream location because it is difficult to know precisely how long it took the contaminant to travel to that location. Two other types of dilution come into play in the performance assessment calculations for Yucca Mountain; these are regulatory and probabilistic in nature.

The responsibility for setting the standards for the release of radioactivity belongs to the U.S. Environmental Protection Agency (EPA). The final standards (EPA 2001b) set limits on annual dose to a "reasonably maximally exposed individual" and limits on radionuclide concentration in groundwater withdrawn in the ten thousand years after disposal (this portion now remanded). The standards also prescribe site-specific conditions on how the Nuclear Regulatory Commission should apply limits in evaluating the Yucca Mountain facility.

I regard portions of these standards to be ill conceived. Of particular concern is the method mandated to calculate the groundwater radionuclide concentration to which an individual would be exposed. This concentration is to be determined by uniformly mixing the quantity of the radionuclide in the groundwater passing the compliance boundary (stipulated as eighteen kilometers to the south of the site, and five kilometers in other directions) in a year into a "representative volume" of three thousand acre-feet per year (an estimate of how much groundwater a farming community would use for irrigation, using methods currently practiced in the Amargosa Valley). But there is no reasonable physical basis to assume such uniform mixing. Farms are spread widely and not served by a centralized water utility. Individual farms will typically have large-capacity irrigation wells and a separate small well for domestic use. Apparently, the EPA (2001a) adopted this unrealistic dilution condition primarily due to an unjustified presumption that it would be too complicated to calculate the concentration expected at individual wells. The DOE has followed the EPA's lead and adopted in its performance assessment calculations a similarly unrealistic assumption about the rate at which contaminated groundwater can be diluted by withdrawing water from the wells.

Does this artificial groundwater withdrawal dilution make a difference? The total volume of water that would infiltrate (say, five millimeters per year) through a completed repository (one thousand five hundred acres) is about twenty-five acre-feet per year. The physical dilution between the site and the compliance boundary will be very small (possibly by a factor of two) because of the steady contaminant release rate, the source's large initial width (say, four kilometers), and the very small amount of transverse mixing that will occur. The EPA-prescribed mixing volume of three thousand acre-feet per year therefore underestimates the concentration of contaminant that will reach human populations by a factor of one hundred. Think of it this way. An individual home water supply well likely would withdraw water from the aquifer at a rate of a few tenths of an acre-foot per year (around one hundred gallons per day per person). That's two orders of magnitude smaller than the contaminated plume volume (which might plausibly be fifty acre-feet per year). This means that it would be much more reasonable to prescribe that the "reasonably maximally exposed individual" drinks water that has a contaminant concentration equal to the maximum concentration in the aquifer at the compliance boundary. In addition, because of the spatial variability effects not accounted for in the DOE analyses, the maximum concentration occurring in the plume at the compliance location could easily be an order of magnitude larger than the DOE calculations. As a result, the mean or median dose of radiation to populated areas could easily be three orders of magnitude larger than the DOE predicts (see figure 14.3).

The EPA standards specify the compliance boundary as an east-west line some eighteen kilometers south of the proposed Yucca Mountain facility and 5 kilometers from the site in all other directions. The primary reason for the much more extensive exclusion zone south of the site in the direction of groundwater flow seems to be the presumption that it would be more difficult to extract groundwater there because of the greater depth to the water table than in region south of the eighteen-kilometer boundary. This reasoning is unconvincing, however, particularly when considering the likely effects of climate change. Paleo-spring deposits and regional groundwater flow modeling work (Czarnecki 1984; D'Agnese et al. 1999) indicate

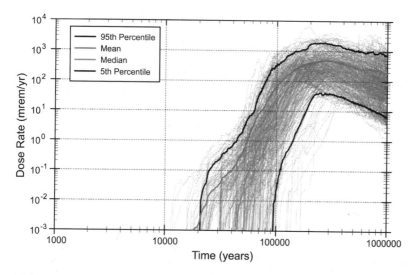

**Figure 14.3**
The DOE's prediction of annual radiation dose to people living in the vicinity of Yucca Mountain over the next million years. From DOE 2001a, figure 4.188.

that the water table in the Yucca Mountain area has in the past been one hundred meters higher than it is now; likely changes in the climate in the future would result in the water table rising again to similar levels. This means that the groundwater could be much more accessible via shallow wells or natural discharge to the surface. The recent drilling by Nye County at the Horsetooth paleo-spring deposits northwest of Lathrop Wells shows the depth to water there to be sixteen to thirty meters. Clearly, groundwater has and will discharge at this location under moderately wetter conditions, and may also discharge in other areas (Forty Mile Wash) north of the eighteen-kilometer boundary. While the DOE performance assessment calculations do consider the increase in flow due to wetter climate, they neglect water table rise and the resulting exposure due to groundwater discharge to the surface. Because the EPA's standards apply only to artificially diluted pumped waters, ultimately a several-kilometers-wide plume could extend tens of kilometers south of the eighteen-kilometer boundary, with radionuclide concentrations several orders of magnitude above the EPA groundwater standards.

Another kind of dilution is introduced through the uncertainty analyses, particularly when the input probability distributions are assigned subjectively without the benefit of real data on which to found those judgments. Such subjectively assigned probability distributions tend to be very wide, reflecting the amalgamation of several experts' unconstrained opinions and diverse biases. The use of such wide distributions is not necessarily conservative because this can result in a lower probability being assigned to a portion of the parameter range that produces adverse consequences. The EPA's prescription that probabilistic results be interpreted by applying the numerical standards to a "reasonable expectation" prescribed to be equal to the mean is troubling. It is difficult to understand how having only an even chance of meeting a health-based standard can be regarded as protective of the public and future generations.

## The Bottom Line

Extremely limited data are available to characterize the movement of contamination away from Yucca Mountain in flowing groundwater. The predictive capabilities of the elaborate models developed to quantify groundwater flow and contaminant transport are severely constrained by the lack of appropriate data needed to determine the large number of unknown parameters used in the models. Key assumptions in the flow modeling and on regulatory dilution do not err on the side of protecting the public. The effort to quantify the uncertainty in these predictions is even more troublesome because it is founded largely on speculative subjective probabilities. The range of outcomes in figure 14.3 is surprisingly narrow.

Is the large uncertainty regarding contaminant transport by flowing groundwater a major concern in terms of the safety of Yucca Mountain? Not if you accept the EPA's ten-thousand-year performance period (now remanded) and the DOE's assumption of perfect integrity of all of the waste containers for more than ten thousand years. Given that the predictions of container integrity are extrapolations of controlled lab tests lasting on the order of months, however, it is doubtful that the uncertainties in such predictions are being realistically portrayed.

Looking beyond the arbitrary ten-thousand-year regulatory period, it seems likely that leakage from the Yucca Mountain facility could ultimately produce a plume several tens of kilometers long and a few kilometers wide in which radionuclide concentrations greatly exceed the current standards for groundwater. Hydrologic and cultural conditions undoubtedly will vary greatly over the time frame of tens or hundreds of thousands of years. What should be done to protect future generations from these potential hazards?

# References

Czarnecki, J.B. (1984) *Simulated Effects of Increased Recharge on the Ground-Water Flow System of Yucca Mountain and Vicinity, Nevada-California.* U.S. Geological Survey. Water-Resources Investigations Report 84–4344.

D'Agnese, F.A., Faunt, C.C., Turner, A.K., and Hill, M.C. (1997) *Hydrogeologic Evaluation and Numerical Simulation of the Death Valley Regional Ground-Water Flow System, Nevada and California.* U.S. Geological Survey. Water-Resources Investigations Report 96–4300.

D'Agnese, F.A., O'Brien, G.M., Faunt, C.C., and San Juan, C.A. (1999) *Simulated Effects of Climate Change on the Death Valley Regional Ground-Water Flow System, Nevada and California.* U.S. Geological Survey. Water-Resources Investigations Report 98–4041.

Department of Energy (DOE) (2000) *Saturated Zone Flow and Transport Model Process Model Report.* CRWMS M&O. TDR–NBS–HS–000001 REV 00 ICN 01.

Department of Energy (DOE) (2001a) *Yucca Mountain Science and Engineering Report.* Office of Civilian Radioactive Waste Management. DOE/RW–0539.

Department of Energy (DOE) (2001b) *FY01 Supplemental Science and Performance Analyses, Volume 1: Scientific Bases and Analyses, Part 1.* TDR–MGR–MD–000007 REV 00. June.

Department of Energy (DOE) (2001c) *FY01 Supplemental Science and Performance Analyses, Volume 2: Performance Analyses.* TDR–MGR–PA–000001 REV 00. July.

Department of Energy (DOE) (2001d) *Probability Distribution for Flowing Interval Spacing.* ANL–NBS–MD–000003 REV 00 ICN 02. Las Vegas, Nevada: CRWMS M&O. ACC: MOL. 20001204.0034.

Environmental Protection Agency (EPA) (2001a) Background Information Document for 40 CFR Part 197 Public Health and Environmental Radiation Protection Standards for Yucca Mountain, Nevada. http://www.epa.gov/radiation/yucca/bid.htm.

Environmental Protection Agency (EPA) (2001b) 40 CFR Part 197 Public Health and Environmental Radiation Protection Standards for Yucca Mountain, Nevada, Final Rule. *Federal Register* 66, pp. 32074–32135.

Fiori, A., and Dagan, G. (2000) Concentration Fluctuations in Aquifer Transport: A Rigorous First-Order Solution and Applications. *Journal of Contaminant Hydrology* 45, pp. 139–163.

Flint, A.L., Flint, L.E., Bodvarsson, G.S., Kwicklis, E.M., and Fabryka-Martin, J. (2001) Development of the Conceptual Model of Unsaturated Zone Hydrology at Yucca Mountain, Nevada. In *Conceptual Models of the Flow and Transport in the Fractured Vadose Zone*, ed. National Research Council. Washington, DC: National Academy Press, pp. 47–85.

Fridrich, C.J., Dudley, W.W., Jr., and Stuckless, J.S. (1994) Hydrogeologic Analysis of the Saturated-Zone Ground-Water System under Yucca Mountain, Nevada. *Journal of Hydrology* 154, pp. 133–168.

Geldon, A.L., Umari, A.M.A., Earle, J.D., Fahy, M.F., Gemmell, J.M., and Darnell, J. (1998) *Analysis of a Multiple-Well Interference Test in Miocene Tuffaceous Rocks at the C-Hole Complex, May–June 1995, Yucca Mountain, Nye County, Nevada.* U.S. Geological Survey. Water-Resources Investigations Report 97–4166.

Geomatrix Consultants, Inc., and TRW. (1997) *Saturated Zone Flow and Transport Expert Elicitation Project.* Contract No. DE-ACX01-91RW00134.

Kapoor, V., and Gelhar, L.W. (1994) Transport in Three-Dimensionally Heterogeneous Aquifers: 2. Concentration Variance for a Finite Size Impulse Input. *Water Resources Research* 30, pp. 1789–1801.

Luckey, R.R., Tucci, P., Faunt, C.C., Ervin, E.M., Steinkampf, W.C., D'Agnese, F.A., and Patterson, G.L. (1996) *Status of Understanding of the Saturated-Zone Ground-Water Flow System at Yucca Mountain, Nevada, as of 1995.* U.S. Geological Survey. Water-Resources Investigations Report 96–4077.

Miralles-Wilhelm, F., and Gelhar, L.W. (1996) Stochastic Analysis of Sorption Macrokinetics in Heterogeneous Aquifers. *Water Resources Research* 32, pp. 1541–1549.

Murphy, W.M. (1995) Contributions of Thermodynamic and Mass Transport Modeling to Evaluation of Groundwater Flow and Groundwater Travel Time at Yucca Mountain, Nevada. *Material Research Society Symposium Proceedings* 353, pp. 419–426.

National Research Council (1996) *Rock Fractures and Fluid Flow: Contemporary Understanding and Applications.* Washington, DC: National Academy Press, 568 p.

National Research Council (2001) *Conceptual Models of the Flow and Transport in the Fractured Vadose Zone.* Washington, DC: National Academy Press, 392 p.

Nye County (2002) Early Warning Drilling Program. Nuclear Waste Repository Project Office, Pahrump, Nevada. http://www.nyecounty.com/ewdpmain.htm.

Talbott, M.E., and Gelhar, L.W. (1994) *Performance Assessment of a Hypothetical Low-Level Waste Facility: Groundwater Flow and Transport Simulation.* U.S. Nuclear Regulatory Commission Report. NUREG/CR-6114, volume 3.

# III

## Thermohydrology

# 15

## Thermohydrologic Effects and Interactions

G. S. Bodvarsson

The radioactivity of high-level nuclear waste releases significant amounts of heat for thousands of years. Heat from stored waste will increase the temperature of the waste packages and, in turn, the temperature of the rock that forms the repository tunnels (known as waste emplacement drifts). The increased temperature of the rocks will induce interactions among the rock, water, and waste that may affect the repository's safety. Despite their complexity, these processes can be predicted closely enough to help forecast the repository's long-term behavior.

These thermally driven processes fall into three categories:

1. Thermohydrologic processes, in which heating affects the water in the rocks
2. Thermohydrologic-chemical processes, in which heating affects the chemistry of the nearby water, the formation or dissolution of minerals in the rocks, and the movement of radionuclides
3. Thermohydrologic-mechanical processes, in which heating affects the opening or closing of rock fractures

Each of these processes could affect repository performance in important ways, such as by increasing water seepage into emplacement drifts, easing the flow of water between emplacement drifts, altering the chemical composition of seeping water, and accelerating the transport of radionuclides from the drift floor through the underlying rock.

To evaluate the potential effect of such phenomena on repository performance, the Department of Energy (DOE) has conducted tests and developed models related to processes that exist under natural conditions, during heating experiments, and after waste emplacement. Specifically, at Yucca Mountain, the DOE is conducting three large-scale heater tests that include a suite of hydrologic, thermal, chemical, and mechanical measurements. It has conducted laboratory tests to evaluate the thermal properties of the rocks and the thermohydrologic, thermohydrologic-chemical, and thermohydrologic-mechanical processes that occur during heating and subsequent cooling. It has also developed models of drift- and mountain-scale behavior of the rock for the design and evaluation of the heater tests as well as for long-term predictions about the effects of these processes on repository behavior.

Existing geothermal reservoirs can serve as natural analogs for some of the thermally driven processes expected to occur in a potential repository. Such reservoirs have been studied in a number of places, including Silangkitang and Kahara-Telega Bodas, Indonesia; Waireakei and Waiotapu, New Zealand; Bulalo, the Philippines; Cerro Prieto, Mexico; and Imperial Valley, California. Data and modeling from these sites corroborate the conceptual and numerical models used for the Yucca Mountain Project (CRWMS M&O, 2002).

No final design has been selected for the potential repository at Yucca Mountain. In the high-temperature design under consideration, the rock would be above the 100°C boiling point of water; in a low-temperature design, rock would be below the boiling point (Buscheck et al. 2003). In general, thermohydrologic and thermohydrologic-chemical effects only become significant when temperatures exceed the boiling point. Thermohydrologic-mechanical effects, on the other hand, do not change dramatically as the temperature goes through the boiling point.

**Heater Tests**

Three large-scale heater tests have been conducted at Yucca Mountain: the Large-Block Test, the Single-Heater Test, and the Drift-Scale Test. The Large-Block Test was conducted on a 3- by 3- by 4.5-meter block excavated from an outcrop at Fran Ridge, near Yucca Mountain (figure 15.1A). Five heaters elevated the block's temperature to 200°C; instrumentation within the block monitored the temperature, water saturation, and mechanical deformation. Prior to heating, air was injected into boreholes that had been drilled into the block. After cooling, the block was taken apart to investigate the geochemical precipitation of minerals. Perhaps the most informative data were collected during rainstorms, when liquid water penetrated into the boiling zone of the block from the top. Models that reproduced these data had to use individual fractures in order to explain the temperature changes observed as the water encountered the boiling zone (Mukhopadhyay and Tsang 2002).

In the Single-Heater Test (figure 15.1B), a five-meter-long heater supplied heat for nine months during 1996 and 1997 (Tsang and Birkholzer 1999). The unit's four-kilowatt output heated the adjacent rocks to a peak temperature of about 180°C. The pretest characterization included various hydrologic, geochemical, and thermal laboratory tests as well as detailed air-permeability testing in more than twenty boreholes within the test bed. During the heating and cooling part of the test, instrumentation in boreholes monitored temperatures, water potential (water content and capillary forces that tend to suck water into unsaturated rock), deformations caused by the thermal expansion and contraction of the rock during heating and cooling, and other parameters.

The Single-Heater Test was designed as a small prototype test for the much larger drift-scale test, which consists of a 47.5-meter long, five-meter diameter drift heated by nine canister heaters and fifty wing heaters (see figure 15.1C). The purpose of the wing heaters, located in horizontal boreholes, is to simulate the effect of heat from an adjacent drift. The total output of the heaters ranged from 140 to 190 kilowatts, which elevated the temperature at the drift wall to about 200°C. Investigators collected data from some thirty-five hundred sensors that are placed in approximately thirty-three hundred meters of boreholes drilled from the heater drifts, the observation drift, and the connecting drift (see figure 15.1C). These sensors monitored the hydrologic, geochemical, thermal, and mechanical behavior of the rock during four years of heating (which ended in January 2002), and will continue to monitor conditions for an additional planned four years of cooling.

Hydrologic data collected include the moisture content of the rock, the fracture gas permeability, and the relative humidity and air pressure within the heated drift. Geochemical data include an analysis of gas and liquid samples as well as the mineralogical-petrologic characteristics of rock samples before and after the test. Thermomechanical data include rock displacement, rock-mass stiffness, and displacement and strain within the concrete liners placed within the drift.

## Modeling of Coupled Processes

The DOE's understanding of thermohydrologic effects is largely based on predictive computer models. When scaling from experiments lasting a few years to thousands of years, modelers must proceed cautiously. Predicting with complete certainty what a repository environment would be like one thousand, five thousand, or ten thousand years from now is impossible. We can, however, establish likely limits to climate variation (based on climate history) as well as the future heat load (which will decrease as the waste decays), and account for such limits within models.

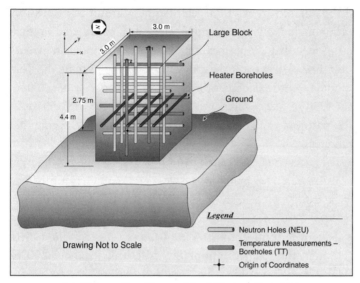

(a) Schematic of Large Block Test at Fran Ridge

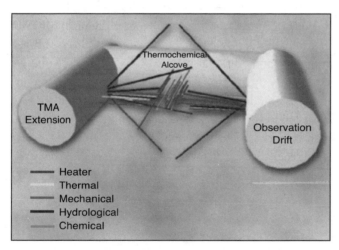

(b) Schematic of Single Heater Test at Alcove 5

**Figure 15.1**
Schematic diagrams of the heater tests conducted at Yucca Mountain: *A*, the large-block test; *B*, the single-heater test; and *C*, the drift-scale test.

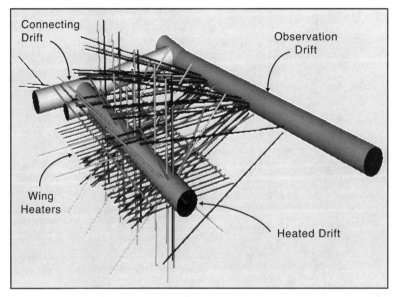

(c) Schematic of Drift Scale Test at Alcove 5

**Figure 15.1** (continued)

## Thermohydrologic Effects

Keeping these considerations in mind, we can use mathematical and numerical models to represent thermally driven processes. The effects of these processes will differ depending on the operating mode (high or low temperature). For the low-temperature operating mode, the temperature of the rock surrounding the waste emplacement drifts remains below the boiling point, and neither the rock matrix nor the fracture network dries out. Although water percolates down to emplacement drifts, capillary forces (which tend to retain the water in the rock) and the conductivity of the fractured rock combine to divert most or all of the percolating water around the drifts. Field tests under ambient conditions have confirmed that water can seep into emplacement drifts only if percolation flux exceeds a certain value defined as the "seepage threshold." The seepage threshold depends also on some important parameters such as the fracture permeability and capillarity (Finsterle et al. 2003).

The main coupled processes predicted to occur at Yucca Mountain for the high-temperature operating mode are shown in figure 15.2. After the waste packages are placed in the repository, heat radiates from the packages to the drift walls and flows

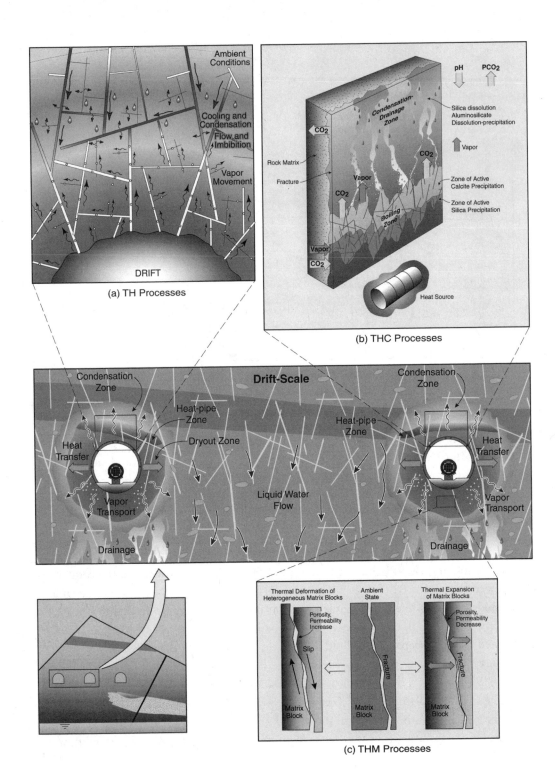

(a) TH Processes

(b) THC Processes

**Drift-Scale**

(c) THM Processes

by conduction into the rock. Once the rocks reach about 100°C, any contained water will begin to boil, and vapor is predicted to migrate away from the drifts. This vapor front is expected to move outward to a maximum distance of about ten meters before condensing in the cooler regions farther from the drifts. This condensed water will flow downward within the fractures; some of the water will be imbibed into and retained by the rock matrix.

Under these conditions, liquid water and vapor tend to flow in opposite directions as a result of the different mechanisms driving gas and liquid flow. This counterflow leads to an efficient transfer of heat. Vapor migrating away from the drifts condenses in the cooler regions of the rock, thereby releasing energy (the latent heat of water). The condensed water drains gradually toward the drift by capillary forces and revaporizes (requiring energy), thereby completing the cycle. This transfer is commonly called a "heat pipe." The rock matrix and fractures next to the drifts are expected to be totally dry for more than one thousand years (see figure 15.2A). This "dryout zone" will expand outward for about one thousand years after emplacement. After that, the dryout zone starts to contract because of the declining heat output from the waste within the waste packages (and the increased precipitation in future, colder climates). After a few thousand years, the fractures and rock matrix will return to near-ambient conditions.

Scientists are concerned that the water mobilized by boiling and the heat-pipe phenomenon may reach and enter drifts during the repository's "thermal period." That period consists of the few hundred years after waste emplacement of boiling or near-boiling temperatures followed by several thousand years of elevated temperatures (Buscheck et al. 2003; Haukwa et al. 2003). The Drift-Scale Test has confirmed both the absence of such seepage and the presence of condensate in boreholes, suggesting the occurrence of boiling and the transport of vapor away from the drift through heat-pipe effects. Computer models calibrated with heater-test data and utilizing heterogeneous rock properties derived from field tests predict no seepage of water during the thermal period under the thermal-loading conditions of the high-temperature operating mode (Haukwa et al. 2002). The models show that for the high-temperature operating-mode case, any water that starts to migrate

**Figure 15.2**
Drift-scale thermally driven coupled processes: *A*, thermohydrologic processes (after Hinds and Bodvarsson 2003); *B*, thermohydrologic-chemical processes (after Sonnenthal and Spycher 2001); and *C*, thermohydrologic-mechanical processes (after Blair et al. 2001).

toward the drifts boils off because of the heat output from the waste. For a low-temperature case, the system always remains below boiling; therefore, no water vaporizes, no dryout zone develops, and seepage is slightly less than that for ambient conditions.

Another concern for both the high- and low-temperature operating modes has been whether the fractured rocks are sufficiently permeable to drain excess water (generated by the condensed steam) away before it enters the drifts (Haukwa et al. 2002). Single-Heater Test and Drift-Scale Test data have verified that the rock has ample drainage capacity. Permeability measurements indicate that the rock has a permeability averaging about ten darcies ($10^{-11} m^2$), which is sufficient to drain orders of magnitude more water than the approximately five millimeters per year now flowing through the rocks. Boiling typically increases water flow by less than an order of magnitude (Haukwa et al. 2002) over preheated (ambient) conditions, and hence it only slightly increases the liquid saturation in the fractures.

Thermohydrologic effects are also expected to affect the transport of radionuclides from the drifts and within the rocks below the drifts, especially for the high-temperature operating mode. The most likely potential pathway for radionuclide transport is by water seeping into emplacement drifts. Models predict that although water will percolate down to emplacement drifts, capillary forces and the conductivity of the fractured rock will combine to divert most or all of the percolating water around the drifts. Simulations of numerous scenarios show that seepage is expected to affect only about 10 percent of the emplacement drifts.

In the absence of seepage, models predict that a zone with practically stagnant water in both fractures and matrix will develop below the drifts (Houseworth et al. 2003). In this case, the migration of the radionuclides into and through the rocks below the drifts will be controlled by diffusion through the matrix, an extremely slow process. These "shadow zone" effects are expected to significantly enhance the performance of the natural barrier at Yucca Mountain. During the thermal period, thermohydrologic effects will only enhance the performance of the shadow zone. Total dryout will occur for thousands of years next to the drifts, with water returning to the fractures next to the drifts after about three thousand years (Bechtel SAIC Company 2001). Thus, no migration of radionuclides can take place until after three thousand years. Radionuclides cannot leave the drifts because no water exists to diffuse into the fractures and matrix surrounding the drifts.

## Thermohydrologic-Chemical Effects

As shown in figure 15.2B, thermohydrologic-chemical effects are primarily related to changes in hydrologic properties stemming from mineral precipitation/dissolution, changes to the chemistry of water seeping into the drifts, and the effects of human-made materials on the radionuclide transport from the drifts. Mineral precipitation and dissolution are greatly enhanced by high temperature and boiling/condensation, and thus the high-temperature operating mode is more affected by these processes. Reactions involving silica, calcite, clay minerals, and other minerals generally proceed faster as the temperature increases. Boiling and condensation also enhance thermohydrologic-chemical effects because boiling concentrates minerals in the remaining liquid, and condensation results in pure water that encourages the dissolution of minerals from fracture coatings and the rock matrix.

Single Heater Test and Drift Scale Test data have indicated that thermohydrologic-chemical effects on hydrologic properties are not significant in the one- to four-year time frame of these tests. Moreover, numerical modeling suggests that thermohydrologic-chemical effects will not result in significant permanent changes (Spycher et al. 2003) under repository conditions of the high-temperature operating mode. In a recent laboratory test (Dobson et al. 2003), however, thermohydrologic-chemical effects involving calcite and silica sealed up a single fracture in about two weeks. Data from geothermal systems also suggest that sealing can occur in a matter of days to years—short time spans in the context of a ten-thousand-year repository (CRWMS M&O, 2002). Because modeling results are not confirmed by laboratory and natural data, this is an area of significant uncertainty in thermal processes.

Water and gas chemistry data from the Single Heater Test and Drift Scale Test have allowed some aspects of the thermohydrologic-chemical models (Spycher et al. 2003) to be validated. Field data and models have both shown an increase in the volume of carbon dioxide, a lowering of the pH (caused by boiling), and slow dissolution of chemicals in the condensate. All of the studies done to date on seepage water chemistry support the conclusion that this water, should it contact the waste packages, will be dilute and noncorrosive. These findings must be further verified, though, because the corrosion rate of the waste packages is sensitive to the concentrations of certain chemicals (such as halogens).

Thermohydrologic-chemical effects on radionuclide transport are difficult to assess because they depend on which human-made materials (e.g., concrete) will be used in the final design. For example, alkaline plumes resulting from the use of significant amounts of concrete in the drifts (perhaps from the lining) will affect hydrologic and transport properties in the rocks below the drifts. Dissolution of the metals within the drifts may result in fluids that encourage ionic exchange with the rock matrix, also resulting in significant changes in hydrologic and transport parameters. Further studies must be conducted to reduce the uncertainties in these processes once a reasonably firm design has been selected.

## Thermohydrologic-Mechanical Effects

Thermohydrologic-mechanical effects can dramatically influence the hydrologic properties of fractures (especially in the high-temperature operating mode) because of the thermal expansion of the rock mass. This expansion could cause fractures to close (reducing permeability) or slip (increasing permeability) (figure 15.2C). Closure is likely to be reversed when the repository cools, but slippage will permanently change the fracture permeability. Reduction in fracture permeability will cause more water seepage into drifts because fracture permeability diverts percolation.

Both the Single Heater Test and the Drift Scale Test measured these effects. During the Single Heater Test, rock was heated and then cooled. Air-permeability data showed that permeability changes were reversed on cooling. The changes in measured fracture permeabilities have been attributed to a combination of increased water saturation in the fractures (thus reducing air permeability) and thermohydrologic-mechanical effects (Tsang et al. 2000). Over the four years of heating in the Drift Scale Test, thermal expansion approached the conditions that will prevail during long-term repository operation. During this test, rock displacement data showed that an initial fifteen-meter separation between two points increased by only ten millimeters, or less than 0.1 percent. At the same time, data from air-injection tests showed a change in fracture permeability of less than an order of magnitude. Both these Drift Scale Test results—the very small changes in displacement and fracture permeability—indicate that the thermohydrologic-mechanical effect on repository performance will be minor (Rutqvist and Tsang 2003).

The numerical modeling of thermohydrologic-mechanical processes at Yucca Mountain was undertaken using data from these tests. The models suggest that the effects of these processes on fracture permeability will be less than two orders of magnitude. The predominant effect will be a reduction in permeability that will mostly be reversed once the repository cools. Some models also suggest that slippage near the emplacement drifts will result in increased fracture permeability, and hence less seepage into the drifts (Rutqvist and Tsang 2003).

The thermohydrologic-mechanical effects on seepage are expected to have only a small impact on repository performance. Because the initial (before heating) fracture permeability is so high, even a two-orders-of-magnitude reduction is not enough to affect water drainage through pillars between emplacement drifts. There is sufficient drainage for about four orders-of-magnitude more water flow than exists now or is ever projected to occur at Yucca Mountain. The thermohydrologic-mechanical results from heater testing and modeling are supported by geothermal analogs, which generally show a modest decrease in fracture permeability with increasing temperature (CRWMS M&O, 2002).

### Summary of Results and Uncertainties

Thermally driven processes in the Yucca Mountain repository can dry out rocks and mobilize water. These processes could also alter the chemistry of liquid and gas phases as well as the chemistry of the rock in the drift. The combined effect will be to influence the amount and chemistry of the water seepage into the drifts, the drainage of water through pillars between the drifts, and the mobilization and migration of radionuclides from emplacement drifts and through the rocks underlying the drifts. After the repository cools, the hydrology and water chemistry will tend to return to preheating conditions.

Based on the available thermal test data, numerical modeling, and data from such natural analogs as geothermal reservoirs, I conclude that:

• Thermohydrologic processes will prevent seepage during the thermal period for the high-temperature operating mode, but will have little effect on seepage for the low-temperature operating mode. For the high-temperature mode, thermohydrologic processes will enhance the performance of the drift shadow zone by causing further dryout and water immobilization in this zone.

• Modeling suggests that thermohydrologic-chemical effects will cause no significant permanent changes in repository conditions. Moreover, modeling studies indicate that seepage water, should it contact waste packages, will be noncorrosive. Thermohydrologic-chemical processes are uncertain for both the high- and (to a lesser degree) low-temperature operating modes, however; this uncertainty stems from insufficient laboratory investigations to assess the importance of fracture sealing for multifractured samples. Similarly, for the high-temperature operating mode, there are significant uncertainties in the chemistry of seepage water, which may have pronounced unknown and unquantified effects on waste package corrosion rates. In addition, it is difficult to know the thermohydrologic-chemical effects on radionuclide transport because these effects depend largely on the engineered barrier materials in the final repository design.

• Thermohydrologic-mechanical effects will be small for both high- and low-temperature operating modes. The thermal expansion of rock will probably reduce fracture permeabilities by one or two orders of magnitude, which is not expected to significantly affect seepage or drainage between the pillars. Furthermore, this process is likely to reverse when the repository cools.

## References

Bechtel SAIC Company. (2001) *FY 01 Supplemental Science and Performance Analyses, Volume 1: Scientific Bases and Analyses.* Department of Energy. TDR–MGR–MD–000007 REV 00 ICN 01.

Blair, S., Carlson, S., Wagoner, J., Wagner, R., Vogt, T., Los, K., and Sun, Y. (2001) *Coupled Thermal-Hydrologic-Mechanical Effects on Permeability Analysis and Models Report.* Bechtel SAIC Company, Las Vegas. ANL–NBS–HS–000037 REV 00.

Buscheck, T.A., Rosenberg, N.D., Blink, J.A., Sun, Y., and Gansemer, J. (2003) Analysis of Thermohydrologic Behavior for Above-Boiling and Below-Boiling Thermal-Operative Modes for a Repository at Yucca Mountain. *Journal of Contaminant Hydrology* 62–63, pp. 441–457.

CRWMS M&O (Civilian Radioactive Waste Management System Management and Operating Contractor) (2002) *Natural Analogue Synthesis Report, Section 11.2.* Department of Energy. TDR-NBS-GS-000027 REV00ICN 02.

Dobson, P.F., Kneafsey, T.J., Sonnenthal, E.L., Spycher, N., and Apps, J.A. (2003) Experimental and Numerical Simulation of Dissolution and Precipitation: Implications for Fracture Sealing at Yucca Mountain, Nevada. *Journal of Contaminant Hydrology* 62–63, pp. 459–476.

Finsterle, S., Ahlers, C.F., Trautz, R.C., and Cook, P.J. (2003) Inverse and Predictive Modeling of Seepage into Underground Openings. *Journal of Contaminant Hydrology* 62–63, pp. 89–110.

Haukwa, C.B., Tsang, Y.W., Wu, Y.-S., and Bodvarsson, G.S. (2003) Effect of Heterogeneity in Fracture Permeability on the Potential for Liquid Seepage into a Heated Emplacement Drift of the Potential Repository. *Journal of Contaminant Hydrology* 62–63, pp. 509–527.

Hinds, J.J., Bodvarsson, G.S., and Nieder-Westermann, G.H. (2003) Conceptual Evaluation of the Potential Role of Fractures in Unsaturated Processes. *Journal of Contaminant Hydrology* 62–63, pp. 111–132.

Houseworth, J.E., Finsterle, S., and Bodvarsson, G.S. (2003) Flow and Transport in the Drift Shadow in a Dual-Continuum Model. *Journal of Contaminant Hydrology* 62–63, pp. 133–156.

Mukhopadhyay, S., and Tsang, Y.W. (2002) Understanding the Anomalous Temperature Data from the Large Block Test at Yucca Mountain. *Water Resources Research* 38, pp. 28–1 to 28–12.

Phillips, O.M. (1996) Infiltration of a Liquid Finger Down a Fracture into Superheated Rock. *Water Resources Research* 32, pp. 1665–1670.

Rutqvist, J., and Tsang, C.F. (2003) Analysis of Thermal-Hydrolic-Mechanical Behavior near an Emplacement Drift at Yucca Mountain. *Journal of Contaminant Hydrology* 62–63, pp. 637–652.

Sonnenthal, E.L., and Spycher, N. (2001) *Drift-Scale Coupled Processes (DST and THC Seepage) Models*. Bechtel SAIC Company, Las Vegas. MDL–NBS–HS–000001 REV 01 ICN 02.

Spycher, N.F., Sonnenthal, E.L., and Apps, J.A. (2003) Fluid Flow and Reactive Transport around Potential Nuclear Emplacement Tunnels at Yucca Mountain, Nevada. *Journal of Contaminant Hydrology* 62–63, pp. 653–674.

Tsang, Y.W., and Birkholzer, J.T. (1999) Predictions and Observations of the Thermal-Hydrological Conditions in the Single Heater Test. *Journal of Contaminant Hydrology* 38, pp. 385–425.

Tsang, Y.W., Huang, K., and Bodvarsson, G.S. (2000) Estimation of the Heterogeneity of Fracture Permeability by Simultaneous Modeling of Multiple Air-Injection Tests in Partially Saturated Fractured Tuff. In *Dynamics of Fluids in Fractured Rocks: Concepts and Recent Advances, Geophysical Monograph 122*, ed. Faybishenko, B., Witherspoon, P.A., and Benson, S. Washington, DC: American Geophysical Union, pp. 99–114.

# 16

## The Near Field at Yucca Mountain: Effects of Coupled Processes on Nuclear Waste Isolation

William M. Murphy

A geologic repository for nuclear waste is commonly divided into two domains: the near field and the far field. The near field is the environment in which the wastes are emplaced and initially isolated. It is distinguished by important material and physical changes due to the wastes and the engineered systems encompassing them. In contrast, the far field of a geologic repository retains characteristics of the natural system with respect to waste isolation. Some near-field changes, such as materials emplacement, will be effectively permanent, and others, such as thermal and hydrologic effects, will diminish with time. The spatial limits of the near field will change with time and will differ according to the processes or conditions under consideration. At Yucca Mountain, the near field will be an extremely complex system requiring a multidisciplinary understanding of many materials and many transient, heterogeneous, and coupled processes.

Thermal, hydrologic, mechanical, chemical, and nuclear processes will modify the near field. These processes will be coupled—that is, some processes will affect others. For example, as radioactive wastes decay they release energy, which heats the near field. Temperature variations change fluid (water and gas) densities and pressures, and cause water to evaporate and condense. The resulting pressure gradients cause water and gas to flow. Excavations and temperature variations cause mechanical deformations of the host rocks. Fluid flow causes heat transport and migration of chemical constituents. Introduced materials such as waste forms, containers, cements, and air interact chemically with the gas, water, and rock of the geologic setting. The dissolution and precipitation of minerals change the environment's hydrologic and sorption properties; such changes may permanently affect radionuclide transport, even after the thermal period. The reaction rates and chemical equilibria among rocks, fluids, and engineered materials vary with temperature and chemical compositions. Radiation affects properties of near-field materials, notably the chemistry of water and gas.

Although detailed predictions of the near-field evolution at Yucca Mountain entail large uncertainties, much can be learned and anticipated through site characterization, field and laboratory experimentation, studies of natural analogs, and modeling. The near field at Yucca Mountain will exhibit complex behavior because of the high thermal loading, steep temperature gradients, temperature excursions above the boiling point of water, gas and liquid (two-phase) flow, chemical instabilities such as chemically reducing container and waste form materials in an oxidizing mountain, and multicomponent chemical reactions among heterogeneous materials.

This chapter paints a mostly qualitative picture of near-field processes for the repository at Yucca Mountain, highlighting important effects and uncertainties, and synthesizing insights from many studies that illuminate the picture. The analysis emphasizes postclosure conditions and processes that will affect long-term radionuclide isolation.

## Materials and Conditions in the Near Field

The natural setting (e.g., rocks, water, and gas) and introduced materials (e.g., wastes, engineered barriers, and construction materials) provide the background and initial conditions for changes in the near-field environment. The repository design determines many important characteristics of the near field. Materials to be emplaced in the geologic system consist mainly of spent nuclear fuel (primarily uranium dioxide) with zirconium alloy cladding, high-level radioactive wastes in a borosilicate glass matrix, alloy containers, metallic structural materials inside waste containers, titanium drip shields, and concrete and metallic ground support materials.

Some materials—such as zirconium alloy cladding, a corrosion-resistant nickel alloy container material called Alloy-22, and titanium drip shields—will corrode at imperceptibly slow rates. Other materials, notably spent nuclear fuel, will—once exposed to natural geochemical conditions—react rapidly relative to the time scale required for isolation of the wastes. Most introduced materials will be chemically unstable in Yucca Mountain, so they will alter eventually to more stable chemical forms. Reaction products of wastes and other engineered materials will permanently modify the near field.

Rocks that will host radioactive wastes at Yucca Mountain are thirteen-million-year-old, silica-rich, fractured, volcanic tuffs composed predominantly of fine-grained, interlocking crystals of alkali feldspar and silica minerals. Secondary minerals including calcite, amorphous silica, and clay minerals occur in fractures. Rocks and fluids throughout the mountain are oxidizing because of their original chemistry and interactions with atmospheric oxygen.

At greater distances from the emplacement zone, the volcanic rocks are variably welded, bedded, vitric (composed of volcanic glass), and altered to secondary zeolite minerals—for example, clinoptilolite. (Zeolites are minerals characterized by crystal structures containing large channels and pores, giving them the ability to reversibly take up water and many chemical species; see chapter 13, this volume.)

Porosity, permeability, and fracture characteristics vary widely among the volcanic strata. The predominant natural alteration of the rocks at Yucca Mountain is the zeolitization of vitric tuffs. This alteration may have occurred shortly following volcanism at a time of elevated temperatures and enhanced fluid circulation. These conditions and the resulting alteration of the rocks at Yucca Mountain provide one natural example of possible induced changes in the near field.

Although waste emplacement at Yucca Mountain will be hundreds of meters above the groundwater table, and the dryness of Yucca Mountain is widely noted, the host rocks contain abundant water. Generally, the rocks at Yucca Mountain above the water table are about 10 percent water by volume (see data in Flint 1998). Zeolites also contain abundant water in their mineral structures. Ambient groundwater is dilute, slightly basic, oxidizing, and dominated by dissolved calcium, sodium, bicarbonate, and silica. Many other components, including chloride, sulfate, and potassium, are present in smaller concentrations (e.g., Yang et al. 1998). Fracture and perched waters tend to be richer in sodium and bicarbonate (like saturated zone groundwaters); matrix waters tend to be richer in calcium, chloride, and sulfate (Browning et al. 2000). The gas phase in the unsaturated zone at Yucca Mountain is air at 100 percent relative humidity with a carbon dioxide content about five times the atmospheric value (e.g., Thorstenson et al. 1998). Radioactive carbon-14 and other data indicate that the flow of gas and water under natural conditions is limited at the depth of the emplacement horizon at Yucca Mountain (Murphy 1995). Nevertheless, the fluids at Yucca Mountain containing bomb-pulse chlorine-36 indicate that some water travels along fast paths from the ground surface through the waste emplacement horizon (see chapter 11, this volume).

## Thermal Effects

The temperature along the natural geothermal gradient increases from about 20°C at the ground surface to about 30°C at the water table seven hundred meters below the crest of Yucca Mountain. Because atmospheric pressure in the unsaturated zone at Yucca Mountain is about 0.85 bar, pure water boils at about 96°C.

The heat output from nuclear waste is initially high, as short-lived radioisotopes (e.g., cesium-137 and strontium-90) decay, but it decreases rapidly over a few hundred years. Heat flow through the near field will occur primarily by conduction, and initial thermal gradients will be steep. Conductive heat flow through repository host rocks will be augmented by convective fluid flow, and by radiation through voids in and around waste packages. Temperatures in the near field around the waste will peak within tens of years and decline thereafter. At greater distances from the waste, temperatures will increase more slowly and reach lower maximum values at longer times.

Maximum near-field temperatures will depend on the spacing and density of waste emplacement, waste characteristics such as age, and postemplacement cooling by ventilation prior to repository closure. Various designs lead to maximum temperatures ranging from near the boiling point of water to over 300°C. A line-loading design, with waste containers arranged in continuous rows in widely spaced parallel drifts, would maintain relatively homogeneous temperatures along the lines and relatively low temperatures in the central sections of the rock pillars between the drifts (Buscheck et al. 2002). Advocates of lower temperature designs argue that uncertainties increase with higher temperatures. Others contend that higher temperatures can be exploited to help isolate wastes by driving water away and maintaining dry conditions in the near field (see chapter 15, this volume).

## Excavation and Thermal Mechanical Effects

The lithostatic pressure at Yucca Mountain is due chiefly to the weight of overlying rocks, with lateral tectonic forces playing a minor role. The stresses due to excavations and heating in the rocks enclosing the emplacement drifts will cause instabilities, leading to rock falls and drift collapse. The rock mass will adjust rapidly to the initial excavation, and artificial rock support will be engineered to maintain safe working conditions during the operational phase of the repository. After waste emplacement, thermal expansion of the host rock will cause the fracture and pore apertures to close, the directions of fluid flow to change, and the directions and magnitudes of stresses to shift. With heating, the maximum principal stress direction will change from approximately vertical to approximately horizontal near the repository horizon, and the location of maximum principal stress will change from drift sidewalls to the roof and floor (Chen et al. 2000).

Cristobalite, an abundant natural silica mineral in the primary host rock of the emplacement horizon, undergoes a spontaneous change in its crystal structure at 220°C (Hatch and Ghose 1991). This transition is accompanied by a volume increase of 5 percent, which would change the mechanical properties of the near field. Other hydrothermal reactions, noted in the following sections, will also cause rock volume and stress changes.

## Thermohydrologic Effects

Thermohydrologic effects will be particularly important and complicated in the Yucca Mountain near field because of the high heat load and the low-pressure, unsaturated hydrologic conditions (for more detail on these effects, see chapter 15, this volume). Heating will dry the near-field environment in an area within meters or tens of meters of the emplacement drifts. Gas pressure gradients due to the vaporization of water and thermal expansion will drive the gas flow away from the heat source. Water will condense from migrating vapor in cooler parts of the geologic repository. Thermal gradients will thus lead to gradients in liquid water saturation, which is the fraction of rock porosity filled with liquid water. In turn, capillary pressures will suck liquid water back toward the heat source. The flow of water vapor away from the wastes and the counterflow of liquid water back toward the heat source will create a heat pipe, transporting energy as latent heat of vaporization and depositing heat energy where the vapor condenses. Because the water flow is affected by gravity as well as suction pressure, the thermohydrologic effects will be asymmetrical above and below the heated waste emplacement zone. The recirculation of gas and liquid water flow will be more likely above the heat source. Drying the waste emplacement area and the drainage of water out of the near-field system will lead to a prolonged period of relatively dry conditions, which will be favorable for waste isolation.

The three-dimensional distribution of heat and water sources along with hydrologic and thermal gradients will lead to heterogeneous near-field fluid flow patterns. Symmetry aids the numerical modeling of thermohydrologic processes, but conditions at the edges of the near field, focused preferential flow pathways, heterogeneous heat sources, and local cold traps will produce near-field flow systems that are far more complex than the average, large-scale behavior. The redistribution of water due to thermohydrologic effects could augment localized perched water systems or produce new ones, even above the emplacement horizon. The focused drainage of water past a relatively cool waste package could promote greater releases of wastes than the estimated average behavior of the whole system.

## Gas-Water-Rock Geochemical Interactions

Near-field heating will cause water evaporation and the release of carbon dioxide dissolved in water to the gas phase. After ventilation ceases, the gas around waste packages will be dominated by water vapor at elevated temperature. Gas flow and the diffusion of air and water vapor will tend to homogenize near-field gases. Gases that are saturated with water and enriched in $CO_2$ will be driven away from the heat source and may reach the ground surface, escaping into the atmosphere through fractures. Such releases could be the earliest chemical manifestations of the repository at the surface and could affect surficial biology.

Near-field heating will change the gas-water-rock equilibria, accelerate the geochemical reactions, and concentrate many dissolved species as water evaporates. Groundwaters in the unsaturated zone are approximately saturated with respect to calcite and strongly supersaturated with respect to quartz. Increasing temperature, increasing pH by $CO_2$ volatilization, and evaporative concentration will promote calcite precipitation in the pores and fractures of the near field. Other secondary minerals, such as clay minerals and zeolites, will form as hydrothermal alteration products of the original feldspars and volcanic glass of the tuff. The extent of these reactions is difficult to predict with precision because of the complexity of coupled thermal-hydrologic-chemical processes. As it evaporates, the residual water will become highly saline, and small volumes of sulfate, carbonate, fluoride, chloride, and nitrate salts will precipitate (e.g., Pabalan et al. 2002). The resulting aqueous concentrations of anions such as chloride, fluoride, and nitrate, which may enhance or inhibit corrosion, will depend on which salts form. The precipitation of salts is likely to occur in the smallest pores where the last residual quantities of water collect. Drying will generally produce small volumes of mineral precipitates because there isn't much dissolved material to begin with. Large-scale changes in mineral dissolution or precipitation would require recirculating fluids or phase changes such as the zeolitization of glass or the conversion of other silica minerals to quartz.

Water that condenses from vapor in cooler parts of the mountain will be chemically aggressive because it will have minimal dissolved constituents except carbonic acid. Initial rapid reactions with rocks, however, will generate less reactive, dilute solutions. The dissolution and alteration of rocks is likely to occur on fracture surfaces where water condenses. As water flowing downward toward the heat source evaporates, minerals may precipitate on fracture surfaces.

As the near-field environment cools, precipitated salts will absorb water (deliquesce) and generate locally concentrated aqueous solutions stable at temperatures above the boiling point of pure water. Salty near-field waters may be corrosive if in contact with engineered materials such as waste containers. Yet fresh water flow through the near field will dissolve the salts, dilute concentrated solutions, and eventually flush the system. Over the course of many thousands of years, the water and gas chemistries will return to conditions similar to those of the unperturbed natural system, with the exception of elements of the introduced engineered and waste materials.

The properties of the aqueous solution will be strongly affected by coupled thermo-hydro-chemical processes and will vary with time. The aqueous solution chemistry will, in turn, control the rates of degradation of the engineered barriers, the rates of the release and concentration limits for radioactive elements, and the geochemical transport of radioactive species. The solution compositions inside the drifts and waste packages will differ strongly at times from natural and hydrothermally conditioned geochemical solutions. Efforts to bound the near-field aqueous solution compositions for Yucca Mountain generally lead to broad ranges of possible compositions.

Although the rocks of Yucca Mountain are chemically stable on a time scale of millions of years, they are susceptible to alteration, particularly by water at elevated temperatures. Volcanic glass in particular is susceptible to alteration. The natural alteration of such glass to zeolites in Yucca Mountain rocks reduces porosity by a large fraction and hydraulic conductivity by many orders of magnitude (see data in Flint 1998). Moreover, zeolites' high ion exchange capacity will affect the transport of certain radioelements—particularly those occurring as aqueous cations. At depths below the water table at Yucca Mountain, the natural mineral alteration assemblages reflect the lower silica activities, and include analcime, kaolinite, and quartz

(Broxton et al. 1987). Predominantly vitric or zeolitic horizons occur at tens or hundreds of meters from the repository heat source at Yucca Mountain, so they may be unaffected by near-field processes.

Silica is the predominant chemical component of the mountain. Many primary minerals are essentially pure silica (cristobalite, quartz, and tridymite), and other primary phases (e.g., alkali feldspars and glass) are composed mostly of silica. Amorphous silica is a common secondary phase, along with silica-rich zeolites and clays. Phase transitions, mobility, and the precipitation of silica in the near field could have large effects on fluid flow and radionuclide transport processes. Natural hydrothermal systems commonly develop zones of intense silica precipitation. Despite the predominance of silica and the potential for silica precipitation reactions to cause near-field changes that affect radionuclide isolation, predicting silica behavior in the near field remains a challenging problem. Uncertainties are compounded by the formation of numerous stable and metastable solids, the slow kinetics of reactions between water and silicate minerals, and the difficulties in characterizing kinetic parameters such as rate constants, nucleation rates, and reactive surface areas. The rates and equilibria depend strongly on varying temperature and uncertain conditions such as solution pH. The basic uncertainty in the behavior of silica highlights the difficulties in predicting coupled thermo-hydro-chemical effects in the near field at Yucca Mountain.

## Seepage

Under natural conditions at Yucca Mountain, episodic, relatively rapid flow occurs in high-permeability fractures, whereas matrix flow is extremely slow. For unsaturated near-field conditions, suction pressures in the rocks surrounding the waste emplacement drifts may divert water around the drift openings. Otherwise, if water saturations increase sufficiently, water may drip or flow into the drifts as seepage. The quantity and chemistry of near-field seepage contacting containers and waste forms has a major effect on potential radionuclide releases. Complex, coupled processes that will affect seepage include the closing of fractures and the redistribution of porosity due to thermal mechanical strain, mineral precipitation in fractures, the dehydration of clay minerals in fractures, the sealing of the rock matrix by mineral precipitates, and the accumulation of water by deliquescence. The difficulty of predicting seepage and its consequences has motivated the introduction of a drip shield in the Yucca Mountain repository design.

## Effects of Introduced Materials

Most engineered materials proposed for Yucca Mountain are chemically unstable in the near-field environment and will alter over time. Metallic waste packages, other metallic materials such as rock bolts, and spent nuclear fuel will oxidize. Although oxygen is in abundant supply, most metals oxidize slowly. Corrosion products such as oxihydroxides of iron and oxidized uranium minerals will have low solubilities and will be geochemically stable at Yucca Mountain.

Natural geologic systems in which uranium dioxide mineral deposits have been oxidized under near-surface conditions provide analogs of the long-term character-istics of the Yucca Mountain near field (e.g., Pearcy et al. 1994) (figure 16.1). These geologic systems and experimental studies provide converging results demonstrating the alteration of uranium dioxide and the formation of a variety of secondary min-erals (e.g., Murphy 2000). In the long term, oxidized alteration products of engi-neered and waste materials, such as iron and nickel oxides as well as uranyl minerals, will dominate portions of the near field. Analogous natural systems com-posed of uranyl minerals and hydrated metal oxides demonstrate geologic stability for periods of millions of years. Certain radionuclides will also be strongly associ-ated with oxidation products of container materials and waste forms. Radioele-ments of concern, such as neptunium and americium, may be incorporated by solid solution in structures of stable uranyl minerals or strongly adsorbed on surfaces of fine-grained metal oxides.

Zones of chemically reducing conditions, ranging in scale from the microscopic to the macroscopic, will develop in and around the waste containers and other engineered materials. A plume of reducing groundwater generated by the oxidation of near-field materials will develop and move along pathways of fluid migration. This transient, localized, reducing plume will affect radionuclide migration in the near field by precluding the migration of certain aqueous radionuclide species that are stable only under oxidizing conditions.

The variety of near-field materials will lead to many different effects. Engineered materials will alter mostly to hydrated oxides; the large increase in volume that this alteration entails will plug porosity. Reactions of groundwater with cement may generate alkaline solutions, augmenting the solubility of silica and the dissolution rate of waste glass. Microscopic pit environments in metals may produce localized acidic solutions. The releases of many radionuclides from the near field will be controlled ultimately by the properties of the alteration products of engineered materials.

**Figure 16.1**
The Nopal I uranium deposit in the Peña Blanca district near Chihuahua, Mexico, has been studied extensively as a natural analog of the Yucca Mountain repository. A concentrated uraninite deposit at this site was hosted in fractured, silica-rich volcanic rocks similar to those at Yucca Mountain. Presently, the deposit is located in hydrologically unsaturated, oxidizing, and semiarid climatic conditions much like those at Yucca Mountain. Uraninite, which is analogous to spent fuel, has been almost completely oxidized to a suite of relatively stable uranyl minerals including schoepite, soddyite, and uranophane. The same sequence of secondary mineral formation has been observed in laboratory experiments designed to mimic the conditions of spent fuel alteration in the near field at Yucca Mountain.

## Radiation Effects

Container materials will largely shield radiation, and radiation will diminish rapidly as the wastes decay. Radiolysis—the chemical breakdown and reorganization of chemical species due to energetic radiation—will occur in the groundwater and gas after the containers fail. Radiolysis can generate highly reactive oxidizing and reducing species or acids, which could increase the rate and extent of corrosion and waste dissolution processes.

## Coupled Processes in the Near-Field Environment

Ideally, the near-field environment would provide conditions for which containers and waste forms are chemically stable and isolated from the flowing groundwater. Geologic site characterization can be used to evaluate geologic and hydrogeochemical conditions conducive to long-term stability and isolation. Yet the introduction of chemically unstable materials and the energy of radioactive decay will substantially alter conditions in the near field at Yucca Mountain through coupled thermal, mechanical, chemical, hydrologic, and radiological processes.

Coupling the near-field processes at Yucca Mountain creates many uncertainties and complexities. Time spans of significance range from the instantaneous to the geologic, while space scales range from microscopic to mountain size. Heterogeneities exist on all scales. Steep gradients or temporal variations in chemical and physical properties lead to unstable systems. Interactions among many solid, liquid, gas, engineered, natural, and radioactive materials involve countless chemical reactions, each of which may be controlled by factors that depend strongly on temperature and chemical variables. Large uncertainties persist in some fundamental areas, such as predictions of silica behavior, the pH of groundwater, changes in the repository rock's porosity and permeability, flow and chemical relations between fracture water and matrix water, the properties of alteration products, the chemical kinetics of mineral alteration, and seepage.

Resolving these uncertainties and understanding the near field at Yucca Mountain and its role in nuclear waste isolation require multifaceted research. Site characterization provides data such as fracture and matrix properties as well as the extent of the natural alteration of glass and zeolites. Experimental measurements provide information on the mechanical, hydraulic, thermodynamic, and kinetic properties of system components. Large-scale experiments of coupled thermal-hydrologic-mechanical-chemical processes, such as the drift-scale heater test, have been conducted at Yucca Mountain. Studies of analogous natural systems, such as uranium mineral deposits in similar geochemical and hydrologic environments and the hydrothermal alteration of analogous rocks, offer insights on time and space scales that are inaccessible to experimental studies. The theoretical modeling of coupled effects on material stability, radionuclide release, and transport is useful to evaluate their consequences. These models, with their inherent uncertainties, will contribute to evaluations of repository safety. Certainty will be impossible to achieve, but great strides have been taken and an adequate understanding of the near field at Yucca Mountain is within our grasp.

## Acknowledgments

I thank Rodney Ewing and Allison Macfarlane for inviting this contribution to *Uncertainty Underground*. Rui Chen, David Bish, and Tom Buscheck provided advice on the technical details of this chapter. Ewing and Mostafa Fayek offered comprehensive and insightful reviews.

## References

Browning, L., Murphy, W.M., Leslie, B.W., and Dam, W.L. (2000) Thermodynamic Interpretations of Chemical Analyses of Unsaturated Zone Water from Yucca Mountain, Nevada. *Materials Research Society Symposium Proceedings* 608, pp. 237–242.

Broxton, D.E., Bish, D.L., and Warren, R.G. (1987) Distribution and Chemistry of Diagenetic Minerals at Yucca Mountain, Nye County, Nevada. *Clays and Clay Minerals* 35, pp. 89–110.

Buscheck, T.A., Rosenberg, N.D., Gansemer, J., and Sun, Y. (2002) Thermohydrologic Behavior at an Underground Nuclear Waste Repository. *Water Resources Research* 38, pp. 10-1–10-19.

Chen, R., Ofoegbu, G.I., and Hsiung, S.M. (2000) Modeling Drift Stability in Fractured Rock Mass at Yucca Mountain, Nevada: Discontinuum Approach. In *Proceedings: Fourth North American Rock Mechanics Symposium*, ed. Girard, J., Liebman, M., Breeds, C., and Doe, T. Rotterdam, Netherlands: Balkema, pp. 945–952.

Flint, L.E. (1998) *Characterization of Hydrogeologic Units Using Matrix Properties.* U.S. Geological Survey. Water-Resources Investigation Report 97–4243.

Hatch, D.M., and Ghose, S. (1991) The $\alpha$-$\beta$ Phase Transition in Cristobalite, $SiO_2$. *Physics and Chemistry of Minerals* 17, pp. 554–562.

Murphy, W.M. (1995) Contributions of Thermodynamic and Mass Transport Modeling to Evaluation of Groundwater Flow and Groundwater Travel Time at Yucca Mountain, Nevada. *Materials Research Society Symposium Proceedings* 353, pp. 419–426.

Murphy, W.M. (2000) Natural Analogs and Performance Assessment for Geologic Disposal of Nuclear Waste. *Materials Research Society Symposium Proceedings* 608, pp. 533–544.

Pabalan, R.T., Yang, L., and Browning, L. (2002) Deliquescence Behavior of Multicomponent Salts: Effects on the Drip Shield and Waste Package Chemical Environment of the Proposed Nuclear Waste Repository at Yucca Mountain, Nevada. *Materials Research Society Symposium Proceedings* 713, pp. JJ1.4.1–JJ1.4.8.

Pearcy, E.C., Prikryl, J.D., Murphy, W.M., and Leslie, B.W. (1994) Alteration of Uraninite from the Nopal I Deposit, Peña Blanca District, Chihuahua, Mexico, Compared to Degradation of Spent Nuclear Fuel in the Proposed U.S. High-Level Nuclear Waste Repository at Yucca Mountain, Nevada. *Applied Geochemistry* 9, pp. 713–732.

Thorstenson, D.C., Weeks, E.P., Haas, H., Busenberg, E., Plummer, L.N., and Peters, C. (1998) Chemistry of Unsaturated Zone Gases Sampled in Open Boreholes at the Crest of Yucca Mountain, Nevada: Data and Basic Concepts of Chemical and Physical Processes in the Mountain. *Water Resources Research* 34, pp. 1507–1529.

Yang, I.C., Yu, P., Rattray, G.W., Ferarese, J.S., and Ryan, J.N. (1998) *Hydrogeochemical Investigations in Characterizing the Unsaturated Zone at Yucca Mountain, Nevada.* U.S. Geological Survey. Water-Resources Investigation Report 98–4132.

# IV

## Waste Package Behavior

# 17

# Waste Package Corrosion

David W. Shoesmith

Retarding the release of radionuclides from the proposed nuclear waste repository at Yucca Mountain will require a system of multiple barriers. An essential component of this system is the metal package that encases the waste. As such, the design and fabrication have been adapted in an attempt to compensate for the uncertainties inherent in natural geologic barriers and for the uncontrollable features of barriers such as the waste form. This approach has led to the perception that the accumulation of uncertainties in performance in all pre- and postwaste package barriers is best dealt with by improving the design and durability of the waste package. In fact, the waste package tends to be viewed as the principal focal point for ensuring overall repository performance. This has contributed to the perception that the natural barriers are inconsequential and has led to both the selection of expensive corrosion-resistant fabrication materials and to design, fabrication, and inspection procedures that could prove complicated because they must be performed remotely. This last requirement is enforced by the safety precautions necessary to prevent the exposure of personnel to the radiation fields on the outside of the packages.

This chapter summarizes the uncertainties inherent in waste package design and fabrication as well as in the nature of the environment it will be exposed to in the repository. It also describes how this environment will influence the long-term degradation of the waste packages. The aim is to demonstrate that the waste package is one in a sequence of important barriers and that its performance is an integral part of the overall vault performance, not merely a means to compensate for geologic deficiencies.

## Waste Package Design and Fabrication

The waste package is designed to contain twenty-one pressurized water reactor spent fuel assemblies within a dual-shell, canlike body and dual-lid design (figure 17.1). The inner shell, fabricated from 316 NG (nuclear grade) stainless steel, gives the package structural rigidity. The outer shell, fabricated from a corrosion-resistant nickel alloy, Alloy-22, provides the corrosion barrier. Similar designs are available for wastes from boiling water reactors, high-level waste from the nuclear weapons complex, and naval spent nuclear fuel. The similarities in these designs are evident in figure 17.2.

The main body of the packages will be fabricated by joining cylindrical segments using longitudinal and circumferential welds (Office of Civilian Radioactive Waste

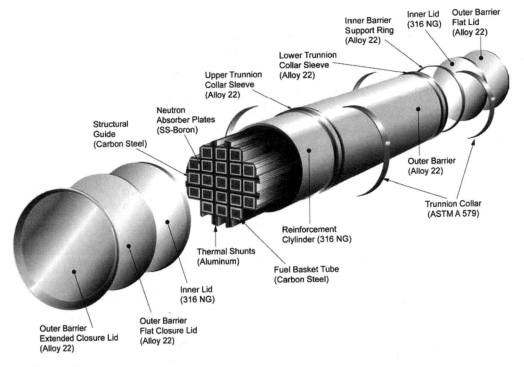

**Figure 17.1**
Schematic showing an exploded view of the waste package designed to contain twenty-one pressurized water reactor spent fuel assemblies. From Office of Civilian Radioactive Waste Management 2002, figure 3.2.

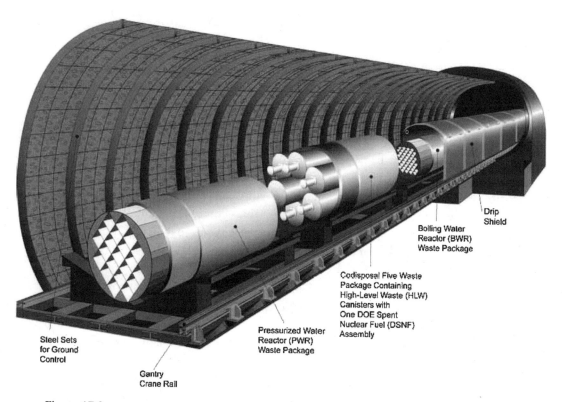

**Figure 17.2**
Schematic showing emplaced waste packages on pallets. From Office of Civilian Radioactive Waste Management 2002, figure 3.3.

Management 2002). This should not pose a significant challenge because it is common industrial practice. As with any similarly manufactured vessel, the sites most likely to fail are the welds. Of particular concern is the possibility of introducing stresses in and around the weld, rendering this area susceptible to stress corrosion cracking on exposure to the aqueous environment. To avoid such cracking, designers of the packages call for a full-scale solution anneal at 1,150°C followed by quenching in water. The quench would place the waste package surface into a state of compressive stress, avoiding the tensile stresses required for the initiation and propagation of cracks. This procedure has been successfully conducted on a full-diameter, quarter-length mock-up. Scale-up should not pose a significant challenge.

## Stress Corrosion Cracking

The key uncertainty in the fabrication process is the performance of the final closure welds on the waste package lids. Because the package will be fully loaded and radioactive when these welds are made, a subsequent full package anneal will not be possible; the high temperatures necessary for such annealing would seriously threaten the integrity of the waste form within the package. Consequently, the final lid closure welds must be done locally, to avoid thermal degradation of the waste form, and remotely, to avoid exposure of personnel to the radiation fields.

To avoid stress corrosion cracking under emplacement conditions, it is essential to minimize the stresses in these welds. The dual-lid design of the waste package provides defense-in-depth against such failure since the probability of appropriately located and aligned flaws and stresses in two closure welds at the same location on the same package is low. The design is complicated, however, and so in practice it will take considerable expertise and experience to achieve the goal of stress-relieved package closure. Numerous welding options are being considered, but the key procedures will be the postweld treatments used to reduce the stresses introduced during the welding process (due to shrinkage during cooling on completion of the welding process). The present plan to alleviate these residual stresses is to use laser-shock peening. This process utilizes a high-power pulsed laser beam to introduce shock waves into the weld surface. These pulses produce compressive stresses to counterbalance the tensile residual stresses introduced during welding. This procedure should produce a compressive surface layer on the outer weld that is a few millimeters thick and a similar layer on the surface of the inner weld.

## Exposure Environment

A major reason for choosing the Yucca Mountain site was that the proposed repository is located well above the water table and, in the arid climate, little water would percolate through the excavated emplacement drifts. Because water is the only viable medium to sustain corrosion processes, this combination of isolation from groundwater and the limited amount of water that would percolate down from the surface would ensure minimal waste package corrosion.

Defining the exposure environment of the waste packages is critical to determining the corrosion processes that may occur and the amount of damage that results. This is a task fraught with difficulties and uncertainties because it involves so many features, ranging from the macroscopic to the microscopic. Examples at the macroscopic scale include the age of the waste (i.e., the time since its discharge from reactor) and the design of the waste vault. These features, in combination, will define the temperature/humidity history of the waste packages—an environmental variable that will strongly affect the corrosion process. Due to the porosity of the rock forming the repository, volcanic tuff, the emplacement drifts within the mountain are an "open" system. The air/water vapor pressure within the drifts will therefore not exceed the ambient atmospheric value. The heat generated by radioactive decay within the waste will guarantee that the temperature on the surface of the waste package will always be greater than, and the relative humidity less than, that of the surroundings. If this surface temperature is sufficiently high, then the relative humidity within the waste drift will be reduced to a very low value since the water will be driven out of the vault to the surrounding rock structure. These dry conditions constitute a benign corrosion scenario, as there is extensive evidence that the dry oxidation of metals and alloys produces insignificant degradation over long periods for temperatures <300°C (OCRWM 2003b).

A vigorous debate has been conducted over whether the repository should have a hot or cool design (Nuclear Waste Technical Review Board 2003). The proposed cool design would maintain the drift wall temperature below the local boiling temperature (96°C) and result in peak waste package temperatures <85°C (Nuclear Waste Technical Review Board 2003). In this scenario, aqueous conditions that could sustain corrosion would occur as soon as the ventilation ceased—that is, within tens to hundreds of years after emplacement. At temperatures of around 85°C, however, corrosion should be extremely slow.

For the hot design, the surface temperature of the waste package would exceed 150°C (Nuclear Waste Technical Review Board 2003), reaching a maximum that would depend on the ventilation scenario and the specifics of the repository design. As a consequence, water vapor would be expelled from the drifts for a few hundred to a thousand years. This design would therefore delay the onset of aqueous corrosion for a considerable period, but eventually allow corrosion to occur at high temperatures when more corrosive environments could be maintained. This could increase the possibility of aggressive localized corrosion that could lead to deep, localized penetrations of the waste package wall. Provided that such localized degradation can be avoided, though, the ongoing testing of Alloy-22 suggests that its corrosion performance is not significantly influenced by temperature (OCRWM 2003b). The material thus appears to be sufficiently robust to withstand the high-temperature design without incurring major corrosion damage.

The cool design would require a wider spacing of both drifts within the repository and waste packages within the drifts—or a considerable extension of the initial ventilation period. This would require the excavation of a much bigger repository. The resolution of this debate requires that the corrosion advantages, if any, gained by adopting the cool design are weighed against the influence of temperature on rock properties and the increased costs of repository construction.

**The Corrosive Environment**

The effect on waste package corrosion of the choice between a hot and cool repository design is shown in figure 17.3. The two schematic profiles denote the temperature-time profiles for hot and cool repositories. Although the period of ventilation is assumed to be the same for both, a much longer ventilation period

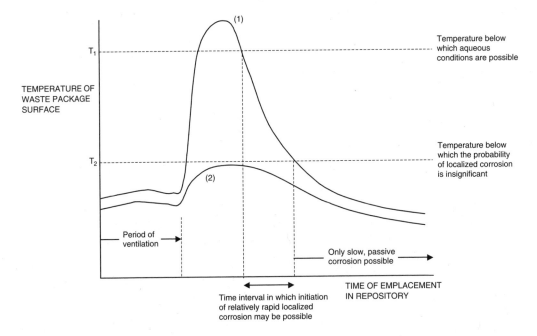

**Figure 17.3**
Schematic illustrating the waste package temperature as a function of time of emplacement in the repository: (1) is for a hot repository, and (2) is for a cool repository.

would be required to ensure a cool repository. Specific features of the repository design will determine the actual values of the temperature-time profiles. Although this schematic uses waste package temperature as the key parameter, it would be equally appropriate to use the relative humidity at the waste package surface. The following discussion assumes that the waste package is not protected by a drip shield (For more on drip shields, see chapter 18, this volume).

A crucial feature of both profiles in figure 17.3 is that after ventilation is complete and the repositories are closed, both designs evolve toward cooler and less corrosive conditions. From the corrosion perspective, the critical difference between the two scenarios is reflected in the two threshold temperatures: $T_1$, the temperature below which aqueous corrosion conditions can be established; and $T_2$, the temperature below which the probability of localized corrosion becomes insignificant. The conditions most able to sustain localized corrosion are a high value of $T_1$, a low value of $T_2$, and an aggressive aqueous environment.

Localized corrosion, in the form of pits, crevices, or cracks, can lead to deep wall penetrations even when there is a relatively small amount of corrosion damage overall. Therefore, to ensure long lifetimes for waste packages, it is essential to demonstrate that the combination of environmental conditions and properties of the waste package's corrosion barrier material will not sustain such localized corrosion.

Figure 17.3 shows that the hot design allows a period within which localized corrosion could occur. One apparently easy way to avoid this possibility is to adopt the cool design. If the more compact hot design is to be adopted, however, then it becomes essential to define the two critical temperatures, $T_1$ and $T_2$. A key difference between $T_1$ and $T_2$ is that $T_1$ is defined by the evolution of the repository environment in response to the waste package temperature; by contrast, $T_2$ is determined by the material's properties in response to the evolution in environment.

The determination of $T_1$ and the chemical nature of the aqueous environment that will then be produced have proven to be difficult. While recent studies have narrowed down the possibilities, considerable uncertainty remains. The time at which aqueous conditions will be established on the drip shield or the waste package will depend on the composition of the groundwater seeping into the emplacement drifts and how it is modified by the hot drip shield/waste package surfaces. A sequence of wetting-evaporative cycles will lead to the development of mineral and salt deposits that readily absorb water and raise its boiling point. These deposits will form predominantly on the drip shield. The humidity and temperature at which aqueous conditions can be durably established will then be determined by the propensity of these deposits to absorb water.

A second potential source of such salts will be the air-borne solids introduced to the repository by convective airflow during ventilation. The drip shield will not prevent their accumulation on the waste package. These particles could lead, via capillary condensation, to aqueous conditions on the waste package by contact with water vapor in the space between the drip shield and the package (figure 17.2).

Whether the waste package becomes susceptible to localized corrosion depends on the composition of the aqueous environment and the corrosion resistance of the material. While the waters in and around Yucca Mountain are dilute, and hence noncorrosive, wetting-evaporative cycles could produce more corrosive concentrated brines. The compositions of these brines and the relative solubilities of the dissolved species in them will be determined by the nature of the deposits, and hence the temperature at which these deposits allow aqueous conditions to be established. While significant uncertainty over the exact composition of aqueous environments remains, a considerable database exists about the geochemistry of similarly concentrated natural waters in arid environments in California and Nevada. These geochemical studies provide an essential reference point against which to evaluate the results and conclusions from laboratory tests. Thanks to tests conducted at Lawrence Livermore National Laboratory, an understanding of the possible environments within Yucca Mountain is evolving (OCRWM 2003a).

The eventual brine composition will depend on the initial composition of the dilute groundwater, as the relative concentrations of dissolved species will dictate which are removed by precipitation and which concentrate in the saturated brine. Chloride-dominated brines would be the most corrosive to the waste package. The temperature at which such brines could sustain aqueous conditions will be determined by the nature of the dominant cations. Calcium and magnesium brines could allow aqueous environments to form at ~160°C (OCRWM 2003a). By contrast, sodium/potassium brines would sustain aqueous conditions only at about 110°C (OCRWM 2003a). The presence in the brines of other anions, such as sulphate and nitrate, significantly reduces their corrosivity. In addition, the level of carbonate is particularly important. A substantial inventory of carbonate would precipitate calcium and magnesium, as well as buffering the pH to relatively neutral conditions. A smaller amount of carbonate could allow the development of either extremely deliquescent calcium and magnesium brines or extremely high pH ($\geq 12$) conditions (OCRWM 2003a).

While these features of evaporative processes may be well understood, significant uncertainties still prevail over the exact environment that will develop on the drip shields and the waste packages. Much of this uncertainty arises from an inability to specify which of the available waters in the mountain will be involved in the evaporative process. These waters include local well waters, waters perched above the repository, and water trapped within the pores of the rock.

Other, more minor uncertainties over the nature of the exposure environment also exist. These include the concentration by evaporation of minor impurities such as lead and mercury, and the presence of microbial colonies. While species such as lead are known to cause cracking of nickel alloys under the extremely hot conditions within operating nuclear reactors (pressurized water at ~300°C), the available evidence suggests such materials will have no effect on corrosion processes under the anticipated repository conditions. Also, although microbial communities exist within Yucca Mountain, laboratory tests give no indication that these microbes have any significant influence on corrosion.

The corrosion models used to assess the repository's performance assume that the drip shields and waste packages will be exposed to nitrate-containing, carbonate-based seepage waters (OCRWM 2003b). These models then conservatively assume that the propensity of sodium nitrate for water will allow an aqueous environment to form at around 120°C ($T_1$ in figure 17.3). While this value seems reasonable, further testing will be required to reliably assess the possibility that more aggressive brines will form at higher temperatures.

## Corrosion Properties of the Waste Package

Given the uncertainties in specifying the exposure environment on the waste packages, the outer shell of the waste package must be fabricated from an extremely corrosion-resistant material. The material selected was Alloy-22, a highly corrosion-resistant alloy of nickel, chromium, and molybdenum. Alloy-22 was only developed in 1981, and until testing in the Yucca Mountain Project began, information on its corrosion performance was sparse. It must be noted, however, that Alloy-22 is the fourth generation of the C-family of nickel-based alloys. Nickel-chromium alloys were first developed in 1898, and each succeeding generation has had an improved corrosion resistance.

Figure 17.3 summarizes the requirements that the material must meet. For the hot repository, if it is impossible to stop the onset of localized corrosion, then it must be demonstrated that the damage is limited enough not to significantly shorten the package's overall lifetime (defined as the time to first through-wall penetration). This assurance would require a considerable database of corrosion damage that would be difficult and time-consuming to accumulate. An easier approach is to demonstrate that the threshold for the initiation of localized corrosion is so high that such corrosion would have a very low probability of occurrence. This condition would be met if $T_2$ exceeded $T_1$. To achieve this last goal, it must be demonstrated that the alloy is not susceptible to localized corrosion in a wide range of environments that encompass all the uncertainties inherent in specifying the actual repository environment. This is a large task, but it is under way (OCRWM 2003b). Like all passive materials, Alloy-22 is protected from corrosion by the presence on its surface of a thin, adherent, and chemically inert oxide film. Localized corrosion begins only after this film is breached—a process that allows the development of extremely acidic and aggressive chemistry at the breakdown site. If this occurs on an open surface, the process would be described as pitting. If it occurs under a mineral deposit or in a gap between, say, the waste package and the support pedestal (figure 17.2), it would be crevice corrosion. If it were to occur in a crack created by tensile stresses, it is stress corrosion cracking.

Testing has shown that pitting is unlikely. Crevice corrosion, however, cannot be so readily dismissed, as crevices exist within the package/pedestal design, and will be present once mineral and salt deposits are formed. Because pits and crevices can act as initiation sites for stress corrosion cracking, this last process must also be considered. Provided the package shell is fully stress-annealed before service (as described above), stress corrosion cracking on the shell is only a remote possibility. Such cracking must be considered in the final closure welds, though.

Tests in concentrated brines at Lawrence Livermore National Laboratory have shown little tendency toward crevice corrosion (OCRWM 2003a). In long-term tests (> two years), no such corrosion occurred at temperatures as high as 90°C; short-term accelerated electrochemical tests revealed no corrosion up to 120°C. The results indicate that crevice corrosion will not be a problem in anticipated repository environments, but more conservative electrochemical testing in chloride-dominated brines, which have a remote chance of occurrence under repository conditions, has shown the process can be stabilized at temperatures around 95°C. The mode of accumulation of corrosion damage remains to be determined.

Given the lingering uncertainties over the possible environment, it would be imprudent to rule out the possibility of crevice corrosion in extremely aggressive brines. Still, as long as the drip shield performs its design function, it is unlikely that such extreme environments will form on the waste package. It is possible that extremely corrosive environments could form due to wetting of solids deposited on the package during ventilation—thus bypassing the protective function of the drip shield and leading to more aggressive corrosion conditions. Yet this process seems unlikely.

Concern over stress corrosion cracking focuses on the two lid closure welds, since for a fully stress-annealed waste package there are few alternative sources of stress. One potential source would be from damage due to the impact of falling rocks, but a properly functioning drip shield should prevent this. Careful handling procedures should avoid the introduction of stresses due to local mechanical damage during emplacement. Once situated in the repository, the waste package will experience none of the dynamic stresses that play a role in many industrial operations during start-up and shutdown operations.

Testing for stress corrosion cracking has been conducted using conservative procedures, which guarantee the onset of crack growth. Crack growth rates were subsequently measured in hot, concentrated brines (up to 110°C) as conditions were made more like those anticipated in the repository (OCRWM 2003b). The observed crack growth rates, while not necessarily zero, were extremely low. These studies contradict the claims that Alloy-22 is immune to stress corrosion cracking and reinforce the need to develop stress mitigation procedures for the closure welds. Nevertheless, the studies demonstrate that a combination of waste package fabrication processes and repository design features (e.g., the use of a drip shield) can make localized corrosion—including stress corrosion cracking—highly unlikely. Continuing studies may succeed in eliminating these processes as credible failure mechanisms.

If this last goal were achieved, the only corrosion mechanism that would still constitute a serious threat to the waste package's integrity would be passive corrosion (figure 17.3). Short-term tests (two to three years) in anticipated repository environments yield extremely low passive corrosion rates; performance assessment models that assume such low rates predict waste package lifetimes well in excess of ten thousand years (OCRWM 2003b). While these extremely low rates are consistent with other short-term laboratory measurements with Alloy-22 and similar materials, they are nonetheless empirical; a firmer understanding of how material properties will evolve over such long time frames is required. Moreover, considerable research is needed to develop a level of understanding that would bestow credibility on performance assessment calculations based on these rates. It is unrealistic to expect that short-term testing will generate this understanding.

In view of these reservations, an extensive research effort is under way at Lawrence Livermore National Laboratory and other government, university, and commercial laboratories. The main goal is to demonstrate that the probability of corrosion accelerating with time in the repository is remote. This will require that in view of the evolution of the repository environment from initially aggressive to eventually more benign, neither the alloy itself nor its protective oxide film degrade significantly with time. If this can be shown, then the use of corrosion rates measured under the aggressive repository environments anticipated in the short term can be confidently extrapolated to longer times.

A major concern is that processes such as dealloying—the loss of key alloying elements (e.g., chromium)—will not degrade the material to a more corrodible form. Aging, for instance, could redistribute key alloying elements to precipitated phases, thereby depriving the main body of the alloy of its resistance to corrosion. It is also necessary to demonstrate that the passive oxide film will not preferentially lose its critical alloying constituents as it oxidizes and dissolves. In addition, one must show that the expansion that results when metal is converted to oxide will not introduce stresses that fracture the oxide film. That process could allow corrosion to bypass the oxide and potentially maintain higher rates of corrosion than presently measured.

Corrosion testing has not shown any evidence of such processes, and the evidence supports claims that passivity will be maintained in the long term. These experiments bolster confidence that long (one hundred thousand years) waste package lifetimes are achievable. Further research is needed, however, to justify present performance assessment predictions.

## Summary

A considerable effort has been expended on both the design and fabrication as well as the mechanical and corrosion testing of the waste package. The key procedure is the performance of the final closure welds. The welding techniques will be those commonly and successfully applied in other industrial situations, but their optimization under remote handling conditions remains to be developed. A major licensing requirement will be the demonstration that the package can be closed with stress-relieved welds that are not susceptible to stress corrosion cracking.

Corrosion testing of the waste package material has shown that it will be extremely resistant to a wide range of anticipated repository environments. Yet significant uncertainties remain over how aggressive the exposure environment may become during the critical early exposure period, making it injudicious to presently rule out the occurrence of localized corrosion processes. While the available data indicate that the waste package material is sufficiently robust to withstand this possibility, further corrosion testing is required to confirm this.

Assuming that the possibility of localized corrosion can be eliminated, the lifetime of the container would be limited by its passive corrosion rate. Short-term tests indicate that these rates are so low that package lifetimes in excess of one hundred thousand years would be achieved. To justify these predictions, a much better fundamental understanding of the science of passivity is required, and a considerable effort is under way to obtain it.

Given the importance of waste package performance to the overall safety case for the repository, these ongoing studies are critical.

## Acknowledgments

I wish to acknowledge many interesting meetings and discussions with Gerald Gordon and Pasu Pasupathi (Framatome Advanced Nuclear Power), Peter Andresen (General Electric Corporate Research and Development), and Gregory Gdowski (Lawrence Livermore National Laboratory). A large portion of the information described here is based on their research and much of the understanding developed in discussions (arguments) with them.

## References

Nuclear Waste Technical Review Board (2003) Transcript of the Meeting on Thermal Aspects of Yucca Mountain Repository Design. May 13–14. http://www.nwtrb.gov/meetings/meetings.html.

Office of Civilian Radioactive Waste Management (OCRWM) (2002) *Yucca Mountain Science and Engineering Report*. Department of Energy. DOE/RW–0539–1.

Office of Civilian Radioactive Waste Management (OCRWM) (2003a) *Technical Basis Document No. 5: In-Drift Chemical Environment Revision 1*. Department of Energy. November.

Office of Civilian Radioactive Waste Management (OCRWM) (2003b) *Technical Basis Document No. 6: Waste Package and Drip Shield Corrosion, Revision 1*. Department of Energy. December.

# 18

# Drip Shield and Backfill

David Stahl

A repository at Yucca Mountain will depend on both the geologic and engineered barriers for safety. Foremost among the engineered barriers is the waste package. Providing a second layer of protection will be a drip shield, essentially an umbrella that covers the waste containers as well as protects them from rockfall and dripping water. An alternative method of protecting waste packages is to fill the empty tunnel space around them with some type of crushed rock, known as backfill. Currently, the Department of Energy (DOE) plans to use drip shields but not backfill in its repository design. This chapter will evaluate the merits of the titanium drip shield concept, and will also briefly discuss the pros and cons of backfill.

The drip shield is one component of the multiple barrier system to be utilized at the proposed repository. The key idea is to engineer the barriers so that they would not fail by the same mechanism. In this case, the drip shield would be made of a titanium alloy, while the waste packages themselves would be made of a different material that would fail by a different mechanism. Thus, if one barrier were to corrode, the other barrier would continue to function.

Several nickel-based alloys were evaluated for the waste packages. Ultimately, the nod went to Alloy-22, a nickel-based alloy that is compatible with titanium by virtue of its close proximity on the galvanic series. Galvanic proximity prevents the development of an electric current between the two materials when they are immersed in a conductive fluid, and thus will not provide a potential mechanism for failure.

Initially, several designs under consideration involved the use of a titanium alloy as part of the waste package container itself, rather than a separate drip shield. These designs were very costly, however, in part because they would have required a significant thickness of titanium alloy to sustain the handling loads. Separating the titanium from the container in the form of a drip shield preserves the advantages of defense-in-depth, while deferring the cost until the time of closure when the drip shields would be inserted.

The drip shield is an inverted "U" fabricated from a fifteen-millimeter-thick rolled plate, which has attached to it vertical stiffeners, lifting lugs, and end members (figure 18.1; OCRWM 2001b). The end members permit subsequent units to interlock with the previous ones, making a continuous barrier to block the ingress of dripping water. Each unit measures about 6.1 by 2.5 by 2.5 meters. Most of the drip shield is to be made of titanium grade 7, with stiffeners on the sides to be made from a stronger alloy, titanium grade 24 or titanium grade 29. One drip shield assembly weighs about 4.2 metric tons.

Prior to the repository's closure, ventilation plus the heat from the radioactive decay of the waste will keep the waste packages dry. Thus, the drip shields are not required during this early period.

To predict the performance of the drip shields, the DOE evaluated the environmental conditions they would need to withstand, from corrosive waters to rockfall conditions to the potential for electrochemical reactions with other materials in the tunnel. The issue is whether the drip shield can perform its function for thousands

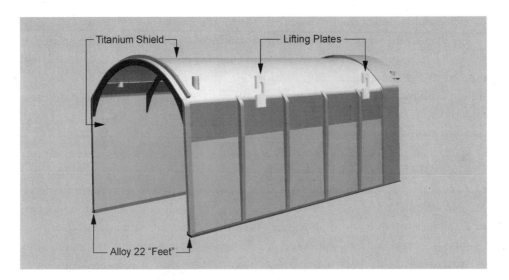

**Figure 18.1**
Representative drip shield. From Office of Civilian Radioactive Waste Management 2001b, figure 2.73.

of years given the potential for aggressive water chemistries and rockfall. These environmental issues are discussed in the next section.

### Environment of the Drip Shield

The environmental conditions that the waste packages and the drip shields experience will evolve over time. The DOE's repository design has two alternatives: a hot option in which the repository environment would heat to over the boiling point of water, and a cooler option where temperatures would remain below boiling. After emplacement, the waste packages will still be fairly warm and the air will be low in relative humidity, due in part to mechanical fans that will provide ventilation. During the first twenty-five years, when fans are operational, the waste package surface temperatures would be about 110°C for the higher-temperature design and about 80°C for the lower-temperature design. When the fans are turned off, the waste package surface temperatures would rise to about 240°C for the higher-temperature design, cooling after three hundred years to about 140°C. (The surface temperatures for the lower-temperature design are proportionately lower.)

After installation of the drip shields, the waste packages will continue to cool and the relative humidity will eventually reach ambient conditions (near 100 percent). Thus, in the far future, moisture is likely to condense onto the drip shield. In addition to the condensation, water may also drip from the ceilings of the emplacement drifts. This water may contain concentrated salts due to evaporation, or the presence of salt-laden moisture or dust brought in by the ventilation system. As a consequence, the DOE has predicted possible corrosion conditions and mechanisms that the drip shields might experience. In doing so, the DOE evaluated two types of water, including water from the J–13 well, which has about 200 parts per million of sodium bicarbonate; and rock pore water, which has about 250 parts per million of chloride-sulfate. The J-13 water is representative of the many waters found in and around Yucca Mountain that flow in the fractures and faults there. The pore water represents water that is contained in pores in the rock. The pore water could potentially contact the drip shield through evaporation-condensation mechanisms.

Knowledge of the water chemistry is critical in determining the rate of corrosion of the drip shields and waste package components. If the water remains dilute, corrosion is unlikely or extremely slow. But high concentrations of aggressive solutes could greatly accelerate corrosion. Therefore, the DOE conducted analytic modeling as well as evaporation tests to estimate the compositions of the waters and salts as evaporation progressed (OCRWM 2001b). The J-13 well water formed salts containing chlorides, nitrates, carbonates, and silicates. Early in the evaporation process, calcium and magnesium precipitated as carbonates; chlorides and nitrates, being more soluble, precipitated only when evaporation was nearly complete.

One noteworthy characteristic of salts is their deliquescent point—that is, the point at which they absorb enough water to form a liquid solution. The deliquescent point is important because it controls when corrosion could begin as the waste package cools; the higher the deliquescent point, the higher the stability of a water film on the drip shield surface and the sooner corrosion could be initiated. Sodium nitrate, formed as a result of the evaporative concentration, has the lowest deliquescent point—about 50 percent relative humidity at 120°C. The result is that a concentrated solution of sodium nitrate has a pH of about twelve, making it very basic. The pore water chemistry formed predominantly chloride and sulfate salts on evaporation. These have a lower deliquescent point; the resulting concentrated solution has a pH of about six (slightly acidic). The conditions resulting from the evaporation of J-13 water were used as a basis for subsequent corrosion testing at Lawrence Livermore National Laboratory as they were considered representative of solutions contacting the waste package and the drip shield. These tests have been expanded to include concentrated pore waters, however.

Another factor in the DOE's environmental analysis for drip shield performance is its electrochemical interaction with other tunnel materials. For instance, the drip shields and the waste packages could interact with the components of the ground support system. Ground support could be as simple as rock bolts and mesh to something more complex such as a steel set structure. Both approaches have advantages and disadvantages. The use of rock bolts and mesh could increase the potential for rockfall, but would reduce the potential for galvanic interactions that would hasten corrosion. On the other hand, the use of steel sets would essentially eliminate rockfall during the operation period and well into the postclosure period, but would increase the potential for galvanic interactions. Currently, the Yucca Mountain Project has taken the approach that stainless steel sheeting with rock bolts will be utilized for the emplacement drifts. This is based on more thorough thermal and stress analyses performed for the openings.

In reality, the response of the drip shield to rockfall will depend on the ground support approach utilized. For the purposes of design, however, it was assumed that the ground support system did not function and that the drip shield would need to withstand the effects of rockfall by itself. This approach requires knowledge of the size and distribution of rocks that could strike the drip shield. A detailed geotechnical study (OCRWM 2001b) led to the assumption that to perform its job, the drip shield would need to withstand the impact of a falling rock weighing less than four metric tons. A finite element analysis indicated that the drip shield required extra stiffening, which was accomplished with the addition of the titanium grade 24 stiffeners.

## Possible Degradation Mechanisms

Due to their exposure to hot, dry air and later to cooler, moist air with the potential for dripping water, the drip shields may be subjected to a wide range of degradation mechanisms (OCRWM 2001b). Any of these mechanisms may lead to the penetration of the titanium over time, and are usually stated in terms of a reduction in the thickness of the material in units of nanometers or millimeters per year. These mechanisms are discussed in more detail below. Shoesmith and colleagues (1997) evaluated the potential for the activation of these mechanisms.

## Dry Oxidation

Dry oxidation involves the reaction of titanium with oxygen in air drier than the threshold for humid air corrosion (usually about 60 percent relative humidity). The DOE obtained rates for the dry oxidation of titanium from the ASM Metals Handbook (1987) and Shoesmith and colleagues (1997). The oxidation rates vary over a range of temperatures, and the DOE explored rates over the temperature range from ambient to 250°C. For the conditions expected at Yucca Mountain, the rates for dry oxidation were found to be quite low, in the range of 0.1 micrometers per year; at that rate, compared to the drip shield thickness of fifteen millimeters, drip shield penetration would not occur for 75,000 years, assuming oxidation on both surfaces. It is recognized that the drip shield may collapse due to oxidation thinning prior to 75,000 years because of rock overburden; however, it is likely that the drip shield will fail much earlier by aqueous corrosion processes.

## General and Localized Corrosion

Aqueous corrosion can take place either generally or locally on the surface of the titanium. The DOE used general corrosion rates for the titanium drip shield based on results from the Lawrence Livermore National Laboratory's Long-Term Corrosion Test Facility. Humid air corrosion is possible when a water film containing aggressive species forms on the surface of the titanium. This may occur if the relative humidity exceeds about 60 percent. Humid air corrosion and general aqueous corrosion thus require similar conditions—such as an appropriate temperature, the presence of water, and aggressive chemistry—for corrosion to be activated. The Yucca Mountain Project conservatively considered both processes together in analyzing their consequences because the relative humidity in the emplacement drifts will approach the ambient relative humidity (near saturation) as the packages cool.

The rates obtained appeared to be independent of temperature between 60°C and 90°C, as well as independent of the chemistry of the test solutions (OCRWM 2001b). The median measured rate was approximately zero, with most measurements for weight-loss samples lying between plus or minus two hundred nanometers per year. The presence of both positive and negative rates indicates that the value is strongly influenced by the presence of surface films—such as a thin resistant titanium oxide layer, as well as silicate deposits—and the process of cleaning the specimens prior to the measurement of weight loss. The DOE is evaluating methods to better define the effect of surface films on corrosion rates.

Localized corrosion can occur either at flaws or defects, creating pits, or under surface deposits or material interfaces, creating crevices. Tests at Lawrence Livermore National Laboratory revealed no evidence for pitting corrosion. Corrosion rates for creviced samples were found to be extremely slow, and as with general corrosion, appeared to be independent of temperature and test solution compositions. As with the uncreviced samples, the median rate was approximately zero, with most of the corrosion rates between plus or minus 350 nanometers per year. Again, the positive and negative rates were likely due to the presence of surface films. In short-term tests, Shoesmith and colleagues (1997) report that crevice corrosion is not observed in titanium grade 7 below 200°C. The highest measured rate in the DOE tests was less than +0.35 micrometers per year and would not lead to the failure of the drip shield during its first ten thousand years (OCRWM 2001b). Based on these data, the life of the drip shield does not appear to be limited by the crevice corrosion of titanium. Testing and model results indicate that the likely corrosion rate for titanium is approximately 25 nanometers per year and the maximum rate is less than 350 nanometers per year.

Fluoride at high concentrations is known to attack titanium. Shoesmith and colleagues (1997) summarized the influence of fluoride on the corrosion of titanium and its alloys. At a fluoride concentration of one hundred parts per million and a temperature of 177°C, the researchers observed no crevice corrosion. The content of fluoride in Yucca Mountain waters is now on the order of one part per million, and the concentration of fluorides is not expected to increase beyond this level significantly even for concentrated solutions, due to precipitation. The presence of fluoride may, however, affect the long-term stability of the passive film and thereby undermine the long-term resistance to corrosion. Recent longer-term (450-hour) studies conducted by the Center for Nuclear Waste Regulatory Analyses (Brossia et al. 2001) on titanium grade 7 have shown that the presence of fluoride above four parts per million in one molar (~35,000 ppm) NaCl solutions significantly accelerates corrosion. Additional current density tests in the passive corrosion region, which contained the potentially inhibiting nitrate and sulfate ions, also indicated increased corrosion rates, at least in the short term. Yet the ionic ratios in these tests differed from those expected at Yucca Mountain. Clearly, the affect of fluoride on the corrosion rate of titanium grade 7 needs further study.

**Stress Corrosion Cracking**

Cracks can develop in susceptible materials if the stress on the material is above a threshold value and an aggressive water chemistry exists on the surface. The major source of stress is rockfall directly on the titanium drip shields due to earthquakes. The resulting load could lead to corrosion cracking and penetration of the drip shields. Because of the drip shields' size, however, these penetrations are not expected to prevent the structures from performing their intended function of diverting water away from the waste packages. The addition of stiffeners to the drip shield should prevent it from being fractured by a rockfall.

Stress corrosion cracks in passive alloys such as titanium grade 7 and Alloy-22 tend to be extremely tight. Analyses of rockfall on the drip shields have produced estimates that the impact of a large (greater than four metric ton) rock would lead to a corrosion crack about one hundred micrometers wide. The crack faces are expected to corrode slowly and eventually fill the crack space with corrosion products. During this period of slow corrosion, a small amount of water may be transported by surface diffusion into the crack and through the drip shield. Nevertheless, the DOE expects that the high thermal conductivity of the drip shield wall will result in a small temperature gradient across it. The slowly flowing water will evaporate on contact with the hot drip shield, and a salt-scale deposit will form over and within the crack on the upper drip shield surface. Such a formation of salt-scale deposits is well documented in seawater environments and heat exchangers. Therefore, stress corrosion cracking will not compromise the function of the drip shield.

### Galvanic Effects

Galvanic interactions can occur between two metals in close proximity, separated by a conducting fluid. The degree of interaction can be evaluated from the two metals' position on the galvanic series. (As noted above, the galvanic series ranks metals or alloys according to their reactivity in specific solutions like seawater; the greater the separation in the series, the greater the interaction.) The drip shield material is very close in the galvanic series to the nickel-base alloy used in the outer layer of the waste packages. Thus, galvanic effects are expected to be negligible.

The drip shield could, however, form a galvanic couple with the carbon steel of the ground support or invert system (such as rock bolts, wire mesh, and the steel liners used in the drift). Corrosion of this carbon steel could accelerate the generation of insoluble iron (ferric) oxides or oxyhydroxides. These compounds do not interact with the titanium and so are not likely to be detrimental. Still, the corrosion of carbon steel could cause trouble through another mechanism. If water is disassociated, the resulting hydrogen could be absorbed by the titanium. If the hydrogen content exceeds a critical value, the result could be hydrogen-induced cracking. To mitigate galvanic coupling between the drip shield and the invert carbon steel, Alloy-22 "feet" at the bottom of the drip shields will separate the drip shields from the invert structure.

### Hydrogen-Induced Cracking

Hydrogen generated from corrosion reactions can embrittle titanium. Generally, the titanium oxide film that readily forms on titanium surfaces acts as an excellent barrier to the transport of hydrogen. In addition, the absorption of significant amounts of hydrogen would not be expected nor has it been observed in chemical systems under conditions similar to those expected at Yucca Mountain. Over the very long time in the repository, even slow hydrogen absorption may be significant. Still, the model developed by the DOE estimated that the hydrogen levels generated are below the threshold for cracking to occur (OCRWM 2001b).

A simple and conservative model has been developed to evaluate the effects of hydrogen-induced cracking of the drip shield (OCRWM 2001b). Using available general corrosion test data, which estimate the amount of hydrogen from the reaction of water with the titanium that is available for absorption into the titanium, the model calculates the hydrogen concentration in the drip shield at ten thousand years after emplacement. The assumption of a corrosion rate of one hundred nanometers per year leads to a calculated hydrogen concentration of 260 parts per million of titanium grade 7. This is far below the critical concentration of hydrogen at which the material will fail, determined by Shoesmith and colleagues (1997) to be one thousand parts per million. The mean estimate of corrosion rate is only twenty-five nanometers per year, which corresponds to an even lower hydrogen level: sixty-five parts per million. In either case, the analysis indicated that hydrogen-induced cracking will not occur.

The DOE accomplished partial model validation using the short-term (two- to ten-year) data available from both the DOE and the literature. The DOE's evaluation, based on the current model, indicates that hydrogen-induced cracking of the drip shield will not occur with a large margin of safety. Additional data on the hydrogen content in corrosion samples are being collected, thereby improving confidence in the model.

### Microbially Influenced Corrosion

Metals corrode when exposed to the right combination of microbes, nutrients (such as carbon, sulfur, and phosphorous), temperature (below the boiling point), and relative humidity (usually above 60 percent; OCRWM 2001b). Microbial growth has been studied utilizing combinations of microbes found at Yucca Mountain as well as different combinations of nutrients. The principal nutrient-limiting factor to microbial growth appears to be the low level of phosphate present (OCRWM 2001b). As the repository cools, temperatures and relative humidities are expected to be in the range required for the growth of microbes. Corrosion tests conducted by Lawrence Livermore National Laboratory (OCRWM 2001b) as well as those noted in Shoesmith and colleagues (1997), however, have produced no observed attack of the titanium alloy by a consortium of Yucca Mountain microbes.

## Radiolysis-Enhanced Corrosion

In the presence of radiation, nitrogen and oxygen in moist air can form aggressive oxidizing species such as nitric acid or hydrogen peroxide. This process, called radiolysis, can enhance corrosion. To simulate the effect of radiolysis, the DOE studies have added hydrogen peroxide to electrochemical polarization tests in various test solutions. (The purpose of such tests is to accelerate corrosion of a test electrode by applying an electrical potential between it and an inert reference electrode.) For Alloy-22, the continued addition of peroxide to J-13 well water led to corrosion potentials well below any threshold where localized corrosion would be expected. Subthreshold corrosion potentials resulted regardless of whether the well water was mildly concentrated (by a factor of ten with 10 percent of the water volume reduced) or very concentrated (by a factor of one thousand). Shoesmith and King (1999) reviewed the effects of radiation on corrosion of Titanium Grade 7 and found no impacts for the radiation doses expected at Yucca Mountain. Hence, it is not expected that radiolysis-enhanced corrosion will be an important degradation mechanism for the drip shield.

## Long-Term Passive Film Stability

The resistance of titanium alloys to general aqueous corrosion depends on the stability of the tenacious titanium oxide film. Data collected by the DOE from the Long-Term Corrosion Test Facility and the polarization studies described above show that the film is stable under the expected Yucca Mountain conditions. The tests were conducted with concentrated solutions noted above in the temperature range of 25°C to 90°C. Still, because these tests have been of relatively short duration (about five years), the long-term stability of these films is uncertain. This uncertainty may be particularly large because the detected corrosion rates are low—less than one micrometer per year, which is near the detection limit by weight-loss measurements.

To better quantify the corrosion rates and further elucidate film stability, new microanalytic techniques are being used. These include atomic force microscopy, X-ray photoelectron spectroscopy, electrochemical impedance spectroscopy, and linear polarization. These techniques can detect corrosion rates in the several nanometer per year range, making them at least one hundred times more accurate than the current weight-loss measurements. The DOE also plans to examine the thicker oxides generated at higher temperatures on aged samples at Lawrence Livermore

National Laboratory. In addition, the DOE is developing an analytic, mechanistically based model for projecting general and localized corrosion behavior and passive film stability. Finally, the DOE can partially validate the estimate of passive film stability by taking advantage of the fact that similar alloys have been utilized in the chemical industry under more aggressive conditions for decades with excellent results.

### Conclusions Regarding Drip Shield Performance

The type of degradation that will limit the life of the drip shield appears to be general aqueous corrosion. Given this mechanism, the DOE estimates that the minimum life of the drip shield is 20,000 years with a mean life of 50,000 years (OCRWM 2001a). This estimate is based on the stability of the passive film on the drip shield's surface, though. Corrosion testing reported in the literature as well as that performed by the DOE shows that this film is stable under conditions (i.e., temperatures up to the boiling point and water films containing salts) expected at Yucca Mountain (OCRWM 2001b; Shoesmith et al. 1997). To provide confidence in the prediction of long-term passive film stability, however, additional work is needed, both to improve the sensitivity of corrosion rate data and further the understanding of the mechanistic basis of film stability. This work will be conducted at Lawrence Livermore National Laboratory or other research entities.

The DOE predicts that the drip shields will begin to fail after about 40,000 years, considering only corrosion and not mechanical degradation processes. The DOE predicts that waste packages will begin to fail after several hundred thousand years. Waste packages will fail first at about 100,000 years by stress corrosion cracking. Breach of the waste packages by general corrosion starts at about 300,000 years. In analyzing waste package corrosion, the DOE assumed that the presence of salt-containing dust particles may result in surface water film with an aggressive water chemistry. Thus, drips on the waste package are not needed for the initiation of waste package corrosion. All that is required is the presence of a water film (which would occur when the relative humidity is above about 50 percent) along with the aggressive chemistry.

Drip shields will not prevent the early initiation of corrosion leading to waste package breach because the condensed water on the waste packages will contain chlorides. Still, drip shields meet their design function in that they provide protection from rockfall and prevent water from dripping onto the waste packages. Moreover, the presence of drip shields delays potential radionuclide releases since any release would be by diffusion and not by the more rapid process of advection.

One could argue that the drip shields should be fabricated of a much cheaper material—for example, stainless steel, which could be coated with a protective ceramic or more corrosion-resistant metallic layer. The stainless steel would be compatible with the Alloy-22 that is used as the waste package outer barrier as well as with the ground support system. While galvanic interactions would not affect the performance of the drip shield, they could lead to faster ground support degradation.

## Backfill

The function of the drip shield is to prevent water from dripping and rocks from falling onto the waste package. Backfill would further fortify the repository by protecting the drip shield itself against rockfall. There is also potential for the reduction of dose to the public as a result of a volcanic event when the emplacement drifts are filled with backfill. The Yucca Mountain Project has considered the use of backfill for several years. A variety of different backfill materials were considered, with the most likely candidate being crushed tuff. While the current plan does not include backfill, the design of the underground facility will not preclude backfill.

Backfill could have several other desirable attributes, such as providing a well-controlled thermal-hydrologic-chemical environment in which the rest of the engineered barriers reside. Backfill also has a potential benefit of reducing the humidity on the waste package surface, thereby delaying the onset of aqueous corrosion. Yet backfill could have negative effects, such as increasing the peak temperature of the spent fuel cladding and accelerating its degradation. Furthermore, there is some doubt about the usefulness of the chemical buffering that would be provided by backfill. Backfill could in fact potentially enhance radionuclide release from the near field.

The DOE performed sensitivity analyses to assess the performance-related impacts of the potential addition of backfill (OCRWM 2001b). The performance assessment examined a base case with and without backfill. The analysis indicated little net effect, positive or negative, of backfill for the period up to one hundred thousand years. This marginal difference in performance coupled with the large expense of installing the backfill at the closure of the repository led the Yucca Mountain Project to reject backfill in the current design.

## Summary and Conclusions

The DOE has selected a titanium alloy to be utilized as a drip shield that would be emplaced just prior to the closure of the repository. This material was chosen since it is extremely corrosion resistant, is compatible with the Alloy-22 waste package outer barrier material, and likely fails by different degradation mechanisms than that alloy. The titanium alloy could nevertheless suffer some attack by fluoride on the passive film that has not been adequately understood, particularly in the long term since the stability of the passive film is the basis for its excellent corrosion performance.

Thus, additional work needs to be performed to provide a better understanding of the long-term response. Further testing and analysis is also needed to provide improved confidence in the DOE's model of hydrogen-induced cracking.

The DOE has begun to examine other material alternatives to the use of titanium as a drip shield. They options could significantly reduce cost. They include the use of stainless steel, which could be coated with a protective ceramic or a more corrosion-resistant amorphous metallic layer. These options deserve further and more detailed investigation.

## References

ASM Metals Handbook (1987) *Corrosion: Corrosion of Titanium and Titanium Alloys*. 9th ed. Vol. 13. Metals Park, OH: ASM International.

Brossia, C.S., Browning L., Dunn, D.S., Moghissi, O.C., Pensado, O., and Yang, L. (2001) *Effect of Environment on the Corrosion of Waste Package and Drip Shield Materials*. Center for Nuclear Waste Regulatory Analyses, San Antonio, Texas. CNWRA 2001–03. September.

Office of Civilian Radioactive Waste Management (OCRWM) (2001a) *Site Suitability Performance Assessment Volume 1*. Department of Energy. TDR–MGR–MD–000007 REV 00.

Office of Civilian Radioactive Waste Management (OCRWM) (2001b) *Yucca Mountain Science and Engineering Report*. Department of Energy. DOE/RW–0539. May.

Shoesmith, D.W., Hardie, D., Ikeda, B.M., and Noel, J.J. (1997) *Hydrogen Absorption and the Lifetime Performance of Titanium Waste Containers*. Atomic Energy of Canada Limited, Pinawa, Manitoba. AECL–11770.

Shoesmith, D.W., and King, F. (1999) *The Effects of Gamma Radiation on the Corrosion of Candidate Materials for the Fabrication of Nuclear Waste Packages*. Atomic Energy of Canada Limited, Pinawa, Manitoba. AECL–11999.

# V

## Waste Forms

# 19

# Zircaloy Cladding

Eric R. Siegmann

More than 40,000 metric tons of spent nuclear fuel are now stored at commercial nuclear plants around the country. Most of this material rests in deep, stainless steel–lined pools that are cooled by recirculating deionized water. The current national strategy calls for transporting the spent fuel to a geologic repository such as the proposed Yucca Mountain site. There, the waste would be sealed in large containers made of durable metals, placed into the mountain, and left for perpetuity. Over tens of thousands of years, the containers and the spent fuel will degrade. The question is when, and at what rate, the radionuclides will be released into the environment.

One of the main factors in predicting the fate of the nuclear waste is the integrity of the spent fuel itself. That integrity depends, in large part, on the material used for the tubes that encase the actual fuel pellets. This cladding—usually a zirconium alloy—prevents the environment from interacting with the fuel pellets and reduces the rate of release of radionuclides into the environment. The early Total System Performance Assessments (TSPA) done by the Department of Energy (DOE) neglected the role of the cladding. The nuclear fuel was assumed to be bare—that is, the fuel pellets were considered to be exposed to water as soon as the metal waste package failed. Thus, all of the fuel in the waste package was immediately available for dissolution. The rate of release of the radionuclides was then determined by the fuel dissolution rate, the solubility of the radionuclides, and the amount of water that passes through the failed waste package. In 1995, the Yucca Mountain Project began to consider the role of the degradation of cladding. Three subsequent performance assessments have considered the impact of the presence of cladding: the 1998 viability assessment (CRWMS M&O 1998), the 2001 site recommendation (McNeish et al. 2000), and the Electric Power Research Institute's (2000) latest TSPA.

This chapter describes the spent fuel produced by commercial nuclear reactors and discusses the various ways that the fuel cladding might degrade once the waste is stored at Yucca Mountain. The chapter also addresses the effect on the performance assessment of including cladding.

## Commercial Reactor Cladding

The light water reactor fuel assemblies consist of a square array of fuel rods connected together by spacer grids. There are two kinds of light water reactors: pressurized water reactors and boiling water reactors. Figure 19.1 shows a typical pressurized water reactor assembly. It is approximately 3.7 meters long and 20 centimeters wide. The rods consist of metal tubing about 1 centimeter in diameter (the cladding) filled with cylindrical ceramic pellets composed of uranium dioxide ($UO_2$); each pellet is about 1.35 centimeters long. The cladding is an alloy consisting mostly of zirconium, alloyed with traces of other metals. For a typical pressurized water reactor generating one thousand megawatts of electricity, the core consists of

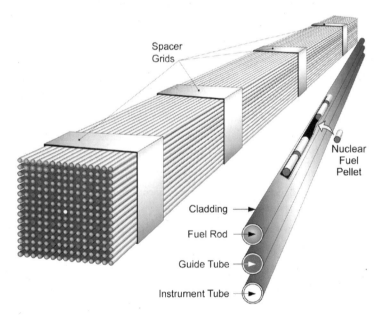

**Figure 19.1**
Representative pressurized water reactor fuel assembly. From Office of Civilian Radioactive Waste Management 2001, figure 3.6.

**Table 19.1**
Composition of nuclear fuel cladding materials

| Component | Zircaloy-2 | Zircaloy-4 | ZIRLO™* | M5™* |
|---|---|---|---|---|
| Zirconium | 97.71 to 98.56% | 97.77 to 98.46% | 97.76 to 97.83% | 98.84% |
| Tin | 1.20 to 1.70% | 1.20 to 1.70% | 0.96 to 0.98% | — |
| Iron | 0.07 to 0.20% | 0.18 to 0.24% | 0.094 to 0.105% | — |
| Chromium | 0.05 to 0.15% | 0.07 to 0.13% | 79 to 80 ppm | — |
| Nickel | 0.03 to 0.08% | <0.0070% | — | — |
| Oxygen | 0.09 to 0.16% | 0.09 to 0.16% | 0.09 to 0.12% | 0.16% |
| Niobium | — | — | 1.02 to 1.04% | 1% |

*ZIRLO is a trademark of Westinghouse Nuclear; M5 is a trademark of Framatome ANP.

about two hundred such fuel assemblies, clamped into the shape of a cylinder roughly 3.7 meters high and 3.7 meters in diameter. The fuel assemblies in a boiling water reactor are narrower and more numerous; in these reactors, unlike the pressurized water plants, the fuel assemblies are enclosed in a metal flow channel.

Some of the first nuclear reactor cores used stainless steel cladding. Zirconium first became available in industrial quantities in 1946. Alloys of zirconium were introduced as nuclear fuel cladding early in the development of light water reactors in the 1950s because this material has excellent corrosion, mechanical, and neutronic properties. When exposed to air, zirconium quickly develops an adhesive layer of zirconium oxide that seals the metal from further corrosion. The material's high cost tends to limit its use to highly corrosive environments within the chemical industry (Yau and Webster 1987). Table 19.1 gives the composition of the different zirconium alloys. Zircaloy-2 is used in boiling water reactors and Zircaloy-4 is used in pressurized water reactors. The mid-1990s brought the introduction of two new and more corrosion-resistant zirconium alloys, called ZIRLO and M5. The new alloys have displaced much of the Zircaloy-4 in U.S. pressurized water reactors.

## Cladding Failure

The importance of cladding is that it is a barrier preventing water from coming into contact with the $UO_2$ waste form. The failure of the cladding means that small cracks or holes occur and permit the outside environment (air, water, or water vapor) to enter the rod and react with the $UO_2$ pellets. The more cladding that fails, the more of the spent fuel that is potentially exposed to contact with water. Such contact leads to the slow degradation of the pellets and the release of radionuclides. Cladding failures may occur either before the fuel bundles arrive at Yucca Mountain, or at the repository due to creep, corrosion, mechanical loading, or hydride formation.

Earlier studies have evaluated cladding failure under repository conditions (Rothman 1984; Pescatore et al. 1990; Henningson 1998). They concluded that most failure mechanisms would not occur. The Environmental Protection Agency (EPA) did a detailed study of cladding degradation in the repository (S. Cohen and Associates 1999) and reached similar conclusions. Einziger and Kohli (1984) and Peehs (1998) also concluded that cladding failure was unlikely. Sanders and colleagues (1992) reviewed the condition of cladding after reactor operation and also the potential for damage from external mechanical loading as a result of transportation accidents. They showed that breakage of the irradiated cladding required very high forces—equivalent to the impact of a drop from higher than nine meters.

The DOE models cladding degradation in the site recommendation as a two-step process. First the cladding fails (becomes perforated). The resulting oxidation of the $UO_2$ fuel pellets and formation of other lower-density minerals (various uranium oxyhydrides) causes the pellet to swell and the cladding to split or "unzip."

**Cladding Failure Prior to Arrival at Yucca Mountain**

Some fuel rods may have failed prior to arrival at the repository. Such breakdowns can occur during reactor operation, pool storage, dry storage, or transportation. Siegmann (2000a) reviewed these mechanisms and concluded that about 0.095 percent, the median value, of the fuel would have failed prior to its arrival at the repository. That value is more conservative than actual fuel reliability after 1990, which experienced a 0.015 percent rod failure rate. The modes of this expected failure distribution are summarized in table 19.2 (Siegmann 2000a). Not all spent fuel has Zircaloy cladding. About 1.1 percent of commercial fuel inventory is clad in stainless steel, which was the material of choice before zirconium alloys became widely available. Because stainless steel is less durable than Zircaloy, the DOE assumes that all of these stainless steel clad rods will have failed (table 19.2).

Ordinary reactor operation will take a small toll on the fuel. The fuel failure rate that the DOE uses to model future performance is based on historical fuel reliabilities, and thus is conservative. In fact, in-reactor fuel reliability studies (Siegmann 2000a) show a significant improvement in fuel performance with time (figure 19.2).

**Table 19.2**
Components of the failed cladding at emplacement in the site recommendation

| Rod failure mode | Site recommendation |
| --- | --- |
| Reactor operation failures | 0.095% (range 0.016 to 1.3) |
| Pool storage | 0.0 |
| Dry storage | 0.045 |
| Transportation (vibration, impact) | 0.01 |
| Stainless steel cladding | 1.1 |
| Stress corrosion cracking (post–reactor operation) | 0.5 |
| Total | 1.75% (range 1.67 to 3.0) |

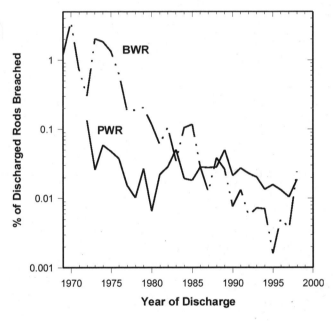

**Figure 19.2**
Fuel reliability as a function of calendar year.

The site recommendation assumed a median failure rate of 0.04 percent (range of 0.016 to 1.23 percent). This value is based on data from 112,000 assemblies, which contained fourteen million fuel rods.

Once the fuel leaves the reactor, its chances of failure continue at a low level. Numerous studies (e.g., Bradley et al. 1981) have determined that no cladding degradation mechanisms are active during pool storage. There is little information available on fuel rod failure occurring during dry storage. The Idaho National Engineering and Environmental Laboratory dry storage program (McKinnon and Doherty 1997) observed an overall failure rate of 0.045 percent (twelve failures of 26,500 rods), mostly due to rod consolidation operations, an experimental process not used in the nuclear energy industry. Some of the nonconsolidated rods are now undergoing metallurgical examination after seventeen years of storage, and no degradation has been observed thus far. During normal shipping of fuel, no failures have been described in the literature. Sanders and colleagues (1992) performed a statistical structural analysis on fuel rods and assemblies. They predicted that if a shipping cask were dropped nine meters (a severe transportation accident), the failure probability would be approximately $2 \times 10^{-4}$ per rod. The TSPA assumed a failure rate of $1 \times 10^{-4}$ during transportation to Yucca Mountain.

## Creep Failure under Repository Conditions

Creep is the slow deformation of a metal under high stresses and temperatures occurring over a period of years. In the repository, the stress of the fuel rod internal pressure could cause an outward ballooning of the cladding. The internal rod pressure is caused by gases produced by both the fission of uranium and the helium injected into the rods during manufacturing. Creep has been identified as the primary degradation mechanism during dry storage. As currently designed, the repository would be relatively cool, and cladding temperatures would generally not rise above 240°C. Therefore, no fuel rods are expected to fail from creep during dry storage or in the repository. Sensitivity studies to assess the effect of one or more variables on the total results gave a more detailed understanding of the likelihood of creep failure (Siegmann 2000b). These studies showed that cladding would start to fail if its temperature rose above 550°C during vacuum drying (the process used to remove traces of water from the fuel when it is removed from the spent fuel pool). Failure would also begin if peak cladding temperatures during either dry storage or repository emplacement exceeded 400°C.

## Corrosion Failure in the Repository

Corrosion is the deterioration of the fuel cladding as a result of chemical or electrochemical reactions with the contents inside a breached waste package. This in-package environment can include a combination of air, water vapor, and groundwater. Each of these components might be modified by the effect of heat and radiation, and by interactions with engineered and natural materials, forming an aqueous solution containing many ionic species.

Three forms of corrosion were considered in the site recommendation: general corrosion, localized corrosion, and stress corrosion cracking. Other types of corrosion, such as nodular corrosion, were ruled out because of the low steam temperatures. For general corrosion, the entire cladding surface is oxidized to $ZrO_2$ from oxygen in the air, water, or water vapor. After an initial corrosion phase, the oxide layer has an inner barrier of columnar zirconia crystallites, located on the metal surface. The corrosion rate is controlled by the diffusion of oxygen through the barrier layer. The corrosion rate is constant with time, but it varies with temperature. The current model for general corrosion, based on Hillner and colleagues (2000), predicts that if the waste package fails after 250 years of emplacement, about 1.3 μm of oxide will form (consuming 0.8 μm of cladding metal) because the repository temperatures will be sufficiently low (<150°C).

In the site recommendation, localized corrosion was modeled by assuming that the general corrosion by fluorine would occur with a low concentration of fluorine (< five parts per million) and was concentrated on a small area of the rod. This fluoride drill model was criticized as not being realistic and not addressing the most probable localized corrosion mechanism, ferric chloride ($FeCl_3$) pitting. A pitting model was developed that used electrochemical theory. The corrosion potential is modeled as a function of ferric ion, chloride ion, and hydrogen peroxide concentration using electrochemical values measured under various concentrations. This is compared with the repassivation potential to predict pitting (and assumed failure). When this model is applied to the current in-package chemistry analysis (including the formation of nitric acid and hydrogen peroxide from radiolysis), no pitting and no cladding failures from localized corrosion are predicted to occur.

Stress corrosion cracking, a chemical reaction at a crack tip (a preexisting flaw in the metal where the stress is concentrated), causes the crack to propagate through the metal. Chloride-induced stress corrosion cracking requires the breakdown of the passive layer to occur. This is the same electrochemical condition at which pitting occurs. Since pitting is not expected, stress corrosion cracking is also not expected.

## Mechanical Failures from External Loads

The cladding may fail from external forces or loads exceeding the strength of the material. The site recommendation included two different failure mechanisms. In one, the fuel fails from impact or mechanical loading against the waste package internals during a large earthquake. In the other scenario, failure would follow from the static weight of rubble that would result from a collapse of the repository tunnel ceiling (McCoy 1999). In the site recommendation, severe seismic events with a probability of occurrence of about one in a million per year were assumed to rupture all the cladding. Other analysis has suggested that after about 200,000 years, when the waste package is no longer protecting the cladding, the cladding would fail from the static loading from rock overburden (McCoy 1999).

## Hydrides

The oxidation of zirconium in water produces hydrogen ($Zr + 2H_2O \rightarrow ZrO_2 + 2H_2$). About 15 percent of this hydrogen is absorbed into the Zircaloy. When the metal becomes saturated with hydrogen, brittle zirconium hydrides are precipitated. These hydrides usually take the form of platelet-shaped grains that are oriented in the circumferential direction such that they do not weaken the material. Five specific hydride failure mechanisms were separately evaluated (Siegmann 2000c): delayed hydride cracking, reorientation of the hydrides into the radial direction, hydrogen embrittlement from waste package corrosion, hydrogen embrittlement from cladding corrosion, and hydrogen embrittlement from galvanic reaction (accelerated corrosion when different metals are electrically coupled). The site recommendation rules out the possibility of both the dissolution of hydrides at temperatures higher than 300°C and the reorientation in the radial direction by the action of hoop stresses (circumferential stress) on cooling. The reason for excluding these two mechanisms is because estimates of stress distributions and the maximum temperature for the cladding do not necessarily capture the uncertainties and variability across the repository. All other mechanisms were also eliminated.

## Cladding Axial Splitting

The cladding that has failed generally has small perforations. This results in small areas of fuel being exposed to the air, water vapor, and groundwater. Oxygen and water can react with the $UO_2$ fuel pellets, yielding uranium oxyhydroxides with lower densities than the $UO_2$. The formation of these alteration phases is postulated to cause swelling that tears the cladding open, exposing more fuel, which leads to more swelling and so forth in a process known as axial splitting or unzipping. Before splitting occurs, the radioisotopes in the gap between the fuel pellets and the cladding are released into the waste package. This release involves 4 percent by weight of the fuel's iodine and 1.4 percent by weight of its cesium. Initially, 0.2 percent of the $UO_2$ in the rod is also released. The estimate of the initial release is based on experiments (Wilson 1990) that measured the releases of radionuclides from damaged cladding under possible repository conditions.

Axial splitting is observed in boiling water reactors and is associated with hydrides cracking in the cladding under anoxic (oxygen-depleted) conditions. In autoclave tests at Argonne National Laboratory, similar splits in cladding were observed over the full length of both samples of failed fuel segments exposed to steam at 175°C. These splits appeared after about 1.5 years. Earlier TSPAs contained an analysis of the axial splitting of the cladding. After two tests showed rapid splitting, efforts to model the splitting were abandoned and instant splitting was assumed. This leaves the fragmented fuel pellets exposed and available for corrosion.

## Effect of Cladding Durability on Performance Assessment Results

For the two most important radioisotopes in the nominal release scenario, technetium and neptunium, commercial spent nuclear fuel supplies over 86 percent and 95 percent of the total repository inventory, respectively. Hence, the performance of the cladding on the spent fuel from commercial reactors directly influences the calculated dose by reducing the release rate relative to that of bare fuel. Figure 19.3 compares the dose for the current cladding model case and the bare fuel case (no cladding). These dose rates are calculated at twenty kilometers from the repository and include all the other barriers, such as sorption and dilution. It should be noted that because few waste packages are predicted to fail, the dose rate is lower than 0.001 millirems per year even beyond 10,000 years.

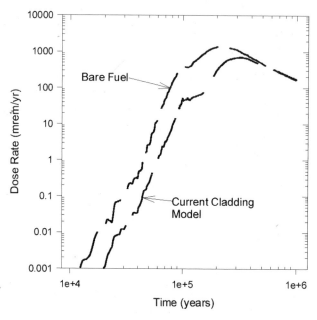

**Figure 19.3**
Comparison of doses using two different cladding models.

In the bare fuel case, the dose is controlled by the failure rate of the waste pack-ages, the solubility limits for certain radionuclides such as plutonium, and the rate of diffusion of the radionuclides through the patches in the waste packages. These effects also contribute to dose reduction in the case with cladding. Overall, the model calculates that for the first 100,000 years, cladding reduces the dose by an order of magnitude as compared with bare fuel. In the site recommendation, tunnel collapse and cladding failure from rock overburden account for the increase in dose after 200,000 years (seen in figure 19.3) because by this time most of the waste packages have corroded and been breached, and the waste package no longer protects the cladding. Nevertheless, after 200,000 years the presence of cladding still reduces the peak dose by 50 percent. This is consistent with the sensitivity studies that show that the peak dose to the critical group is controlled by the frac-tion of cladding that is calculated to have failed.

The various components of the cladding model were analyzed in sensitivity studies as summarized in figure 19.4. The base case refers to the cladding model described

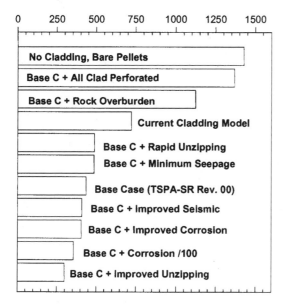

Peak Dose (mrem/yr)

**Figure 19.4**
Sensitivity of dose from various cladding model components. The base case represents the TSPA-site recommendation case. The current cladding model was used in the Supplemental Science and Performance Analyses.

and used in the site recommendation (rev. 00). Cladding degradation is modeled in two steps: perforation (failure) and axial splitting. Sensitivity studies show that significant dose reduction is achieved from unfailed fuel being unavailable for dissolution (compare the "base case + all perforated" to "bare fuel" in figure 19.4). After 200,000 years, due to the weight of the rock overburden, most of the cladding starts to fail and the peak dose approaches the "all cladding perforated" case. The details of the unzipping mechanism and $UO_2$ dissolution rate are less important, probably because the mechanism currently considered is rapid relative to the TSPA time scale (compare the "base case + rapid unzipping" to the "base case"). Sensitivity studies of different corrosion models and improved seismic risk models show little effect on the calculated dose. All of these sensitivity studies predict peak doses in excess of the EPA standards (fifteen millirems per year), but the peak dose occurs after 200,000 years, well after the regulatory period (now remanded) of 10,000 years.

## Conclusions: Uncertainties in the Cladding Model

Zircaloy cladding has been extensively studied over the last fifty years. The aim of these studies has been the improvement of the fuel performance in reactors. Studies of the behavior expected for twenty years of dry storage exist, but for disposal the research effort is scarce. This becomes an important issue for high burn-up fuel as well as many variations in the stockpile of spent fuel. Such variations are found in the DOE-owned spent fuels that exist on the weapons complex. They include cladding chemical composition (e.g., low-tin Zircaloy), tube processing (e.g., annealing treatments), dual wall cladding (e.g., an inner zirconium barrier to control pellet-clad interactions, a failure method associated with power changes in boiling water reactors), among other fabrication changes. The nuclear fuel suppliers continue to study cladding. This work is reported in the American Society for Testing and Materials' meeting on zirconium in the nuclear industry, the American Nuclear Society's topical meeting on light water reactor performance, the *Journal of Nuclear Materials*, and other technical publications. Zirconium research for the chemistry industry is reported at meetings of the National Association of Corrosion Engineers and published in the journal *Corrosion*. The Argonne National Laboratory is performing a series of creep tests as part of the dry storage research program. In France, the Commissariat a L'Energie Atomique has a similar research program.

In the TSPA–site recommendation, the greatest uncertainty is in the extrapolated behavior of cladding for the extremely long period (more than ten thousand years) under repository conditions. As shown by the sensitivity analyses, the most important issue is the prediction of the fraction of cladding that fails as a function of time. The greatest uncertainty in this prediction lies with the potential effects of localized corrosion. The measured rates of general corrosion are extremely slow. The uncertainty in the estimates of localized corrosion comes from the difficulty in estimating the probability of local chemical conditions that may become extremely acidic; such an acidic environment could contain chemical species that accelerate corrosion (e.g., $FeCl_3$ that can cause pitting). These chemical and pH variations may be caused by a combination of microbial reactions, radiolysis due to radiation, and corrosion of waste package internals. The additional modeling of localized corrosion using recent test results by Cragnolino and colleagues (2001) and studies of microbial colonies at the repository (Jolley and Screiber 2001) are ongoing. These efforts will be combined with the results of models of the in-package chemistry analysis (Brady 2001) in order to predict the probability of failure by pit penetration. This modeling has been completed for inclusion in the Yucca Mountain repository license application.

There is considerable uncertainty in the details of the cladding axial splitting model (anoxic axial splitting versus wet axial splitting), but the details of the mechanism would not affect the performance assessment results. Wet axial splitting has not been observed with failed fuel after twenty years in spent fuel pools, and anoxic axial splitting has been observed within one and a half years in Argonne National Laboratory fuel segment tests. Instant splitting is currently being assumed.

In summary, the degradation of the Zircaloy cladding is of limited importance because it has only a small effect on the safety of the repository. For the first 100,000 years after closure, the inclusion of cladding as a barrier appears to reduce the dose by about an order of magnitude as compared with the calculated release from bare fuel. The peak dose, which occurs after two hundred thousand years, is reduced by 50 percent. Zirconium metal and Zircaloy cladding have been extensively studied in both the chemical and nuclear industries, and there is a sufficient technical basis for their inclusion in repository modeling to conduct performance assessment computations.

## References

Bradley, E.R., Bailey, W.J., Johnson, A.B., Jr., and Lowry, L.M. (1981) *Examination of Zircaloy-Clad Spent Fuel after Extended Pool Storage.* Pacific Northwest Laboratory. PNL–3921.

Brady, P. (2001) *In-Package Chemistry for Waste Forms.* CRWMS M&O, Las Vegas. ANL–EBS–MD–000056 REV 00.

Cragnolino, G., Brossia, C.S., Dunn, D.S., and Greene, C.A. (2001) General and Localized Corrosion of Zircaloy under High-Level Radioactive Waste Disposal Conditions. Paper presented at the tenth International Conference on Environmental Degradation of Materials in Nuclear Power Systems–Water Reactors, Lake Tahoe, Nevada, August 5–9.

CRWMS M&O (Civilian Radiactive Waste Management System Management and Operating Contractor) (1998) *Total System Performance Assessment–Viability Assessment (TSPA–VA) Analyses Technical Basis Document.* Department of Energy, Las Vegas.

Einziger, R.E., and Kohli, R. (1984) Low-Temperature Rupture Behavior of Zircaloy-Clad Pressurized Water Reactor Spent Fuel Rods under Dry Storage Conditions. *Nuclear Technology* 67, pp. 107–123.

Electric Power Research Institute (EPRI) (2000) *Evaluation of the Candidate High-Level Radioactive Waste Repository at Yucca Mountain Using Total System Performance Assessment Phase 5.* Electric Power Research Institute, Palo Alto, California. EPRI TR-1000802.

Henningson, P.J. (1998) *Cladding Integrity under Long-Term Disposal.* Framatome Technologies, Lynchburg, Virginia. 51-1267509-00.

Hillner, E., Franklin, D.G., and Smee, J.D. (2000) Long-Term Corrosion of Zircaloy before and after Irradiation. *Journal of Nuclear Materials* 278, pp. 334–345.

Jolley, D.M., and Screiber, J.D. (2001) Modeling of In-Drift Microbial Communities at Yucca Mountain, Nevada. In *Proceedings of the 9th International Conference on High-Level Radioactive Waste Management.* La Grange Park, IL: American Nuclear Society.

McCoy, J.K. (1999) *Breakage of Commercial Spent Fuel Cladding by Mechanical Loading.* CRWMS M&O, Las Vegas. CAL-EBS-MD-000001 REV 00.

McKinnon, M.A., and Doherty, A.L. (1997) *Spent Nuclear Fuel Integrity during Dry Storage: Performance Tests and Demonstrations.* Pacific Northwest Laboratory. PNNL–11576.

McNeish, J.A., et al. (2000) *Total System Performance Assessment for the Site Recommendation.* CRWMS M&O, Las Vegas. TDR–WIS–PA–000001 REV 00, ICN 01.

Office of Civilian Radioactive Waste Management (2001) *Yucca Mountain Science and Engineering Report.* Department of Energy. DOE/RW–0539. May.

Peehs, M. (1998) Assessment of Dry Storage Performance of Spent LWR Fuel Assemblies with Increasing Burn-Up. Coordinated Research Program on Spent Fuel Performance Assessment and Research, Erlangen, Germany: Bereich Energieerzeugung. April 20–24.

Pescatore, C., Cowgill, M.G., and Sullivan, T.M. (1990) *Zircaloy Cladding Performance under Spent Fuel Disposal Conditions Progress Report, May 1–October 31, 1989.* Brookhaven National Laboratory. BNL 52235.

Rothman, A.J. (1984) *Potential Corrosion and Degradation Mechanisms of Zircaloy Cladding on Spent Nuclear Fuel in a Tuff Repository.* Lawrence Livermore National Laboratory. UCID–20172.

Sanders, T.L., Seager, K.D., Rashid, Y.R., Barrett, P.R., Malinauskas, A.P., Einziger, R.E., Jordan, H., Duffey, T.A., Sutherland, S.H., and Reardon, P.C. (1992) *A Method for Determining the Spent-Fuel Contribution to Transport Cask Containment Requirements.* Sandia National Laboratories. SAND90–2406.

S. Cohen and Associates (1999) *Effectiveness of Fuel Rod Cladding as an Engineered Barrier in the Yucca Mountain Repository.* S. Cohen and Associates, McLean, Virginia.

Siegmann, E.R. (2000a) *Initial Cladding Condition.* CRWMS M&O, Las Vegas. ANL–EBS–MD–000048 REV 00 ICN 01.

Siegmann, E.R. (2000b) *Clad Degradation: Summary and Abstraction.* CRWMS M&O, Las Vegas. ANL–WIS–MD–000007 REV 00 ICN 01.

Siegmann, E.R. (2000c) *Clad Degradation: FEPs Screening Arguments.* CRWMS M&O, Las Vegas. ANL–WIS–MD–000008 REV 00 ICN 01.

Wilson, C.N. (1990) *Results from NNWSI Series 3 Spent Fuel Dissolution Tests.* Pacific Northwest Laboratory. PNL-7170.

Yau, T.L., and Webster, R.T. (1987) Corrosion of Zirconium and Hafnium. In *Corrosion,* 13, ASM Handbook. Materials Park, OH: ASM International, pp. 707–721.

# 20

# Spent Fuel

Jordi Bruno and Esther Cera

For nations that use nuclear power, the fate of the used nuclear fuel is a major concern. The United States has chosen direct geologic disposal of spent fuel after once-through burn up in a nuclear reactor. This strategy avoids reprocessing—the reclamation of plutonium from the spent fuel—which is an important means of preventing the proliferation of fissile material. Among the other countries that favor the direct geologic disposal of spent fuel are Canada, Sweden, Finland, and Spain. Uranium dioxide ($UO_2$), the main component of spent fuel (>96 percent), has the same chemical composition and structure as uraninite or pitchblende—minerals that are abundant in natural uranium deposits. Hence, there is considerable information available on the long-term behavior of $UO_2$ in these deposits.

The proposed nuclear waste repositories will provide either a reducing environment, which is characterized by the absence of oxygen, or an oxidizing environment, where oxygen is always present. The stability of the major component of the spent fuel, $UO_2$, is very different in these two environments. Under reducing conditions, $UO_2$ is stable and insoluble; in the presence of oxygen, it destabilizes and becomes soluble. The proposed repository at Yucca Mountain is an oxidizing regime because it is located approximately three hundred meters above the water table. This has critical implications for the stability and behavior of the spent fuel.

Five kinds of radioactive waste would be stored at Yucca Mountain (DOE 2001):

• Spent fuel from commercial nuclear power plants

• Spent fuel and vitrified high-level waste from the production of nuclear weapons, including naval spent fuel, spent fuel from the Fort Saint Vrain reactor, irradiated debris from the Three Mile Island II reactor, and uranium metal fuel

• Spent fuel from research reactors operated by the Department of Energy; (DOE)

• High-level waste from reprocessing facilities generated before 1982
• Small quantities of spent fuel from foreign reactors for which the United States has maintained ownership.

At the end of 1999, the United States had accumulated about forty thousand metric tons coming from commercial spent nuclear fuel. This amount could more than double by 2035 if all currently operating plants complete their initial forty-year license period (CRWMS M&O 2000). According to these estimates, about 60 percent of commercial spent fuel assemblies generated by 2040 will be from boiling water reactors and 40 percent from pressurized water reactors (CRWMS M&O, 2000). In addition, as much as thirty-four metric tons of surplus weapons-grade plutonium will be fabricated into uranium-plutonium fuel and irradiated in commercial reactors. Use of this so-called mixed-oxide fuel will be limited to only a few specific commercial reactors (CRWMS M&O 1999).

**Physical and Chemical Properties of the Spent Fuel**

The diverse types of spent fuel that the DOE is planning to dispose of exhibit a wide range of physical characteristics, such as length, cross-section, weight, and cladding. Table 20.1 summarizes the dimensions and weights of boiling water reactor assemblies, and table 20.2 provides similar information for pressurized water reactor assemblies (CRWMS M&O 1999).

Most commercial spent fuel assemblies would arrive at the potential repository undamaged and suitable for immediate disposal. The fuel assemblies are rods composed of $UO_2$ compacted into cylindrical pellets eight to ten millimeters in diameter and nine to fifteen millimeters long. The assemblies are constructed from noncorrosive cladding materials (mainly Zircaloy and stainless steel), and they will be transported in waste packages (see chapter 17, this volume).

The radioactivity in uranium dioxide comes from the isotopes of the uranium itself as well as uranium's radioactive decay products. The rate at which a radioactive isotope, or radionuclide, decays is given by its half-life—the time it takes for half of the initial amount to transform into another element. Half-lives vary from fractions of a second to billions of years. After about ten half-lives, about 99.9 percent of a radionuclide is gone. Uranium-238 and uranium-235 have half-lives of 4.5 and 0.7 billion years, respectively. Natural uranium consists of mostly uranium-238 (99.3 percent) with only 0.7 percent of uranium-235, but uranium-235 is the

**Table 20.1**
Design basis dimensions and weight for boiling water reactor assemblies

| Assembly group | Length (inches) | Section (inches) | Weight (pounds) | Primary cladding | % of BWR assemblies | Total BWR assemblies |
|---|---|---|---|---|---|---|
| Big Rock Point | 81.6–84.8 | 6.50–7.21 | 457–591 | Zircaloy-2 | <1 | 524 |
| Misc. shutdown reactors | 95.0–141.4 | 4.00–6.31 | 276–480 | Zircaloy-2 | 1 | 1,615 |
| General Electric | 171–178 | 5.24–6.07 | 556–725 | Zircaloy-2 | 99 | 164,800 |
| All BWR assemblies | 81.6–178.0 | 4.00–7.21 | 276–725 | | 100 | 166,939 |

**Table 20.2**
Design basis dimensions and weight for pressurized water reactor assemblies

| Assembly group | Length (inches) | Section (inches) | Weight (pounds) | Primary cladding | % of BWR assemblies | Total PWR assemblies |
|---|---|---|---|---|---|---|
| Misc. shutdown reactors | 111.7–140.2 | 6.27–8.60 | 437–1612 | 304 Stainless steel | 2 | 2,460 |
| Westinghouse, Babcock & Wilcox and others | 146.0–173.5 | 7.76–8.64 | 1096–1705 | Zircaloy-4 | 84 | 105,500 |
| Combustion Engineering & South Texas | 176.8–201.2 | 8.03–8.53 | 1430–1945 | Zircaloy-4 | 14 | 16,900 |
| All PWR Assemblies | 111.7–201.2 | 6.27–8.64 | 437–1945 | | 100 | 124,860 |

isotope that fissions or splits, giving off energy and neutrons in the process. Many power reactors, including the light water type used in the United States, use a fuel slightly enriched in uranium-235, up to about 4 percent.

At the end of the reactor operation, about 95 to 99 percent of the spent nuclear fuel still consists of $UO_2$. The rest of the material consists of fission products, activation products, and actinides. During the operation of a nuclear power reactor, nuclei of uranium-235 fission to form new elements, known as fission products. Most of the fission products have short half-lives (approximately thirty years), but others that are formed in significant amounts have long half-lives: 200,000 years for technetium-99 and 1.7 million years for iodine-129, for instance. Examples of other fission products are iodine-131, cesium-134, cesium-135, cesium-137, and strontium-90. Because of their relatively short half-lives, fission products increase the radioactivity of the fuel. Also in spent fuel are activation products—the result of the interaction of neutrons from the fission process with metals of the fuel assembly, and include the radioactive isotopes of nickel, cobalt, and others elements.

The actinides are formed when the heavy elements uranium and plutonium absorb neutrons without splitting. These elements become heavier than uranium due to the "absorbed" neutrons, and they are called the transuranic elements. The most important are plutonium-238, -239, -240, -241, and -242; neptunium-237; and the isotopes of americium and curium: americium-241, americium-243, curium-243, and curium-244. As a result of natural disintegration, the amount of these long-lived radionuclides may increase after the spent fuel is removed from the reactor operation. This phenomenon is called ingrowth. The most abundant actinide is plutonium, which may compose up to 1 percent of the total mass of the spent fuel.

The amount of radionuclides contained in the spent fuel dictates the thermal output, its chemical and radiotoxicity, and the persistence over time of these effects. The radionuclide inventory depends partly on the type and amount of the initial radioactive elements contained in the fuel. Another crucial factor is the history of the fuel within the reactor—that is, the burn up. The burn up is particularly important in defining the types and amounts of long-lived isotopes that will be produced. Table 20.3 shows the inventory at different times after discharge of the nuclear reactor for fuel in a pressurized water reactor with a typical burn up of thirty-three megawatt-days per kilogram of uranium (Johnson and Shoesmith 1988).

Figure 20.1 shows the relative activity of Swedish spent fuel, with an average burn up of thirty-eight megawatt-days per kilogram of uranium, as a function of time. The

Table 20.3

Calculated radionuclide inventory for pressurized water reactor used fuel with burn up of thirty-three megawatt-days per kilogram of uranium, in milicuries per kg uranium

| Time since discharge, years | | | | |
|---|---|---|---|---|
| Nuclide | 10 | 100 | 1,000 | 10,000 |
| *Major fission and activation products:* | | | | |
| H-3 | 464 | 3 | | |
| C-14 | 1.55 | 1.53 | 1.38 | 0.46 |
| Ni-59 | 5.15 | 5.15 | 5.11 | 4.72 |
| Ni-63 | 652 | 331 | 0.38 | |
| Se-79 | 0.41 | 0.41 | 0.40 | 0.37 |
| Kr-85 | 4,850 | 14.4 | | |
| Sr-90 | 57,200 | 6,710 | | |
| Zr-93 | 1.93 | 1.93 | 1.93 | 1.92 |
| Nb-94 | 1.28 | 1.28 | 1.24 | 0.91 |
| Tc-99 | 13 | 13 | 13 | 12.6 |
| Pd-107 | 0.11 | 0.11 | 0.11 | 0.11 |
| Sn-126 | 0.78 | 0.78 | 0.77 | 0.72 |
| I-129 | 0.032 | 0.032 | 0.032 | 0.032 |
| Cs-135 | 0.345 | 0.345 | 0.345 | 0.345 |
| Cs-137 | 82,100 | 10,300 | | |
| Sm-151 | 331 | 166 | 0.16 | |
| Totals | 150,312 | 17,551 | 24.9 | 22.2 |
| Listed isotopes | | | | |
| All isotopes | 307,000 | 33,900 | 27.6 | 24.9 |
| *Actinides and their decay products:* | | | | |
| Ra-226 | | $2.66 \cdot 10^{-5}$ | $3.12 \cdot 10^{-3}$ | 0.134 |
| Th-230 | | 0.001 | 0.017 | 0.168 |
| U-233 | | $1.7 \cdot 10^{-4}$ | $3.22 \cdot 10^{-3}$ | 0.048 |
| U-234 | 1.2 | 1.6 | 2.03 | 1.99 |
| U-235 | 0.02 | 0.02 | 0.02 | 0.02 |
| U-236 | 0.26 | 0.26 | 0.27 | 0.35 |
| U-238 | 0.318 | 0.318 | 0.318 | 0.318 |
| Np-237 | 0.32 | 0.42 | 1.0 | 1.18 |
| Np-239 | 17.1 | 16.9 | 15.6 | 6.7 |
| Pu-238 | 2,330 | 1,150 | 1.08 | |
| Pu-239 | 313 | 312 | 305 | 237 |
| Pu-240 | 527 | 526 | 478 | 184 |
| Pu-241 | 77,600 | 1,020 | | |
| Pu-242 | 1.76 | 1.76 | 1.72 | 1.69 |
| Am-241 | 1,690 | 3,750 | 893 | |

**Table 20.3** (continued)

| Nuclide | Time since discharge, years | | | |
|---|---|---|---|---|
| | 10 | 100 | 1,000 | 10,000 |
| Am-242 | 7 | 4.6 | 0.1 | |
| Am-243 | 17.1 | 16.9 | 15.6 | 6.7 |
| Cm-242 | 5.7 | 3.8 | 0.1 | |
| Cm-243 | 16.6 | 1.86 | | |
| Cm-244 | 1,320 | 42.1 | | |
| Cm-245 | 0.39 | 0.39 | 0.36 | 0.17 |
| Cm-246 | 0.11 | 0.11 | 0.10 | 0.03 |
| Totals | | | | |
| Actinides | 83,855 | 6,854 | 1,714 | 440 |
| Grand total | 390,855 | 40,754 | 1,742 | 465 |

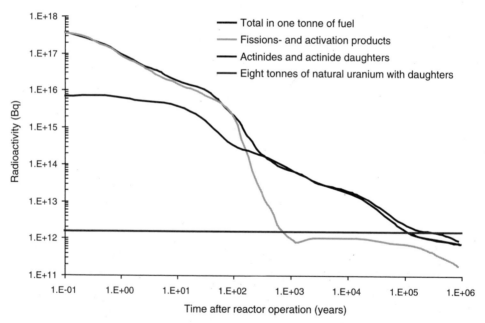

**Figure 20.1**
Activity of a boiling water reactor fuel with a burn up of thirty-eight megawatt-days per kilogram of uranium. Source: Säkerhets Redovisning 97 1999. Bq stands for becquerel = disintegration per second.

relative contributions of fission products, activation products, and actinides are also shown. Similar decay curves could be drawn for the various U.S. spent fuel types.

## Distribution of the Radionuclides in Spent Fuel

Radionuclides are mostly embedded in the $UO_2$ matrix, although they are not homogeneously distributed. During reactor operation, the fuel temperature reaches 1700°C. This heat causes the lighter elements (and some gases) to diffuse to the surface of the $UO_2$ pellets. On contact with groundwater, the lighter radionuclides that have preferentially migrated to the surface, such as iodine and cesium, are preferentially released. This is because these elements are not only physically available but also soluble in water. Some of these elements diffuse into the gap between the fuel pellets and the metal cladding. Consequently, their radioactive isotopes will be important contributors to the initial radioactive dose. Some of the remaining radionuclides generated during reactor operation are homogeneously mixed in the $UO_2$ matrix (forming solid solutions), such as the actinides (plutonium, americium, and neptunium) and the lanthanides (cerium and europium).

Some of the fission products exist as small metallic inclusions. Among these elements are technetium, palladium, ruthenium, and molybdenum, with half-lives ranging from 3,500 to 6.5 million years. These inclusions are insoluble in water and are an important constituent of spent nuclear fuel's radiotoxicity during a repository's middle years (more than one thousand years). Figure 20.2 shows a micrograph

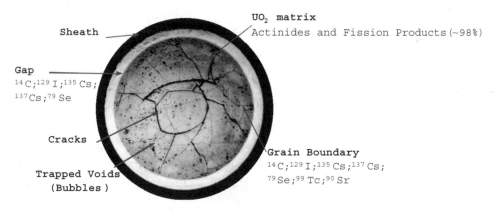

**Figure 20.2**
Cross-section of a spent fuel pellet.

of a used fuel pellet, indicating the distribution of fission products, activation products, and actinides.

## Thermal Output of Spent Fuel

Spent fuel arriving at the repository from commercial reactors is expected to have a wide range of thermal output. The heat generated from the different types of spent fuel will affect the physical and chemical evolution of the repository over time. This is particularly true if the spent fuel is disposed of shortly after discharge from the reactor, when it is hot from the energy produced by the decay of very short-lived radionuclides.

The key factors used to determine the thermal output of the spent fuel are the burn up, the number of years it has spent out of the reactor, and the initial amount of fissile material (uranium-235 or plutonium) in the spent fuel. The variable heat output given by these factors will affect the corrosion process.

## How Will Spent Fuel Behave in Yucca Mountain?

Spent nuclear fuel is a polycrystalline ceramic that is not soluble in water when kept under the physical and chemical conditions that prevent the transformation of $UO_2$ to other solid phases. Because most radionuclides are embedded in the $UO_2$ matrix, the potential release and migration of radionuclides from the repository depends on the stability of that matrix.

The key physical parameters controlling the stability of spent fuel are temperature and contact with water. The temperature depends on the heat output, which is a function of the repository design and fuel burn up. Figure 20.3 shows the strong temperature dependence of the stability of the spent fuel matrix. The amounts of water and the timing of its contact with the spent fuel depend mainly on the hydrologic properties of the repository site (see chapters 10 and 11, this volume). The main chemical parameters controlling the stability of spent fuel are the acidity, oxidation potential, and chemical composition of the contacting waters (see figures 20.4 and 20.5). For a given pH, the solubility of the $UO_2$ increases dramatically with the increasing oxidation potential (figure 20.4). Also, if the acidity of the seeping water is high, the $UO_2$ becomes soluble even under reducing conditions. Hence, the spent fuel matrix is more stable under reducing conditions and at neutral or slightly alkaline (high pH) conditions.

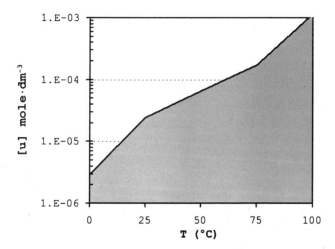

Figure 20.3
Effect of the temperature on the solubility of the spent nuclear fuel matrix ($UO_2(s)$) Eh = 0, groundwater composition: J-13 well water.

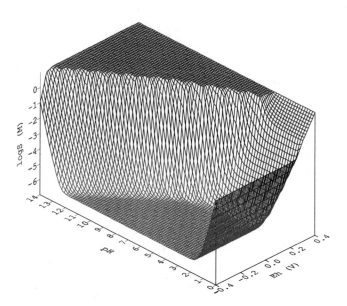

Figure 20.4
Surface plot showing the $UO_2(s)$ solubility dependence on pH and Eh, T = 25°C.

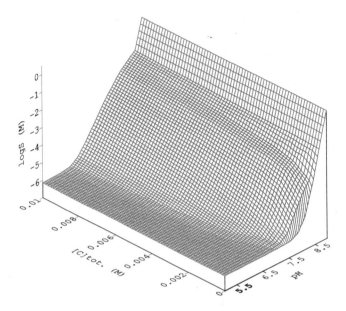

**Figure 20.5**
Surface plot showing the solubility of $UO_2(s)$ as a function of carbonate content in the ground water and pH, Eh = 0, T = 25°C.

The chemical constituent of the contacting water that most affects the stability of the $UO_2$ is carbonate ($CO_3^{2-}$). The higher the concentration of carbonate in the water, the less stable the $UO_2$ (figure 20.5). This effect arises because aqueous carbonate complexes of uranium in its oxidized form ($U^{6+}$) destabilize the spent nuclear fuel matrix. The greater the carbonate content in the groundwater, the higher the solubility of the spent fuel matrix.

These chemical parameters are mainly controlled by the geochemical conditions of the repository and the chemical evolution of water in the immediate vicinity of the spent fuel. At Yucca Mountain, the groundwater is slightly alkaline (pH = 7.4), with a moderate bicarbonate content. This is typical of the composition of water from Yucca Mountain's J-13 well, a standard groundwater from the site (table 20.4).

Any groundwater at the repository horizon will be heated by the thermal output of the spent fuel assemblies. Figure 20.6 illustrates the calculated temperature regime during the initial times of the Yucca Mountain repository. These estimates show that the vicinity of the spent fuel assemblies will be hot (50°C to 250°C).

These warmer temperatures will change the composition of the groundwater around the repository due to evaporation, mineral precipitation, and dissolution

**Table 20.4**
Major element composition of J–13 well water

| Component | Concentration (ppm) | Component | Concentration (ppm) |
| --- | --- | --- | --- |
| Na | 45.8 | HCO3 | 128.9 |
| K | 5.04 | SiO2(aq) | 61.0 |
| Ca | 13.0 | SO4 | 18.4 |
| Mg | 2.01 | F | 2.18 |
| pH = 7.41 | | Cl | 7.14 |

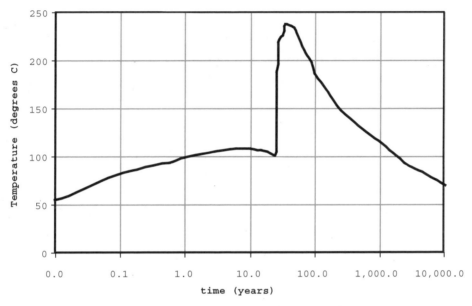

**Figure 20.6**
Temperature of the waste package outer barrier surface as a function of time, 21-pressurized water reactor design basis with twenty-five years of ventilation, without backfill (taken from CRWMS M&O 2000).

processes. Hence, these waters will become more concentrated in salts. Because the repository sits above the water table, atmospheric oxygen and carbon dioxide will be dissolved in these waters, increasing both their oxidation potential and carbonate content. Because of this chemical composition, the Yucca Mountain groundwaters that come into contact with the fuel after corrosion of the container will accelerate the fuel's oxidation and dissolution. The higher temperatures will also have an impact: the warmer the water, the faster the corrosion. The combination of oxygenated, hot, and concentrated fluids leads to the rapid degradation of the fuel.

Another important process that may affect the integrity of the spent fuel matrix and enhance the release of the radionuclides is the radiolysis of water due to the effect of radiation. The main result is the generation of chemically reducing and, more important, oxidizing species, such as $O_2$ and $H_2O_2$. Nevertheless, the Yucca Mountain repository is already under oxidizing conditions; oxygen will be present even without the radiolysis of water. There are a large number of experimental results concerning the oxidation and dissolution of spent nuclear fuel in the presence of atmospheric oxygen (Gray et al. 1993; Eriksen et al. 1995; Shoesmith et al. 1998; Torrero 1998; Bruno et al. 2000; references therein). All of these investigations indicate that atmospheric oxygen has a stronger oxidizing effect than do the products of water radiolysis. There is a large uncertainty, however, regarding the combined effects of temperature and oxidizing conditions on the degree of alteration of the spent fuel matrix.

## Alteration Processes of the Spent Fuel Matrix

The dissolution of the spent fuel matrix in contact with air follows a pattern that geologists and geochemists have studied for years when investigating the alteration of uranium deposits in contact with air (McKinley et al. 1988; Isobe et al. 1992; Finch and Ewing 1992; Murphy and Pearcy 1992; Chapman et al. 1993; Blomqvist et al. 2000). Furthermore, there have been numerous laboratory studies devoted to understanding the main processes affecting the stability of the $UO_2$ matrix under repository conditions (Grandstaff 1976; Stroes-Gascoyne 1985; Forsyth and Werme 1992; Wronkiewicz et al. 1992; Bruno et al. 1995; de Pablo et al. 1999; references therein). From all this information, we can propose the following general reaction scheme for the alteration of the spent fuel matrix under the oxidizing conditions at Yucca Mountain.

In the presence of air and water, the $UO_2$ surface is readily oxidized to $U(6^+)$ containing oxides:

$$UO_2 + x/2 \ O_2(gas) \rightarrow UO_{2+x} \tag{20.1}$$

This process oxidizes the uranium from a 4+ to 6+ state. The degree of oxidation depends on the temperature and the amount of water in contact with the spent nuclear fuel. The limit on the oxidation of $UO_{2+x}$ before dissolution is about $x = 0.33$, at ambient temperatures. The extent of the surface oxidation prior to dissolution at higher temperatures is not well-known, however. The higher the x value, the greater the proportion of $U(6^+)$ in the structure.

Independent of the degree of oxidation, the next step is the dissolution of the oxidized $U(6^+)$ as a result of the interaction with the carbonate molecules present in the Yucca Mountain groundwater and resulting from the water-rock interaction processes at higher temperatures:

$$UO_{2+x} + 2xCO_3^{2-} + 2xH^+ = xUO_2(CO_3)_2^{2-} + (1-x)UO_2 + xH_2O \tag{20.2}$$

Reaction (20.2) results in the dissolution of the oxidized surface of $U(6^+)$. The dissolution of the oxidized surface exposes more $UO_2$ to oxygen, and the cycle of oxidation and dissolution continues. This process is interrupted only when the concentration of dissolved $U(6^+)$ is high enough to trigger the precipitation of $U(6^+)$ phases on the surface of the corroding fuel.

Normally, the first phases to precipitate are $U(6^+)$ oxides and hydroxides, such as schoepite. This process can be approximated by the reaction:

$$UO_2(CO_3)_2^{2-} + (y+1) \ H_2O \Leftrightarrow UO_3 \cdot yH_2O + 2 \ CO_3^{2-} + 2 \ H^+ \tag{20.3}$$

Reactions (20.1) and (20.2) are in principle irreversible: they can only proceed in the direction of the arrow. These are nonequilibrium processes, governed by the rate at which a given reaction proceeds. Reaction (20.3), however, is an equilibrium process; even at low temperatures, it proceeds sufficiently fast in both directions. Regarding reaction (20.2), there is a large degree of uncertainty about the rates and mechanisms controlling the precipitation of U(6+) hydroxides. Key unknowns include the degree of oversaturation of U(6+) required to initiate the precipitation and the role that other surfaces play in the process.

The next step in the oxidative alteration of the spent fuel matrix depends strongly on the chemical composition of the contacting fluids. The sequence of alteration that is most commonly seen in natural systems (Finch and Ewing 1992) as well as laboratory studies (Wronkiewicz et al. 1992) is the conversion of the $U(6^+)$ oxides and hydroxides to silicates and phosphates, depending on the groundwater composition. One of the most common uranium silicates found in nature and laboratory experiments with $UO_2$ is uranophane, ideally $Ca[(UO_2)(SiO_3OH)]_2(H_2O)_5$. The reaction can be written:

$$2\ UO_3 \cdot yH_2O + 2\ H_4SiO_4 + Ca^{2+} + (3 - y)\ H_2O \Leftrightarrow$$
$$[(UO_2)(SiO_3OH)]_2(H_2O)_5 + 2\ H^+ \tag{20.4}$$

This process will proceed toward the formation of uranophane, which is more stable in the expected geochemical conditions of the groundwater at Yucca Mountain (see chapter 16, this volume). Nevertheless, the decrease of silicate or calcium concentrations with time in these waters might drive the reaction to the left.

These reaction steps—including their mechanisms, rates, and the stabilities of the uranium phases that form—are well-known at low to moderate temperatures (up to 90°C). Geochemical models that describe these processes have been tested against an extensive amount of information available from natural systems.

## Behavior of the Trace Elements Contained in the Spent Fuel

The alteration and dissolution of the spent fuel leads to the release of radionuclides contained in the $UO_2$. An important issue for repository safety is how these radionuclides will behave as the matrix corrodes and alters. Trace elements in natural rocks can be used to approximate the behavior of radionuclides.

Studies of natural uranium deposits have provided important insights into the behavior of trace elements during the alteration of $UO_2$ (Murphy and Pearcy 1992; Chapman et al. 1993; Louvat et al. 2000; Blomqvist et al. 2000). These sites exist in a variety of geochemical conditions, from the reducing environment of the Oklo uranium deposits in Gabon, to the oxidizing environment of the Peña Blanca deposit in Mexico. The geochemical modeling of the behavior of trace elements in these well-characterized natural systems has provided crucial data on the main processes affecting radionuclide migration in a repository system (Bruno et al. 2002).

Some radionuclides can be retained in secondary phases that are formed during the alteration process (Jensen and Ewing 2001). Laboratory tests have documented that strontium and cesium in particular are retained in secondary alteration phases (Finn et al. 1996). Actinides and lanthanides are not as easily incorporated into these new phases. These new phases may increase the mobility of these radionuclides, affecting the expected radiation dose to humans.

In addition, the process in reaction (20.1) that causes the oxidation of the $UO_2$ matrix will also change the oxidation state of other elements. Notably, neptunium—which is in the ($4^+$) redox state as $NpO_2$ and similar to $UO_2$ in the spent fuel matrix—will be oxidized to $Np(5^+)$ according to the reaction:

$$NpO_2 + 0.25 \ O_2(gas) + H^+ \rightarrow NpO_2^+ + 0.5 \ H_2O \qquad (20.5)$$

$Np(5^+)$ is soluble in water. Recent studies have shown that instead of remaining in water, neptunium metal may be incorporated into the structure of schoepite (Buck et al. 1998). Clearly, there are still some open questions concerning the extent and reversibility of the incorporation of $Np(5^+)$ into uranium phases.

A similar process occurs for the radionuclides contained in the matrix as metallic inclusions such as technetium, palladium, ruthenium, and molybdenum. As the matrix is oxidized according to reaction (20.1), these elements oxidize and become more soluble in water. This transformation is particularly critical for an element like technetium, which is almost insoluble and therefore not mobile in metallic form, but that becomes soluble and mobile in the oxidized form, as $Tc(7^+)$. The general redox process is:

$$Tc + 1.75 \ O_2(gas) + 0.5 \ H_2O \rightarrow TcO_4^- + H^+ \qquad (20.6)$$

The incorporation of $Tc(7^+)$ into the structure of uranium secondary phases should be feasible by the substitution of the silicate and phosphate anions. These substitutions, however, will result in a destabilization of the structures given by the underbonding of the $U(6^+)$ positions. Thus, significant substitution cannot occur in uranyl phases formed as alteration products of the corroded spent fuel (Chen et al. 2000).

## Which Alteration Sequence Is Expected in Yucca Mountain?

The preceding discussions might suggest that the hydrogeologic setting at Yucca Mountain is not optimal for the stability of $UO_2$ in the spent nuclear fuel matrix. The situation is further complicated by the high temperatures that will prevail during the respository's first thousand years. The stability of the waste matrix critically depends on the access of water and oxygen to the repository, and the lifetime of the metallic container under these conditions.

Once the oxygen-carrying fluids come into contact with the spent fuel, the sequence of events will be:

• The lighter and more soluble radionuclides (iodine-129 and cesium-137) will be washed away from the surfaces of the spent fuel pellets.

• The spent fuel will be oxidized; uranium, some actinides (especially neptunium), and the metallic inclusions will be oxidized and released into the groundwater solution near the waste packages.

• As uranium reaches a certain concentration in solution, it will precipitate as U(6$^+$) oxides and hydroxides, such as schoepite. Some of the actinides and lanthanides (thorium, plutonium, americium, cerium, and probably neptunium) will coprecipitate with the U(6$^+$) oxide; the oxidized metals (technetium, ruthenium, and molybdenum) will not.

• As the alteration sequence proceeds, uranium silicates will form that may or may not incorporate some of the actinides and other elements. The rest of the radionuclides would be scavenged by sorption processes on the iron corrosion products and other phases present in the geochemical environment of Yucca Mountain.

Although this sequence of events can be predicted with high confidence, there is more uncertainty about the degree of alteration and the amount of radionuclides that will be mobilized during this process. The main unknown is what the temperature of the repository will be when water comes into contact with the spent fuel matrix. This also affects the uncertainty about the chemical composition of the waters intruding into the repository, which strongly influences how much the matrix is altered and which phases are formed. Further research is required to develop an understanding of the geochemical evolution of the Yucca Mountain groundwater system at higher temperatures; such an understanding will be needed to estimate experimentally the alteration rates and mechanisms of spent fuel under these conditions. This has critical consequences for the processes that control the scavenging of radionuclides by the alteration products of spent fuel. This is one of the key areas of uncertainty with regard to the consequences of the oxidation of spent fuel at high temperatures.

### Uncertainty in the DOE's Modeling of Spent Fuel Behavior

To evaluate the performance of a potential monitored geologic repository at Yucca Mountain, a Total System Performance Assessment (TSPA) has been conducted

(CRWMS M&O 1998). The performance assessment model is supported by nine process models. One of these process models, the waste form degradation model, accounts for the behavior of the waste; a summary of the technical basis of this model is given in CRWMS M&O (2000).

The waste form degradation model incorporates eight submodels and analyses. Of these eight components, three are directly related to the alteration of the spent fuel: the radioisotope inventory model, which estimates the quantity of the radionuclides most important to human dose in the disposed spent nuclear fuel; the spent fuel matrix degradation model, which determines the rate of the alteration and dissolution of the commercial spent nuclear fuel matrix as well as the included radionuclides (this model describes how, and how fast, the matrix is altered according to the processes described above); and the dissolved radioisotope concentration model, which evaluates the dissolved concentration of radionuclides released from the wastes, based on the solubilities of the phases expected to precipitate under the geochemical conditions in the repository.

The radionuclide inventory will depend mainly on the type and history of the waste. The radionuclide inventory has been calculated using radionuclide concentrations in commercial spent fuel assemblies, defense spent fuel, and high-level waste canisters.

The model for the degradation of commercial spent fuel has been based on the results obtained from experimental measurements of the dissolution rate of fresh and spent fuels. These measurements have been taken under both saturated and unsaturated conditions, and using flow-through, batch, and drip reactors, with J-13 water (references in Stout and Leider 1998). Isotopes of five different elements (cesium, technetium, iodine, strontium, and uranium) were considered as a measure of dissolution in the flow-through tests. The data used represent more than ten years of experiments with about nine different fuel types.

The dissolved radionuclide concentration model provides an estimate of the amount of initial radioactivity released from the waste, but the amount of radionuclides dissolved in the incoming fluid will be limited by the precipitation of secondary solid phases. Such precipitation could occur either as a pure solid phase (i.e., with the radionuclide as the dominant element of the solid) or coprecipitated (i.e., with trace amounts of the radionuclide present in a solid phase that consists primarily of other materials). One of the main inadequacies of the model is that it considers only pure solid phases. This approach can lead to an overestimation of the concentration of some radionuclides in solution in the expected repository conditions. A model that accounted for coprecipitation should give more realistic values.

One area of uncertainty is the role of the secondary alteration phases in the retention of radionuclides—particularly those that present a larger radiological risk in the long-term behavior of the repository (technetium-99 and neptunium-237). An understanding of this phenomenon would be enhanced by dedicated experiments to ascertain some of the evidence that has been brought forward from the study of natural systems as well as the application of structural and chemical analogies.

The limited amount of thermodynamic data available and the incomplete knowledge about which solid phases will precipitate under repository conditions are important limitations for the solubility calculations. The combination of approaches used in the DOE models can lead to wide ranges of calculated solubilities. An effort to limit these ranges would entail experimental programs to determine the solubility limits of some of these key radionuclides under the conditions expected in the repository.

Other significant sources of uncertainty in the thermodynamic calculations are the inherent limitations of the databases used and the lack of the availability of data for a given radionuclide.

The overall model combines kinetic processes with thermodynamic constraints, including the main geochemical variables of the system. Yet the model does not take into account the evolution of the isotopic inventory in the original $UO_2$ pellet, the solution, and the precipitated material. These processes can be crucial to define concentrations available for transport for some of the key radionuclides (Cera et al. 2001).

The model describes the oxidation and dissolution of the matrix followed by the precipitation of schoepite ([20.1], [20.2], and [20.3]). It does not, however, take into account the subsequent alteration mechanisms to silicate and phosphate phases, despite laboratory and natural data. The models that reproduce these processes have not been incorporated into the base-case study of the TSPA but they have been used for sensitivity analyses intended to point toward the future development of these models.

## What We Still Need to Know

From a geochemical perspective, the environment for the disposal of spent nuclear fuel at Yucca Mountain is far from ideal. Nevertheless, evidence from laboratory and natural systems indicate that the alteration of spent fuel under the conditions expected at Yucca Mountain will tend to constrain the release of most of the radionuclides contained in spent fuel. The major uncertainty arises from insufficient knowledge about the coupled thermal and chemical processes that will occur in the waste package—in particular, the influence of the high temperatures on the rates of the formation and fate of the alteration phases. This uncertainty makes it difficult to establish how fast the alteration of the spent fuel will proceed. It also clouds predictions of the consequences that this alteration will have on the release and subsequent migration of radionuclides. There are large uncertainties associated with the amount and composition of the fluids that will come into contact with the spent fuel as well as the portion of the spent fuel surface that will be exposed to these chemical interactions. Conservative estimates of these parameters overemphasize their consequences.

There is an urgent need to support the calculations made in the TSPA with experimental and theoretical developments that will constrain these uncertainties. This is particularly critical in the areas concerning the chemical composition of the reacting fluids at high temperatures, their consequences on the alteration degree of the spent nuclear fuel matrix, and the subsequent release of radionuclides from the waste form.

## References

Blomqvist, R., Ruskeeniemi, T., Kaija, J., Ahonen, L., Paananen, M., Smellie, J., Grundfelt, B., Pedersen, K., Bruno, J., Pérez del Villar, L., Cera, E., Rasilainen, K., Pitkänen, P., Suksi, J., Casanova, J., Read, D., and Frape, S. (2000) *The Palmottu Natural Analogue Project; Phase II: Transport of Radionuclides in a Natural Flow System at Palmottu.* (Finland) European Commission Report EUR 19611 EN.

Bruno, J., Casas, I., Cera, E., Ewing, R.C., Finch, R.J., and Werme, L.O. (1995) The Assessment of the Long-Term Evolution of the Spent Nuclear Fuel Matrix by Kinetic, Thermodynamic, and Spectroscopic Studies of Uranium Minerals. *Material Research Society Symposium Proceedings* 353, pp. 633–639.

Bruno, J., Cera, E., Eklund, U.-B., Eriksen, T.E., Grivé, M., and Spahiu, K. (2000) Experimental Determination and Chemical Modelling of Radiolytic Processes at the Spent Fuel/Water Interface. *Radiochimica Acta* 88, pp. 513–519.

Bruno, J., Duro, L., and Grivé M. (2002) The Applicability and Limitations of the Geochemical Models and Tools Used in Simulating Radionuclide Behaviour in Natural Waters: Lessons Learned from the Blind Predictive Modelling Exercises Performed in Conjunction with Natural Analogue Studies. *Chemical Geology* 190, pp. 371–393.

Buck, E.C., Finch, R.J., Finn, P.A., and Bates, J.K. (1998) Retention of Neptunium in Uranyl Alteration Phases Formed during Spent Fuel Corrosion. *Material Research Society Symposium Proceedings* 506, pp. 87–94.

Cera, E., Merino, J., and Bruno, J. (2001) Release of Radionuclides from Spent Fuel under Repository Conditions: Mathematical Modelling and Preliminary Results. *Material Research Society Symposium Proceedings* 663, pp. 459–468.

Chapman, N.A., McKinley, I.G., Shea, M.E., and Smellie, J.A.T. (1993) *The Poços de Caldas Project: Natural Analogues of Processes in a Radioactive Waste Repository*. Amsterdam: Elsevier, 603 p.

Chen, F., Burns, P.C., and Ewing, R.C. (2000) Near-Field Behavior of 99Tc during the Oxidative Alteration of Spent Nuclear Fuel. *Journal of Nuclear Material* 278, pp. 225–232.

CRWMS M&O (Civilian Radioactive Waste Management System Management and Operating Contractor) (1998) *Total System Performance Assessment Viability Assessment (TSPA-VA) Analyses Technical Basis Report*. Department of Energy. B00000000-01717-4301-00001 REV 01. November.

CRWMS M&O (Civilian Radioactive Waste Management System Management and Operating Contractor) (1999) *1999 Design Basis Waste Input Report for Commercial Spent Nuclear Fuel*. Department of Energy. B00000000-01717-5700-00041 REV 01. December.

CRWMS M&O (Civilian Radioactive Waste Management System Management and Operating Contractor) (2000) *Waste Form Degradation Process Model Report*. Department of Energy. TDR-WIS-MD-000001 REV 00, ICN 01. July.

de Pablo, J., Casas, I., Giménez, J., Molera, M., Rovira, M., Duro, L., and Bruno, J. (1999) The Oxidative Dissolution Mechanism of Uranium Dioxide: I. The Effect of Temperature in Hydrogen Carbonate Medium. *Geochimica et Cosmochimica Acta* 63, pp. 3097–3103.

Department of Energy (DOE) (2001) *Yucca Mountain Science and Engineering Report*. Office of Civilian Radioactive Waste Management. DOE/RW–0539. May.

Eriksen, T.E., Eklund, U.-B., Werme, L.O., and Bruno, J. (1995) Dissolution of Irradiated Fuel: A Mass Balance Study. *Journal of Nuclear Materials* 277, pp. 76–82.

Finn, P.A., Hoh, J.C., Wolf, S.F., Slater, S.A., and Bares, J.K. (1996) The Release of Uranium, Plutonium, Cesium, Strontium, Technicium, and Iodine from Spent Fuel under Unsaturated Conditions. *Radiochimica Acta* 74, p. 65.

Finch, R.J., and Ewing, R.C. (1992) The Corrosion of Uraninite under Oxidizing Conditions. *Journal of Nuclear Materials* 190, pp. 133–156.

Forsyth, R.S., and Werme, L.O. (1992) Spent Fuel Corrosion and Dissolution. *Journal of Nuclear Materials* 190, pp. 3–19.

Grandstaff, D.E. (1976) A Kinetic Study of the Dissolution of Uraninite. *Economic Geology* 71, pp. 1493–1506.

Gray, W.J., Thomas, L.E., and Einziger, R.E. (1993) Effects of Air Oxidation on the Dissolution Rate of LWR Spent Fuel. *Materials Research Society Symposium Proceedings* 294, pp. 47–54.

Isobe, H., Murakami, T., and Ewing, R.C. (1992) Alteration of Uranium Minerals in the Koongarra Deposit, Australia: Unweathered Zone, *Journal of Nuclear Materials*, 190, pp. 174–187.

Jensen, K.A., and Ewing, R.C. (2001) The Okélobondo Natural Fission Reactor, Southeast Gabon: Geology, Mineralogy, and Retardation of Nuclear Reaction Products. *Geological Society of America Bulletin* 113, pp. 32–62.

Johnson, L.H., and Shoesmith, D.W. (1988) Spent Fuel. In *Radiaoctive Waste Forms for the Future*, ed. Lutze, W., and Ewing, R.C. New York: Elsevier, pp. 435–698.

Louvat, D., Michaud, V., and von Maravic, H. (2000) *Oklo Working Group: Proceedings of the Second Joint EC-CEA Workshop on OKLO, Phase II*. European Commission Report EUR 19116 EN. Helsinki. June 16–18.

McKinley, I.G., Bath, A.H., Berner, U., Cave, M., and Neal, C. (1988) Results of the Oman Analogue Study. *Radiochimica Acta* 44–45, pp. 311–316.

Murphy, W.M., and Pearcy, E.C. (1992) Source-Term Constraints for the Proposed Repository at Yucca Mountain, Nevada, Derived from the Natural Analog at Peña Blanca, Mexico. *Materials Research Society Symposium Proceedings* 257, pp. 521–527.

Shoesmith, D.W., Sunder, S., and Tait, J.C. (1998) Validation of an Electrochemical Model for the Oxidative Dissolution of Used CANDU Fuel. *Journal of Nuclear Materials* 257, pp. 89–98.

Säkerhets Redovisning 97 (1997) SKB (Swedish Fuel and Waste Management Co.), Stockholm.

Stout, R.B., and Leider, H.R. (1998) *Waste Form Characteristics Report*. CD-ROM Version. Lawrence Livermore National Laboratory. UCRL–ID–132375.

Stroes-Gascoyne, S., Johnson, L.H., Beeley, P.A., and Sellinger, D.M. (1985) Dissolution of Used CANDU Fuel at Various Temperatures and Redox Conditions. *Materials Research Society Symposium Proceedings* 50, pp. 317–326.

Torrero, M.E., Casas, I., de Pablo, J., Duro, L., and Bruno, J. (1998) Oxidative Dissolution Mechanism of Uranium Dioxide at 25°C. *Mineralogical Magazine* 3, 62A, pp. 1529–1530.

Wronkiewicz, D.J., Bates, J.K., Gerding, T.J., and Veleckis, E. (1992) Uranium Release and Secondary Phase Formation during Unsaturated Testing of $UO_2$ at 90°C. *Journal of Nuclear Materials* 190, pp. 107–127.

# 21

## Glass

Werner Lutze

High-level radioactive waste from the burn up of seventy thousand metric tons of heavy metal may be disposed of in the Yucca Mountain geologic repository (Nuclear Waste Policy Act 1982; Nuclear Waste Policy Amendments Act 1987). The Department of Energy (DOE) has determined that spent fuel (accounting for 65,333 metric tons of heavy metal) and solidified high-level radioactive waste (accounting for 4,667 metric tons of heavy metal) will be shipped for disposal (DOE 2002). The solidified waste (8,315 canisters) will be mostly glass that contains high-level waste from the reprocessing of spent fuel and the separation of plutonium for nuclear weapons. In this chapter, I talk about glass as a waste form.

Uranium and plutonium in spent nuclear reactor fuel can be recovered by reprocessing. This requires the dissolution of the irradiated fuel as well as the chemical separation of uranium and plutonium from fission products, neptunium, and transplutonium elements generated in the fuel during reactor operation. The aqueous phase, from which uranium and plutonium have been separated, is high-level waste. Liquid high-level radioactive waste is stored in tanks at the reprocessing facility prior to solidification and disposal in a geologic repository. The preferred solidification process is vitrification—that is, conversion of the waste into glass.

In the United States, liquid high-level waste originated almost entirely from the reprocessing of fuel irradiated for the production of plutonium for nuclear weapons. This waste is referred to as defense high-level waste. Waste compositions vary widely depending on the fuel composition and the chemical separation process. Processes in the 1940s and 1950s used chemical precipitation to separate plutonium, thereby producing wastes containing high concentrations of nonradioactive chemicals. Uranium became part of the waste. Later, this waste was processed again to recover uranium. In the 1950s, solvent extraction processes were developed to recover uranium and plutonium, resulting in less volume and simpler compositions of the high-level waste. The fuel dissolution process created a highly acidic waste, which had to be neutralized to minimize corrosion of the carbon steel tanks in which the waste was stored. For this purpose, large quantities of sodium hydroxide were added.

Reactor fuel has been reprocessed at three sites owned by the DOE: the Hanford site in Washington State, the Savannah River site in South Carolina, and the Idaho National Laboratory. Fuel was also reprocessed at a fourth site, located in West Valley, New York, and owned by the state of New York. The DOE is responsible for retrieving and disposing of the high-level waste at West Valley.

Table 21.1 shows inventories and some characteristics of the high-level waste in the United States. The total radioactivity of the defense waste in the United States amounts to 460 million curies. By the time this waste will be disposed of in a repository, currently planned to begin in 2010 and lasting several decades, the activity will have decreased naturally to less than 230 million curies. It will probably take several decades before all the waste at Hanford is converted into a disposable form.

**Table 21.1**
Inventory and characteristics of liquid high-level waste in the United States

| Site | No. of tanks | Volume liters (gallons) | Characteristics | Total activity Bq* |
|---|---|---|---|---|
| Hanford | 177** | 208(54) million | pH ≈ 12, highly inhomogeneous waste, sludge, salt cake, liquid | $7.4 \times 10^{18}$ |
| Savannah River Site | 51 | 125(33.4) million*** | pH ≈ 14, liquid and sludge | $8.9 \times 10^{18}$ |
| Idaho National Engineering Laboratory | 11 +7 calcine vaults | 5.3(1.4) million; 3.8(1.0) million calcine | pH ≈ 0, sludge, liquid, calcine | $1.9 \times 10^{16}$ $8.9 \times 10^{17}$ |

*1 Becquerel = 1 transmutation per second; $3.7 \times 10^{10}$ Bq = 1 Curie
**149 single-shell tanks and 28 double-shell tanks
***Vitrification in progress

## Making a Glass Waste Form for Geologic Disposal

Some of the wastes at Hanford have been stored in tanks for more than fifty years. Over the decades, wastes have separated into solid and liquid phases; another characteristic is the chemical inhomogeneity of the solid masses. More than one-third (67) of the 177 tanks (table 21.1) have leaked 1.5 million curies of radioactive liquid into the ground (Gephart and Lundgren 1998). To alleviate leakage, all 149 single-shell tanks have been treated to reduce their contents of liquid. After treatment, these tanks now contain three types of material: liquid (sodium hydroxide and sodium nitrate/nitrite solution near saturation), salt cake (crystalline mixture of salts), and sludge (viscous, amorphous, insoluble material). A fourth type of material—slurry (solid, mostly crystalline particles suspended in a liquid)—is found in the double-shell tanks.

Making a disposable waste form requires recovery of the waste from the tanks, quantitative characterization, and chemical treatment prior to solidification. These processes are particularly complicated at Hanford. At the Savannah River site, waste inventories are less heterogeneous and chemical compositions vary less between tanks. At the Idaho National Laboratory, a large fraction of the liquid waste has been dried at an elevated temperature and stored on-site as a solid (calcine) in steel bins. Calcine is not considered a disposable waste form, as it is fairly soluble in water. Treatment of the calcine will be necessary prior to disposal.

A large number of candidate waste forms have been developed worldwide for high-level waste. Most of these waste forms have been described and reviewed by Lutze and Ewing (1988). Only two types of materials, glasses and ceramics, passed U.S. and international peer reviews and waste form selection procedures (e.g., DOE 1981). Donald and colleagues (1997) published a more recent review of these two types of waste forms. Using the report (DOE 1981) compiled by the Hench Committee, the DOE chose borosilicate glass for the defense waste at Savannah River. Borosilicate glass is any silicate glass that contains more than 5 percent by weight of boron oxide ($B_2O_3$). Meanwhile, the DOE has applied this choice to all its sites except Idaho, where a decision has not yet been made for the treatment of the calcine.

Current practice requires that high-level waste be mixed with glass-forming additives, primarily silica ($SiO_2$) and $B_2O_3$, melted, and poured into metallic containers. On cooling, the melt freezes and forms a glass. In this process, most of the radioactive elements become part of the glass structure; some are just dispersed in the glass. If there is insufficient alkali in the waste, sodium oxide ($Na_2O$), lithium oxide ($Li_2O$), or both are added. Other oxides are frequently added to fine-tune material properties important for the melting process. The lower the production temperature (typically 1150°C versus 1450°C for container or flat glass), the lower is the loss of volatile radioactive material and the lower is the corrosion of melter components. Some additives, such as zirconia ($ZrO_2$), are selected to improve chemical durability—the most important property of glass in the repository.

The industrial-scale vitrification of high-level waste was developed in France in the 1970s. French and British companies operate commercial reprocessing and vitrification plants. They reprocess domestic and foreign spent fuel from nuclear power reactors, vitrify the high-level waste, and return the glass. There are two noncommercial vitrification facilities in the United States: the West Valley Demonstration Project and Savannah's Defense Waste Processing Facility. The high-level waste at West Valley has already been vitrified and the melter has been shut down. The glass is contained in 255 cylindrical canisters, each 62 centimeters in diameter and about 3 meters long. These "glass logs" are stored on-site awaiting disposal in a geologic repository. The Savannah facility started radioactive operation in 1996. About fifteen hundred glass logs ($\approx$ two thousand seven hundred metric tons of glass) had been produced by early 2004. A total of some six thousand logs will be produced to convert all the high-level waste at Savannah into borosilicate glass. At the current pace, vitrification will continue for at least two decades.

The vitrification of waste at Hanford is expected to commence in 2011. Given the large tank inventory of waste, we can assume that it will take several decades until all of the high-level waste at Hanford is available for geologic disposal. This estimate may change, and work may be completed earlier if more efficient processing technologies become available or alternative strategies are developed and pursued.

Expressing the total amount of U.S. vitrified waste in a number of glass logs to be disposed of is difficult—in part because not all glass logs are the same size. The canister to be used at Hanford will be 4.5 meters long (versus the 3-meter-long ones at Savannah River), holding 1.08 cubic meters of glass (Ohme et al. 1996). Another factor affecting the number of glass logs is waste loading—that is, the mass fraction of waste in the glass. Waste loading depends on the waste composition and how the waste is pretreated. Pretreatment separates radioactive from nonradioactive elements and reduces the volume of glass. A current estimate of the total number of glass logs to be generated at all the DOE sites is about twenty thousand, roughly half of which will go to Yucca Mountain.

### Integrity of the Borosilicate Glass Waste Form

Except for Russia, which has produced phosphate glass for decades, borosilicate glass has become the only waste form for liquid high-level radioactive waste in the world. The advantages of borosilicate and phosphate glasses compared with other waste forms include the large throughput, low melting temperature, and high solubility for most of the waste constituents, yielding fairly homogeneous and chemically durable products. Only a few fission products, such as ruthenium, rhodium, and palladium, are insoluble in glass. They form small crystals embedded in the glass. These crystals can act as nucleation sites; that is, they can enhance the formation of other crystalline phases (such as spinel) during melting and on cooling of the glass.

Several effects and processes can degrade the properties of high-level waste glass prior to and during disposal. Note that property changes are of no concern in terms of repository safety as long as they do not reduce the chemical durability of glass in water. That's because the only way for radionuclides to escape from the repository into the biosphere (aside from human intrusion and catastrophic effects such as earthquakes, volcanism, and meteoritic impact) is by corrosion of the glass and transport in water.

Glasses can be ranked by their chemical durability. Between two glasses with the same specific activity (radioactivity per unit mass), the glass with the higher chemical durability reacts more slowly with water and releases less radionuclides per unit of surface area than the glass with the lower chemical durability.

Due in part to inadequate or missing data and a lack of understanding of the glass corrosion process, uncertainty prevails regarding how well different glasses would perform in the repository over geologic periods of time. Even a complete understanding of all glass properties would not allow us to predict with accuracy the rate at which a high-level waste glass will corrode and release radionuclides. The inherent problem is that the corrosion rate of glass depends on the flow rate and chemical composition of the water in the repository—variables that are difficult to predict accurately over geologic periods of time.

High-level waste glasses are subject to heat and radiation. At high temperatures, crystallization is the main concern. High temperatures prevail only for short periods of time, however. After pouring into a canister, the glass cools within weeks (or sooner) from about 1150°C to below the glass transition temperature $T_g$, a temperature below which crystallization becomes immeasurably slow. For waste glasses, $T_g$ is typically around 500°C. In the Yucca Mountain repository, temperatures below $T_g$ but above ambient will prevail in the glass for several centuries as a result of the radioactive decay of strontium-90 (with a half-life of twenty-nine years) and cesium-137 (with a half-life of thirty years). Over geologic periods of time, the glass will be subject to damaging alpha radiation from the waste.

Heat and radiation have four principal effects on glass properties:

• Temperature gradients in the glass (hotter in the middle and cooler on the surface) cause thermal stress, which supports the formation and growth of cracks. Cracks may increase the surface area accessible to water.

• Crystallization in the glass may lower its chemical durability.

• Alpha and beta decay transform elements into others, thereby changing the chemical composition and damaging the glass structure. Alpha particles end up as helium in the glass. Gas bubbles may form, the density of glass may change, and decay energy may be stored in structural defects. These radiation effects may degrade the chemical durability of glass.

• Radiolysis (that is, the formation of highly reactive species, called radicals, in water) can enhance glass corrosion.

The following addresses how these effects result in the deterioration of the chemical durability of a nuclear waste glass.

Temperature gradients induced by cooling full-size glass logs have led to surface area increases on the order of ten times the geometric surface area of a glass cylinder. If the total surface area is available to the attack of water, the glass corrosion rate, measured in grams of glass reacted per day, increases by the same factor as the surface area—and so does the amount of radionuclides released.

The crystallization of glasses has been studied extensively. Lutze and Ewing (1988) as well as Donald and colleagues (1997) have summarized this work. Time-temperature-transformation diagrams have been established for many glasses, allowing quantitative predictions of crystal yields and thus of the stability of the vitreous state on cooling of the melt to below $T_g$. Typically, high-level waste glasses contain up to a few percent by weight of crystalline material. Only a few crystalline phases have a deleterious effect on a glass's chemical durability, mainly those lowering the concentration of silicon in the glass. Proper tailoring of the glass composition can improve the stability of the vitreous state. As mentioned above, crystallization becomes negligible as soon as the temperature drops below $T_g$. The maximum temperature in the glass in a repository at Yucca Mountain will be several hundred degrees lower than $T_g$. In this temperature range, the effects of phase transformation on the chemical durability of the glass are not expected.

Many studies have measured the stored energy, accumulation and release of helium, density changes, and chemical transmutation in glass waste forms (Lutze and Ewing 1988). More recently, Weber and colleagues (1997) have summarized the published work on radiation effects in high-level waste glasses. The influence of these effects on chemical durability is small or insignificant. Some work has been conducted to determine the change of chemical durability as a function of radiation dose. Burns and colleagues (1981) studied the full range of doses delivered to a glass by alpha decay up to $1.7 \times 10^{19}$ decay events per cubic centimeter using plutonium-238. A defense waste glass would reach this dose in ten million years. The glass corrosion rate was enhanced by a factor of three compared with that of the nonirradiated glass. This result is in good agreement with what other authors have found in different experiments and with different glasses.

The effect of radiolysis on glass corrosion is also small (Lutze and Ewing 1988). Radiation effects on glass corrosion have usually been studied at room temperature in diluted aqueous solutions where the corrosion rate is controlled by the composition of the glass. In a repository, however, it is the composition of the solution—particularly its silica concentration—that is likely to control the corrosion rate. Hence, short-term experiments in pure water or diluted solutions do not yield the significance of radiation damage in glass when corroded under repository conditions. Experiments under silica saturation conditions would be more realistic and could reveal changes in the long-term corrosion rate resulting from simultaneous internal alpha irradiation and radiolysis. Respective corrosion experiments should be conducted for ten years and longer.

There are different ways to look at the release of matter from a glass as a result of corrosion in water. Here, I consider radionuclides as released when they are no longer part of the pristine glass; that is, when they are either dissolved in water or contained in newly formed solid phases on the surface of the glass or elsewhere. The release (or corrosion) rate is a quantitative measure of the release of radionuclides from the glass. In a closed system (no flowing water), this rate is a function of the glass composition, water composition, pH, and temperature. Typically, the rate decreases with time at constant temperature, but corrosion does not cease. In flowing water, the corrosion rate varies with the water flow rate up to a maximum flow rate, above which the corrosion rate is constant.

In a geologic repository, engineered barriers (canisters and potentially others) around the glass are designed to prevent contact between the glass and the water or the moisture. The time to failure of engineered barriers is the critical variable determining the earliest possible release of radionuclides after contact between the glass and the water. It doesn't matter whether this time is one thousand or ten thousand years; the inventory of long-lived radionuclides will be essentially the same as at the time of disposal. These radionuclides include selenium-79 (half-life = 6.5 × $10^5$ years), technetium-99 ($2.11 \times 10^5$ years), cesium-135 ($2.3 \times 10^6$ years), uranium-238 ($4.5 \times 10^9$ years), neptunium-237 ($2.1 \times 10^6$ years), and plutonium-239 ($2.4 \times 10^4$ years). The one exception is the much more rapidly decaying radionuclide americium-241, with a half-life of a mere 433 years. The inventory of neptunium-237 will have actually increased, due to the decay of americium-241. The release of radionuclides with half-lives longer than ten thousand years is essential to the calculation of radiation doses, which might be received by humans living in the vicinity of the repository.

In the repository at Yucca Mountain, glass logs will be disposed of together with spent fuel. The radionuclide inventory of spent fuel in the repository will be more than one hundred times higher than that of the glass logs; spent fuel will therefore account for most of the heat generation. The release of radionuclides originating from spent fuel is likely to mask releases from the glass. This is due not only to the relatively small radionuclide inventory in the glass but also to the glass's chemical durability, which under oxidizing conditions (those expected for the repository at Yucca Mountain) is higher than that of spent fuel. Nevertheless, for the sake of completeness, I will examine the release of radioactivity from borosilicate glass.

If the time to failure of the canisters filled with glass is longer than five hundred to one thousand years, the two heat-generating radionuclides strontium-90 and cesium-137 will have decayed to insignificant levels, and the temperature in the repository will have returned to close to its prerepository value. Then, the glass corrosion rate will depend on the glass composition, water composition, flow rate, and effects from engineered barrier materials. Mass transfer from the glass into the repository environment can be lessened or even blocked by corrosion products. The corrosion products of barrier materials can change the water composition by adsorbing or releasing silica. The glass composition is a variable because the waste listed in table 21.1 will be immobilized in glasses of a different composition. All glasses to be disposed at Yucca Mountain, however, must meet the stability criteria laid out in the waste acceptance product specifications (DOE 1996).

In a simplified way, the glass corrosion rate $R$ can be described mathematically by the expression $R = k \cdot (1 - a)$ with $k$ being the forward rate of dissolution and $a$ the silica concentration in the solution divided by a silica saturation concentration

(Grambow 1985). The forward rate $k$ depends on the temperature and the pH. Clearly, $R$ is highest—that is, the glass corrodes fastest—if $a = 0$.

At Yucca Mountain, glass corrosion is expected to occur in the presence of small amounts of slowly flowing or temporarily stagnant water. This water will most likely be nearly saturated with respect to silica. Hence, $a$ will not be 0 and the glass will not corrode at the highest possible rate. If the flow rate of water is small enough, the silica released from the corroding glass will concentrate in solution— that is, $a$ will increase and $R$ will decrease. Yet it is a characteristic of glass in general that corrosion will continue at a low rate, until the glassy state disappears, even at $a = 1$. There is no chemical equilibrium between glass and an aqueous solution. A diffusion-controlled alteration process in the pristine glass beneath a corroded surface layer, not covered by the simplified rate equation above, can explain this phenomenon.

**Regulatory Issues**

The glass product to be disposed of in a deep geologic repository must meet certain regulatory criteria. For Yucca Mountain, these are the Waste Acceptance Product Specifications for Vitrified High-Level Waste Forms (WAPS) (DOE 1996).

The WAPS require that the chemical durability of a radioactive glass be better than that of a reference glass. The reference glass is the "environmental assessment benchmark glass" (see Jantzen 1993). This same WAPS criterion further requires chemical durability to be measured by a specific test procedure, the Product Consistency Test developed by the Savannah River Technology Center (American Society of Testing and Materials 2002). The Product Consistency Test measures the concentration of major soluble elements (e.g., sodium, lithium, and boron) released from crushed glass into distilled water after seven days at 90°C. By normalizing the respective concentrations in solution to those in the glass, different glasses can be compared and ranked in terms of chemical durability. As an example, in terms of normalized boron release, Pyrex kitchenware glass is about forty times more durable than the environmental assessment glass (author's unpublished data). On the other hand, state-of-the-art high-level waste borosilicate glasses compare well with Pyrex glass in terms of chemical durability, indicating that it is not a challenge to meet the minimum chemical durability set by the environmental assessment glass. The Product Consistency Test measures the short-term chemical durability.

Models to calculate the degradation of high-level waste glass in the Yucca Mountain repository are based on an expression closely related to the one given above: $R_i = k \times S \times I_i$, where $R_i$ is the release rate of radionuclide $i$, $S$ is the glass surface area contacted by water, and $I_i$ is the inventory of radionuclide $i$ in the glass. The effect of the silica concentration ratio $a$ on $R_i$ is considered by modifying $k$. Details can be found in *Technical Basis Document No. 7* (DOE 2004).

## Evaluation of the Remaining Uncertainties

Vitrified defense high-level radioactive waste (borosilicate glass) along with commercial and DOE spent reactor fuel will be disposed of together at Yucca Mountain. The waste forms will be contained in the same type of waste package, but several different designs would be needed. The waste package is the sealed and tested disposable container consisting of the barrier materials and internal components in which the spent fuel and high-level waste would be placed (DOE 2002). One can assume, in a first approximation, that waste package failure is independent of what is inside. In this case, the release of radioactivity from glass and spent fuel is equally likely. Assuming that canister failure occurs at a predisposal temperature in the repository, the release rates of the long-lived radionuclides from glass and spent fuel depend on specific activity as well as the chemical durability of the two waste forms. Glass has a lower specific activity than spent fuel and, under oxidizing conditions, it also has the lower corrosion rate. Hence, the radioactivity coming from the glass will be small relative to spent fuel, also keeping in mind the repository's total inventory of 63,000 metric tons of heavy metal of high burn-up commercial spent fuel (plus 2,333 metric tons of heavy metal of low burn-up DOE fuel), versus a "glass equivalent" of only 4,667 metric tons of heavy metal of low burn-up fuel (DOE 2002).

For any uncertainty associated with the long-term performance of glass to be of concern, we have to assume the unrealistic event that glass is the major source of radioactivity release in the repository. In this event, the main uncertainty with glass lies in the prediction of its long-term performance in the presence of water. Prediction requires knowledge of the long-term corrosion rate for each glass composition. Experimental corrosion rates cannot simply be extrapolated over long periods of time because the corrosion mechanism may not remain constant. The long-term corrosion rate depends on silica concentration in the water, flow rate, water diffusion in the glass, corrosion products, and mass transfer, all of which may vary over the millennia. The more accurate the flow rate, the easier it is to evaluate the significance of other variables, and the accuracy of predictions of radionuclide release from glass increases.

## References

American Society of Testing and Materials (2002) *Standard Test Methods for Determining Chemical Durability of Nuclear, Hazardous, and Mixed Waste Glasses and Multiphase Ceramics: The Product Consistency Test (PCT).* C–1285–02.

Burns, W.G., Hughes, A.E., Marples, J.A.C., Nelson, R.S., and Stoneham, A.M. (1981) Report AERE–R 10189.

Department of Energy (DOE) (1981) *The Evaluation and Review of Alternative Waste Forms for Immobilization of High-Level Radioactive Wastes: Report No. 3, Alternative Waste Form Peer Review Panel.* Office of Civilian Radioactive Waste Management. DOE/TIC–11472.

Department of Energy (DOE) (1996) *Waste Acceptance Product Specifications for Vitrified High-Level Waste Forms (WAPS).* Office of Civilian Radioactive Waste Management. EM–WAPS Rev. 02.

Department of Energy (DOE) (2002) *Final Environmental Impact Statement for a Geologic Repository for the Disposal of Spent Fuel and High-Level Radioactive Waste at Yucca Mountain, Nye County, Nevada.* Office of Civilian Radioactive Waste Management. DOE/EIS–0250. Volume 1, Chapter 2. http://www.ocrwm.doe.gov/documents/feis_a/index_v1.htm.

Department of Energy (DOE) (2004) *Technical Basis Document No. 7: In-Package Environment and Waste Form Degradation and Solubility Revision 1.* Office of Civilian Radioactive Waste Management. July. http://www.ocrwm.doe.gov/technical/tbd.shtml.

Donald, I.W., Metcalfe, B.L., and Taylor, R.N. J. (1997) The Immobilization of High-Level Radioactive Wastes Using Ceramics and Glasses. *Journal of Materials Science* 32, pp. 5851–5887.

Gephart, R.E., and Lundgren, R.E. (1998) *Hanford Tank Clean Up: A Guide to Understanding the Technical Issues.* Report PNL–10773.

Grambow, B. (1985) A General Rate Equation for Nuclear Waste Glass Corrosion. *Material Research Society Symposium Proceedings* 44, pp. 15–27.

Jantzen, C.M. (1993) *Characterization of the Defense Waste Processing Facility (DWPF) Environmental Assessment (EA) Glass Standard Reference Material, Revision 1.* Westinghouse Savannah River Company, Aiken, South Carolina, WSRC–TR–92–346.

Lutze, W., and Ewing, R.C. (1988) *Radioactive Waste Forms for the Future.* Amsterdam: North Holland, 778 p.

Nuclear Waste Policy Act (1982) Public Law 97–425. January 7.

Nuclear Waste Policy Amendments Act (1987) Excerpt from Public Law 100–203. December 22.

Ohme, R.M., Manuel, A.F., Shelton, L.W., and Slaathaug, E.J. (1996) *Revised Tank Waste Remediation System Privatization Process Technical Baseline.* Hanford. Report WHC–SD–WM–TI–774.

Weber, W.J., Ewing, R.C., Angell, C.A., Arnold, G.W., Cormack, A.N., Delaye, J.M., Griscom, D.L., Hobbs, L.W., Navrotsky, A., Price, D.L., Stoneham, A.M., and Weinberg, M.C. (1997) Radiation Effects in Glasses Used for Immobilization of High-Level Waste and Plutonium Disposition. *Journal of Materials Research* 12, pp. 1946–1978.

# 22

## Storing Waste in Ceramic

William L. Bourcier and Kurt Sickafus

Not all the nuclear waste destined for Yucca Mountain is in the form of spent fuel. Some of it will be radioactive waste generated from the production of nuclear weapons. This so-called defense waste exists mainly as corrosive liquids and sludge in underground tanks. An essential task of the U.S. high-level radioactive waste program is to process these defense wastes into a solid material—called a waste form. An ideal waste form would be extremely durable and unreactive with other repository materials. It would be simple to fabricate remotely so that it could be safely transported to a repository for permanent storage. What's more, the material should be able to tolerate exposure to intense radiation without degradation. And to minimize waste volume, the material must be able to contain high concentrations of radionuclides.

The material most likely to be used for the immobilization of radioactive waste is glass. Glasses are produced by the rapid cooling of high-temperature liquids, such that the liquidlike nonperiodic structure is preserved at lower temperatures. This rapid cooling does not allow enough time for thermodynamically stable crystalline phases (mineral species) to form. In spite of their thermodynamic instability, glasses can persist for millions of years (see chapter 21, this volume).

An alternate to glass is a ceramic waste form—an assemblage of mineral-like crystalline solids that incorporate radionuclides into their structures. The crystalline phases are thermodynamically stable at the temperature of their synthesis; ceramics therefore tend to be more durable than glasses. Ceramic waste forms are fabricated at temperatures below their melting points and so avoid the danger of handling molten radioactive liquid—a danger that exists with the incorporation of waste in glasses.

The waste form provides a repository's first line of defense against the release of radionuclides. Along with the canister, it is the barrier in the repository over which we have the most control. When a waste form is designed, the atomic environment of the radionuclides is chosen to maximize the chemical durability. Elements such as zirconium and phosphorous can be included in the waste form that react with and make some radionuclides less soluble, and therefore less likely to be released.

The long-term performance assessment of radionuclide containment requires the development of models for each part of the barrier system. It is almost certainly easier to model the corrosion and alteration of waste forms than it is to develop coupled hydrologic, chemical, and geophysical models of radionuclide transport away from a repository. Thus, much time and effort has been spent optimizing the chemical durability of both glass and ceramic waste forms for radionuclide containment (Ewing 2001). This has not been an easy task. Three problems in particular posed the greatest challenges.

The first is that radionuclides decay, transmuting into daughter elements that may have different chemical properties. These new elements might degrade the existing mineral by making it unstable. A good waste form that works well for uranium may work poorly for lead, its final decay product.

The second problem is that the radioactive decay itself damages the solid over time. Radioactive decay is an energetic process in which ejected particles and the recoiling nucleus disrupt the surrounding atoms. A single alpha-decay event can displace thousands of atoms in the surrounding volume. We know from laboratory measurements that radionuclides are more easily released from radiation-damaged structures than from materials that do not sustain radiation damage.

The third problem is that radioactive waste, particularly the high-level waste from the reprocessing of spent nuclear fuel to extract plutonium and uranium, contains a variety of elements with widely varying chemistry. The waste form must incorporate the radionuclides as well as the nonradioactive elements, such as silicon and sodium, that are present in the waste stream as a result of waste processing.

A number of ceramic waste forms have been developed that minimize these problems and provide a potentially useful host for radionuclides (Lutze and Ewing 1988a). For ceramics, the mineralogy can be tailored to the waste stream by selecting solid mineral phases with structural sites that can accommodate the waste elements as well as the newly formed radioactive decay elements. Radiation damage can be minimized by selecting mineral phases that allow atoms to renew or regain their original crystalline structure—a process known as annealing. For example, actinide phosphate minerals anneal more readily than actinide silicate minerals.

Despite the superior thermodynamic stability of crystalline materials, borosilicate glasses have become the preferred waste forms. One reason is that the processing technologies associated with this glass are believed to be easier to adapt to handling highly radioactive material. In addition, borosilicate glass is relatively insensitive to variability in the composition of the waste stream (for detailed reviews of waste form development, see Hench et al. 1981; Lutze and Ewing 1988b; Donald et al. 1997). The United States decided in 1982 to use borosilicate glass as the waste form for storing high-level defense waste at the Department of Energy's (DOE) Savannah River site, and in 1990 to use borosilicate glass as the waste form for high-level waste at Hanford, Washington. The other U.S. site having a large amount of high-level waste—the Idaho National Engineering Laboratory—is still in the process of choosing a waste form. Until recently, the DOE was considering a titanate-based ceramic waste form for the disposal of fissile plutonium and uranium from dismantled nuclear weapons. Work on this waste form was suspended early in 2002 (DOE 2002), and there are no current plans to include ceramic waste forms in the U.S. high-level waste repository.

## Ceramic Corrosion

Corrosion refers to the reaction of the ceramic with water. Mechanisms of corrosion include the selective leaching of certain elements, the dissolution of material, and the breaking of chemical bonds by water molecules (hydrolysis). Any of these reactions may result in the release of radionuclides from the waste form in a repository.

Except for some gaseous radioactive decay products, neither ceramic nor glass waste forms will release radionuclides under dry conditions. Radionuclides will be released only through water contact. In addition, the movement of radionuclides through ceramic is so slow that they cannot simply migrate to the surface and escape—even at the highest anticipated repository temperatures. To release radionuclides, corrosion must be able to penetrate into the ceramic.

Waste forms are subjected to numerous tests to determine their corrosion rates. Generally, the waste form is immersed in water and the concentrations of released species are measured over time. The temperature and composition of the water solution, chosen because of the anticipated repository conditions, are varied to gauge their effects on the release rates. The measured release rates allow estimates to be made of the release rates from a waste form in a repository. An additional goal is to use the test results to develop a mechanistic understanding of waste form corrosion, including the identification of the rate-limiting corrosion mechanisms. This understanding allows defensible extrapolations of short-term test results to long-term predictions of waste form corrosion rates in a repository.

## Radiation Damage

Any prediction of the long-term durability of a ceramic waste form must also consider the effect on the material of radiation damage. The type of radiation that most compromises the ceramic waste form's durability is due to alpha decay. Each alpha decay gives rise to the emission of an alpha particle and the recoil in the opposite direction of a heavy nucleus. As a result, the local environment surrounding the alpha decay site is full of displaced ions and alpha particles. Radiation damage may be accompanied by a swelling of the material of 15 volume percent or more. The swelling causes cracking, the flaking off of thin layers, and an increase in the surface area. All these processes enhance the radionuclide release.

Experimental work has shown that materials damaged by alpha decay can corrode ten to a hundred times faster than undamaged materials (Wald and Weber 1984; Weber et al. 1998). In these tests, the ceramics are either doped with a short half-life radionuclide to accelerate the radiation damage or bombarded with high-energy ions to simulate the effects of radiation.

Radiation-damaged crystals can and do repair themselves through annealing, however. If diffusion rates are high enough, ions in the material tend to migrate back and resume their original structure. Whether the damage can be repaired depends on the temperature (higher temperatures allow faster annealing) and the rate of further damage. If alpha damage piles on top of previous damage that has not been repaired, the entire structure may become amorphous—the so-called metamict state sometimes found in natural minerals that contain radionuclides.

In addition, some mineral structures are better able to accommodate radiation-induced defects into their structures. Sickafus and colleagues (2000) have shown that fluorite-structured solids such as zirconates are less perturbed by defects introduced by radiation than comparable pyrochlore structures. Such knowledge allows for the tailoring of both the durability and radiation tolerance of potential ceramic waste forms.

## Natural Analogs: The Long-Term Data

Fortunately, information is available to help predict the long-term performance of waste forms. Both ceramics (i.e., minerals and rocks) and glasses occur naturally, and are millions or, for some ceramics, billions of years old. These materials have compositions and structures similar to those of the proposed nuclear waste forms. The minerals in rocks sometimes contain uranium and thorium that decay and over time generate radiation damage. Therefore, the durabilities of these minerals are thought to be good indicators of the long-term durabilities of comparable radiation-damaged nuclear waste forms (Ewing and Jercinovic 1987).

The major uncertainty in the use of natural analogs is that the environmental history of the samples is often poorly known. Although radiometric dating may supply an age, no information is available to tell anything about either the time-temperature history of the sample or the amount of time spent in contact with water (Lumpkin 2001). It is known, however, that these mineral phases have the durability needed to survive in a variety of geologic environments for millions of years.

## Corrosion of Ceramic Waste Forms in a Repository

As the preceding discussion makes clear, there are significant uncertainties in using laboratory and natural analog data to make long-term predictions of ceramic waste form performance in a repository. There are two main issues.

The first source of uncertainty involves the limitations of laboratory corrosion data. The range of anticipated repository conditions, in terms of temperature and fluid chemistry, is extensive. Because of this, data on waste form corrosion do not cover the entire range of potential repository conditions. This problem is amplified by the underlying uncertainty of the model predictions of repository conditions that the waste form will experience.

Most laboratory tests necessarily have durations of only a few years at most, whereas repository lifetimes are on the order of ten thousand to one hundred thousand years. If the rate-limiting reaction mechanism for the waste form has not been identified, there is no mechanistic basis with which to extrapolate laboratory test results to long time periods. There is currently no consensus in the waste form community that the correct rate-limiting mechanisms have been identified for any of the existing waste forms. Natural analog data, while reassuring, have always

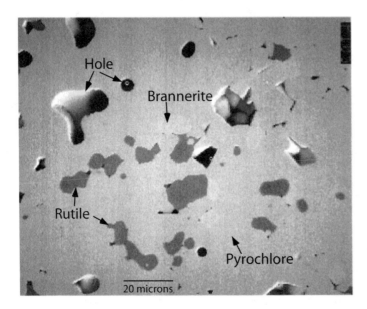

**Figure 22.1**
Scanning electron microscope image of titanate ceramic.

been problematic because of the lack of knowledge of the geologic history and in particular the duration of the water contact of the natural analog material.

The second uncertainty is the effect of radiation damage. Based on corrosion tests of irradiated materials, radiation damage is known to significantly accelerate ceramic corrosion. Of particular concern is the possibility that the swelling that accompanies the accumulation of radiation damage may break apart the ceramic along the crystal's grain boundaries and greatly increase the exposed surface area (see figure 22.1). Since the release rates of radionuclides are generally proportional to the waste form surface area, radiation damage could lead directly to an increase in radionuclide release. It is difficult to quantify this effect from tests of irradiated materials because their radiation damage is induced rapidly and may not be representative of real ceramic waste forms. So while it is known that radiation damage will lessen the ceramic's durability, the magnitude of the effect is unknown.

## A Ceramic Waste Form for Plutonium Immobilization

One ceramic waste form was, until recently, being considered for the Yucca Mountain repository (DOE 2002): the pyrochlore-based titanate ceramic. This waste form was developed as part of the DOE's effort to immobilize as much as fifty metric tons of surplus weapons-usable plutonium and uranium from dismantled nuclear weapons. The waste stream is primarily plutonium and uranium metal. Yet it also includes plutonium and uranium alloys, impure oxides, unirradiated fuel residues, and other materials generated during fissile material processing.

The selected ceramic waste form is titania-based ($TiO_2$) and contains roughly 10 weight percent plutonium. The solid phases are mainly pyrochlore, with lesser amounts of brannerite and rutile (table 22.1). The pyrochlore structure is essentially a titanium oxide framework containing two atomic sites that readily accommodate plutonium and uranium. Minor components of the waste stream generally substitute into one or more of the three mineral phases. Figure 22.1 shows a scanning electron microscope image of a section through the ceramic.

The titanate ceramic is fabricated using a cold press and sinter technique. Fine-grained reactive oxide precursors and the waste oxides are mixed and pressed into pellets at thirteen to twenty megapascals (130–200 atmospheres), and then heated at 1350°C for four hours in a nonreactive gas. The process has been automated for remote operation in a glove box.

Before selecting this waste form, the DOE considered several other ceramics, including cubic zirconia ($ZrO_2$), zircon ($ZrSiO_4$), uraninite ($UO_2$), gadolinium-zirconate ($Gd_2Zr_2O_7$), monazite ($CePO_4$), apatite ($Ca_5(PO_4)_3OH$), and perovskite ($CaTiO_3$). Summaries of the properties of these minerals can be found in Ewing and colleagues (1996) as well as Ewing (1999). The titanate was chosen for four reasons. First, extensive data are available from previous work to develop titanate waste forms for other waste streams. Second, uranium and plutonium have high solubilities in titanate (>10 percent by weight). Third, a simple fabrication method is available for titanate's synthesis. Finally, the titanates have high durability in water. Susceptibility to radiation damage was not a primary consideration. Ceramics were

**Table 22.1**
Solid phases present in the titanate ceramic waste form being developed for fissile element immobilization

| Mineral | Formula | Composition in ceramic |
|---------|---------|------------------------|
| Pyrochlore | $ABTi_2O_7$ | (Ca, Gd, Pu) (Hf, Pu, U, Gd) $Ti_2O_7$ |
| Rutile | $TiO_2$ | (Ti, Hf) $O_2$ |
| Brannerite | $UTi_2O_6$ | (Hf, Pu, U) $Ti_2O_6$ |

chosen over glasses for several reasons, including their much lower corrosion rates (Cochran et al. 1997).

Two key issues were considered in developing the titanate waste form. First, the waste form must pose an effective barrier that would prevent terrorists from having access to the plutonium in the waste. In 1994, the National Academy of Sciences recommended the use of the "spent fuel standard." This standard mandates that the retrieval of fissile elements from the waste should be at least as difficult as retrieving such elements from spent reactor fuel (National Research Council 1994). For this reason, the ceramic waste will be placed inside canisters of high-level waste; this "can-in-canister" approach is illustrated in figure 22.2. High-level waste contains the highly radioactive elements cesium-137 and strontium-90, which supply a source of penetrating radiation not present in the ceramic waste form. The high-level waste provides the radiation barrier to prevent human access.

The second issue is criticality. The waste consists primarily of fissile plutonium-239 and uranium-235, which could potentially become sufficiently concentrated to provide critical masses and hence accelerated nuclear reactions. Two scenarios deserve consideration. The first is that criticality arises during waste form production and within the waste form itself. The second is that criticality occurs during storage. This could happen if, for example, plutonium or uranium is leached from the waste form, transported, and deposited in a localized area. A similar process occurs in nature and is responsible for the formation of localized concentrations of

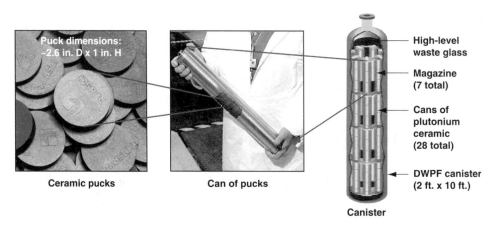

Figure 22.2
Ceramic can-in-canister concept.

metals in ore bodies, such as gold or uranium. Calculations of the transport and deposition of fissile elements in a repository system needed to evaluate the potential for criticality rely on coupled chemical-thermal-hydrologic models of repository evolution. There is considerable uncertainty in the interpretation of the results of these calculations, as described in chapters 11 12, and 14 (this volume).

Because of these concerns, two neutron absorbers, hafnium and gadolinium, were added to the ceramic. Both of these elements capture neutrons without fissioning and therefore greatly reduce the chance for criticality for both scenarios, as long as these neutron absorbers remain with the plutonium or uranium.

## Corrosion Tests of the Titanate Waste Form

The titanate waste form has been subjected to a variety of corrosion tests to measure its chemical durability under repository-relevant conditions. These tests show that as long as the ceramic has not been damaged by radiation, radionuclide release rates are extremely slow (Roberts et al. 2000). For many tests, in fact, it is not clear that anything at all is being released because the measured concentrations are too near the background levels to distinguish a signal (Bakel et al. 1999). An examination of the surfaces of reacted ceramics using electron microscopy usually show no reacted layer or, in a few cases, a layer only a few nanometers thick (see figure 22.3).

Figure 22.4 shows that at the measured corrosion rates, the maximum rates of plutonium and uranium release would, even after a million years, result in the penetration and potential release of radionuclides of only about a one-millimeter-thick layer of ceramic. At this rate, only a small fraction of the total radionuclide inventory could be released over the lifetime of the repository.

The tests that were designed to determine the rate-limiting mechanism for corrosion of the titanate were inconclusive. The reaction rates were so slow that it was impossible to quantify the influence of changing environmental parameters and thus to deduce the rate-limiting mechanism.

Over time, titanates will accumulate radiation damage due to the presence of alpha emitters. For alpha decay in the pyrochlore ceramic, the average alpha particle will travel about ten micrometers and the average recoil nucleus will travel about fifty nanometers (Weber et al. 1998). Each alpha particle produces about one hundred atomic displacements and each recoil nucleus produces about one thousand displacements. This self-radiation damage will cause the ceramic to become

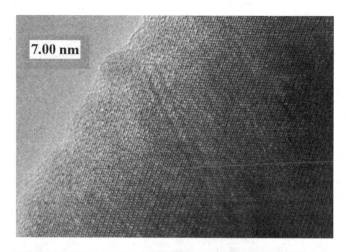

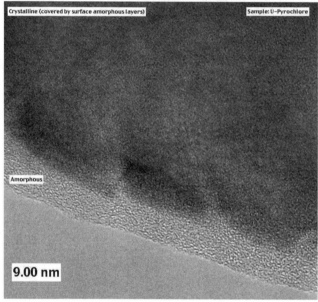

**Figure 22.3**

Transmission electron microscope images of unreacted (*upper*) and reacted (*lower*) pyrochlore samples. The lower sample was reacted for twenty-eight months in a pH 4 solution at 250°C and shows a thin amorphous alteration layer (photos courtesy of David Chamberlain, Argonne National Laboratory).

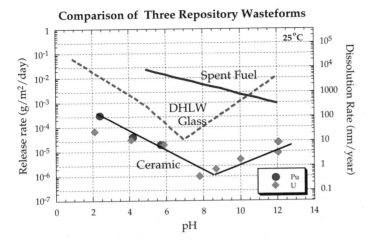

**Figure 22.4**
Corrosion rates of three repository waste forms: spent fuel, defense high-level waste (borosilicate) glass, and the titanate ceramic (Roberts et al. 2000).

amorphous in as little as one thousand years (Muller and Weber 2001) depending on the temperature evolution of the repository.

Even after sustaining radiation damage, however, pyrochlores may be able to retain their radionuclides for a long time. Evidence of this capability comes from observations of natural titanium-rich radioactive pyrochlores (betafites). A study of betafites from nine locations including completely amorphous samples showed that although the minerals had lost or exchanged cations in their "A" sites (see table 22.1), the minerals were able to retain virtually all of their uranium and thorium in their "B" sites. This was true even in specimens as old as 1.4 billion years (Lumpkin and Ewing 1985; Lumpkin 2001).

The long-term effect of radiation damage on the corrosion rate of titanate ceramic is unknown. To account for radiation damage in the performance assessment, the measured reaction rates were increased by a factor of thirty above the measured values for undamaged ceramics (DOE 2001; Shaw et al. 2001). This factor is an average value based on previous measurements of radiation damage effects on the corrosion rates of related ceramics.

## The Titanate Waste Form in the Repository

It is useful to compare the relative uncertainties of the performance predictions of ceramic waste relative to the waste forms currently destined for the repository—

borosilicate glass and spent fuel. Glass waste forms are thermodynamically unstable and will always seek to achieve a lower energy state by restructuring into crystalline forms at some rate that is difficult to predict. Ceramic degradation rates do not suffer from this uncertainty. They are already in a low-energy ordered state and do not transform. Ceramics have a known solubility that can be used to provide a conservative upper limit to their solubilities and therefore the radionuclide release rates. Glasses have no solubility limit that can be used with certainty to provide these conservative estimates.

Both glass and ceramic waste forms can be engineered to be compatible with the predicted oxidation state of the repository. For Yucca Mountain, this will be a relatively oxidized state—that is, some free oxygen will be present. Compatible oxidation states between the waste form and the local environment will avoid the potential for the occurrence of oxidation reactions between the site groundwater and the waste form. Such reactions are known to degrade the waste form and release radionuclides (see chapter 20, this volume). These types of reactions are often enzymatically catalyzed by microbes, which could significantly increase the reaction rate. Most spent nuclear fuel is composed dominantly of reduced uranium—that is, uranium that will react with oxygen to form new uranium solids and aqueous species. In a hot, humid oxidizing repository such as is expected for Yucca Mountain, the uranium will oxidize and, in the process, radionuclides are likely to be released. The rate of this process is highly uncertain, but it must be estimated in order to predict the radionuclide release rates from the spent fuel. Neither glass nor ceramic waste form corrosion rate predictions suffer from this uncertainty.

## Summary

The waste form is one component of the repository that we have a great deal of control over. It can be engineered to be durable in site groundwaters and resistant to radiation damage. Ceramic waste forms are preferable to borosilicate glass due to their greater intrinsic durability. Both glass and ceramic are preferable to spent fuel because they will not undergo oxidation reactions that are likely to release radionuclides. Long-term predictions of waste form performance are limited by both the uncertainties in the extrapolation of short-term tests to long time periods and the inadequacies of our mechanistic models of waste form corrosion. An even more severe uncertainty may be that the calculations of waste form corrosion rely on and are limited by the uncertainties in the results of the thermohydrologic model predictions of repository conditions over time.

## References

Bakel, A.J., Merz, C.J., Hash, M.C., and Chamberlain, D.B. (1999) *The Long-Term Corrosion Behavior of Titanate Ceramics for Pu Disposition: Rate-Controlling Processes.* Argonne National Laboratory Report. ANL/CMT/CP-99412.

Cochran, S.G., Dunlop, W.H., Edmunds, T.A., MacLean, L.M., and Gould, T.H. (1997) *Fissile Material Disposition Program: Final Immobilization Form Assessment and Recommendation.* Lawrence Livermore National Laboratory. UCRL–ID–128705. October 3.

Department of Energy (DOE) (2001) *Defense Spent Nuclear Fuel (DSNF) and Other Waste Form Degradation Abstraction.* Argonne National Laboratory. ANL-WIS-MD-000004 REV01 ICN 01. December 20.

Department of Energy (DOE) (2002) Secretary Abraham Announces Administration Plan to Proceed with Plutonium Disposition and Reduce Proliferation Concerns. Press Release. http://www.energy.gov/engine/content.do?PUBLIC_ID=12974&BT_CODE=PR_PRESSRELEASES&TT_CODE=PRESSRELEASE.

Donald, I.W., Metcalfe, B.L., and Taylor, R.N.J. (1997) Review: The Immobilization of High-Level Radioactive Wastes Using Ceramics and Glasses. *Journal of Materials Sciences* 32, pp. 5851–5887.

Ewing, R.C. (1999) Nuclear Waste Forms for Actinides. *Proceedings of the National Academy of Sciences, USA* 96, March, pp. 3432–3439.

Ewing, R.C. (2001) The Design and Evaluation of Nuclear-Waste Forms: Clues from Mineralogy. *Canadian Mineralogist* 39, pp. 697–715.

Ewing, R.C., and Jercinovic, M.J. (1987) Natural Analogues: Their Application to the Prediction of the Long-Term Behavior of Nuclear Waste Glasses. In *Materials Research Society Symposium Proceedings*, vol. 84, ed. Bates, J.K., and Seefeldt, W.B. Pittsburgh: Materials Research Society, pp. 67–83.

Ewing, R.C., Weber, W.J., and Lutze, W. (1996) Crystalline Ceramics: Waste Forms for the Disposal of Weapons Plutonium. In *Disposal of Weapon Plutonium*, ed. Merz, E.R., and Walter, C.E. Boston: Kluwer Academic Publishers, pp. 65–83.

Hench, L.L., Charles, R.J., Cooper, R.C., Ewing, R.C., Hutchins, J.R., Readey, D.W., VerSnyder, F.L., and Widerhorn, S.M. (1981) *The Evaluation of Candidate High-Level Waste Forms, Report No. 3.* Department of Energy, Springfield, Virginia. USDOE/TIC–11472.

Lumpkin, G.R. (2001) Alpha-Decay Damage and Aqueous Durability of Actinide Host Phases in Natural Systems. *Journal of Nuclear Materials* 289, pp. 136–166.

Lumpkin, G.R., and Ewing, R.C. (1985) Natural Pyrochlores: Analogs for Actinide Host Phases in Radioactive Waste Forms. In *Materials Research Society Symposium Proceedings*, Vol. 44, ed. Jantzen, C.M., Stone, J.A., and Ewing, R.C. Pittsburgh: Materials Research Society, pp. 647–654.

Lutze, W., and Ewing, R.C. (1988a) *Radioactive Waste Forms for the Future*. Amsterdam: North-Holland, 778 p.

Lutze, W., and Ewing, R.C. (1988b) Silicate Glasses. In *Radioactive Waste Forms for the Future*, ed. Lutze, W., and Ewing, R.C. Amsterdam: North-Holland, pp. 1–159.

Muller, I., and Weber, W.J. (2001) Plutonium in Crystalline Ceramics and Glasses. *Materials Research Society Bulletin* 26, pp. 698–706.

National Research Council (1994) *Management and Disposition of Excess Weapons Plutonium*. Washington, DC: National Academy Press, 275 p.

Roberts, S.K., Bourcier, W.L., and Shaw, H.F. (2000) Aqueous Dissolution Kinetics of Pyrochlore, Zirconolite, and Brannerite at 25, 50, and 75 Degrees C. *Radiochimica Acta* 88, pp. 539–543.

Shaw, H.F., Ebbinghaus, B.B., Bourcier, W.L., and Gray, L. (2001) *Plutonium Immobilization Project Input for Yucca Mountain Total Systems Performance Assessment (Rev 4)*. Lawrence Livermore National Laboratory. PIP-01–004. January.

Sickafus, K.E., Mineruinu, L., Grimes, R.W., Ualdez, J.A., Ishimaru, M., Li, F., McClellan, K.J., and Hartmann, T. (2000) Radiation Tolerance of Complex Oxides. *Science* 289, pp. 748–751.

Wald, J.W., and Weber, W.J. (1984) Effects of Self-Radiation Damage on the Leachability of Actinide-Host Phases. In *Nuclear Waste Management Advances in Ceramics, Vol. 8*, ed. Wicks, G.G., and Ross, W.A. Westerville, OH: American Ceramic Society, p. 71.

Weber, W.J., Ewing, R.C., Catlow, C.R.A., delaRubia, T.D., Hobbs, L.W., Kinoshita, C., Matzke, H., Motta, A.T., Nastasi, M., Salje, E.K.H., Vance, E.R., and Zinkle, S.J. (1998) Radiation Effects in Crystalline Ceramics for the Immobilization of High-Level Nuclear Waste and Plutonium. *Journal of Materials Research* 13, pp. 1434–1484.

# Part III
## Coping with Uncertainty

# 23

## The Path to Yucca Mountain and Beyond

Luther J. Carter

The legitimacy of the Yucca Mountain Project has been clouded for more than a few scientists by a belief that narrowing the search to Yucca Mountain was blatantly political, arbitrary, and without a sound technical basis. The congressional override of Nevada's veto of the site did not overcome the distrust because many saw this as a predictable act of political expediency by the majority at the expense of a politically weak host state.

But the performance standard for the Yucca Mountain Project is now likely to be high enough that, if it is ultimately met in a rigorously contested licensing proceeding before the U.S. Nuclear Regulatory Commission (NRC), most fair-minded critics may concede that the project design is, at last, convincingly robust. A decision tightening the standard was the ruling by the U.S. Circuit Court of Appeals for the District of Columbia (July 9, 2004) requiring, in effect, that radiation protection for the public must continue not just for the ten thousand years specified by the Environmental Protection Agency (EPA) but for a vastly longer time. Indeed, the protection would have to extend over the hundreds of thousands of years when radiation doses will be at their peak. The implications of this ruling for the project design will be discussed later.

This chapter presents several major arguments:

• That the Nuclear Waste Policy Act of 1982 and its attendant circumstances made it virtually inevitable that Yucca Mountain would become the site chosen.

• That the suitability of Yucca Mountain or any other specific repository site must be evaluated against a proof of safety embracing adaptive strategies of waste packaging and emplacement that allow for a credible calculation of long-term performance—and that the Yucca Mountain Project, properly pursued, can meet this test.

• That the Yucca Mountain Project should be placed under stronger direction in a new, independent federal corporation and given new assurance of essential funding from the existing user fee on nuclear-generated electricity and the Nuclear Waste Fund.

• That a central interim surface (or near-surface) storage facility for older spent fuel should be established for the sake of a compelling mix of political, technical, and safety objectives, including a more technically prudent pacing of the geologic repository project. Success in providing both interim storage and final geologic isolation may be a prerequisite for an international network of centralized storage centers and geologic repositories that could strengthen the nuclear nonproliferation regime.

The Circuit Court of Appeals decision of July 2004 has significantly altered the context in which the Yucca Mountain Project must henceforth be considered. Lawsuits had been brought against the Yucca Mountain Project by Nevada, Las Vegas, and Clark County, and the Natural Resources Defense Council together with various other environmental groups. These suits challenged the legality of the U.S. Department of Energy's (DOE) site-selection criteria, the EPA's radiation protection standards, the Nuclear Regulatory Commission's (NRC's) licensing criteria, and even the constitutionality of the congressional action that imposed upon a single state an unwanted nuclear waste repository. The court rejected all the claims except the one challenging the EPA standard limiting the period of compliance for radiation protection to ten thousand years. Here, the EPA was found to have failed to honor its mandate under the Energy Policy Act of 1992 to adopt radiation protection standards "based upon and consistent with" the findings and recommendations of the National Academy of Sciences.

In 1995, a panel of the academy's National Research Council had found "no scientific basis for limiting the time period of the individual risk standard to 10,000 years or any other value." It noted, moreover, that compliance assessment for most physical and geologic aspects of repository performance is feasible on a timescale consistent with the stability of the geologic regime, which is reckoned at on the order of a million years (National Research Council 1995). Peak radiation doses, attributable to the extremely long-lived actinides neptunium-237 and plutonium-242 as well as the fission product iodine-129, will not occur until after 100,000 years or longer (CRWMS M&O, 1998).

The Department of Justice is not appealing the Circuit Court of Appeals decision, and after conferring with the EPA and the DOE, is leaving it to the EPA to rework its radiation protection standard to the court's satisfaction (Tetreault 2004). In August 2005, the EPA issued in draft for public comment a standard establishing a two-tiered standard, with a 15-millirem per year maximum dose for the first ten thousand years and a 350-millirem per year maximum dose for the hundreds of thousands of years to come thereafter. The 350-millirem maximum is itself lower by orders of magnitude than the peak doses reported by the Yucca Mountain Project in past TSPAs. But a 15-millirem standard for the entire period of hazard would not be out of reach for a properly designed repository and the state of Nevada will surely be demanding nothing less.

In a lead editorial, the *New York Times* has observed that to require protection out to hundreds of thousands of years is "so outlandishly stringent it may not be achievable" (Roadblock at Yucca Mountain 2004). But while this may square with almost everyone's intuitive belief, it happens to be wrong. Paradoxically, the uncertainty as to radiation safety actually may be greatest during the initial ten thousand-year period when the integrity of the container is in play (Pigford 1999). Moreover, new approaches to waste emplacement may hold promise of safe repository performance over virtually the entire trajectory of the radiation hazard.

### The Inevitability of Fetching up at Yucca Mountain

The Nuclear Waste Policy Act of 1982 provided for what came to be called a "first round" and "second round" of site screening and selection. The first round had to do with choosing an initial repository site from among the nine sites available at the time the act was passed; the screening was not expressly limited to those sites, but the act prescribed a schedule so accelerated as to afford no chance to look beyond them. But the second round, overlapping with the first one, proceeded from a blank slate to identify a site for a second repository from among some 235 crystalline rock bodies in the eastern half of the United States. The screening was to be progressively narrowed by applying guidelines and weighted criteria arrived at by the DOE in collaboration with the potential host states (Carter 1987a; pp. 402–413).

Both rounds were to fail politically. The second round caused anxiety across the upper Midwest, northern New England, and the Southeast. The Reagan administration, worried that this issue might tip control of the Senate to the Democrats, canceled the second round in May 1986. At the same time, the first round screening was narrowed to only three sites. One was a bedded salt site in an extremely rich agricultural county (Deaf Smith) in the west Texas panhandle, which originally had been among seven salt sites under investigation. The other two finalists were each in volcanic rock on federal land in the West: the basalt site at Hanford in Washington State, and the welded tuff site at Yucca Mountain in Nevada.

The Deaf Smith site was almost certainly unobtainable given the high place of Texas in the political firmament and the fear that a nuclear waste repository might contaminate the Ogallala Aquifer. The Hanford basalt site enjoyed a locally sympathetic political base, but it presented technical problems great enough to raise questions as to whether a repository could be safely built there at all. For instance, high internal stresses in the basalt would pose a danger of "rock bursts" when the basalt was exposed to atmospheric pressure (National Research Council 1983). Also, given the high water pressure in the extremely prolific deep aquifer, the deep mining and tunneling would pose a risk of catastrophic flooding (U.S. Geological Survey 1983).

Political support for the siting of the first repository, always shaky, deteriorated even further from the cancellation of the selection of a second-round site. But other factors also contributed to the mounting controversy about siting the first repository and to the narrowing of the search to Yucca Mountain. These included the mounting cost of site characterization, Hanford's last place in the site rankings, and, yes, the convenience of looking to a remote desert site on federal land in a politically weak state.

Knowing that accusations of political expediency could be expected, Congress would have been wiser to have narrowed the choice to Yucca Mountain not in one fell swoop but by two steps taken about a year apart. Initially, Congress could have chosen the site tentatively subject to a review by the National Research Council of the National Academy of Sciences and the Natural Academy of Engineering, or, more particularly, the council's Board on Radioactive Waste Management. Inasmuch as the board's mandate would have been to judge only whether Congress was making a technically defensible choice of a site where no repository would be built without extensive exploratory testing by the DOE and intensive licensing review by the NRC, its finding would quite predictably have been positive. Nonetheless, that would have made the choice of Yucca Mountain less easily characterized by Nevada as a decision of raw expediency (Carter 1987b).

## Geologic Isolation and the Challenge of Long-Term Safety

Twenty-two years ago when the Nuclear Waste Policy Act was passed, government policymakers, whether in Congress or the executive branch, had no idea that the isolation of long-lived, heat-generating waste in a geologic formation was so technically daunting. Even many technical people in nuclear waste management were to find (and appear, even now, to still find) the concept that project design must be driven and shaped by a robustly believable safety case to be elusive.

The bill that the chair of the Senate Energy and Natural Resources Committee introduced in 1981 allowed only five years for screening sites and obtaining an NRC construction license for the site chosen. Yet the DOE's assistant secretary for nuclear energy applauded this ridiculously compressed schedule and said it could be met. Even as finally enacted, the Nuclear Waste Policy Act allowed less than seven years for site screening and selection and issuance of a construction license. The siting, licensing, and building of a repository was assumed to be a straightforward engineering task, less difficult, if anything, than doing the same for a nuclear power plant (Carter 1987a, pp. 195–230).

Yet some earth scientists had warned that the scientific foundations of geologic disposal were much weaker than had been supposed. In 1978, a U.S. Geological Survey paper warned of a variety of poorly understood "interactions among the mined opening of the repository, the [heat-generating] waste, the host rock, and any water that the rock may contain" (Bredehoeft et al. 1978).

The geologic medium of principal interest at the time was salt, so it was that the earliest findings of disturbing uncertainties pertained to salt. Experience in building the Waste Isolation Pilot Plant in New Mexico for defense transuranic waste was to bear out the need for caution. The tunnels excavated from the bedded salt at the Waste Isolation Pilot Plant, with no heat source present, were found by actual measurement to tend to undergo closure from the tendency of salt as a plastic medium to "creep." Calculations by a Sandia National Laboratory scientist indicated that if forty canisters of heat-generating high-level waste were placed in vertical holes in the floor of an eighteen-foot-high Waste Isolation Pilot Plant chamber, the chamber would lose almost half its height within ten years, with the floor creeping up and the ceiling creeping down (Carter 1987a, pp. 188–191). (Although salt creep alone does not disqualify salt as a repository host rock, it poses a formidably complicating factor if, as called for by the Nuclear Waste Policy Act, a spent fuel/high-level waste retrieval option must be maintained for up to a century.)

## An Adaptive Strategy and the Oxidizing Environment

The Yucca Mountain Project has from the start presented its own compelling need for an inventive and adaptive design strategy. The repository horizon inside the mountain is about 300 meters below the ridgetop and some 240 meters above the water table in what geologists know as the vadose or unsaturated zone. Water fills up to 80 percent of the pores in the rock, but because it does not saturate the rock, the water typically moves quite slowly, tending to be held in place by capillary tension (CRWMS M&O 1998).

Yucca Mountain offers major advantages over sites beneath the water table in that the repository, accessible by ramps from outside the mountain, would be relatively dry and amenable to monitoring and inspection for centuries. But oxygen is present and the mountain breathes. Hence, the unsaturated zone is an oxidizing environment where canisters and spent fuel may undergo rapid corrosion if they become wet or damp, as from high humidity or water dripping from the ceiling.

In 1998, I wrote an article with Thomas H. Pigford of the University of California at Berkeley for the *Bulletin of Atomic Scientists* on "Getting Yucca Mountain Right." The article concluded that the DOE effort was mistaken to focus on such expedients as drip shields and the use of containers lacking credible calculations of long-term performance (Carter and Pigford 1998). It also concluded that the department was wrong to stick to a schedule that might not allow time for a design effort supported by robustly defensible performance assessments (Carter and Pigford 1998). The DOE's goal to have the repository receiving waste by 2010 was unrealistic from the start.

The court ruling voiding the EPA's 10,000-year period of compliance for radiation protection makes it all the clearer that a license application should await inventive and convincing new approaches to the central question at Yucca Mountain of how to cope with the oxidizing environment.

A central aim must be to achieve proof of safety. As Pigford has emphasized, "proof of safety relies on defensible predictions using bounding calculations based on known phenomena. The strategy rests on designs amenable to such calculation. If all the science can't be understood, that's all right as long as a reliable prediction can be made that the safety limit is not exceeded" (personal communication).

Two examples of promising innovative design strategies cited in our work (Carter and Pigford 1998) are reprised here. First, a *capillary barrier* can be created to keep water from reaching the waste containers by covering the sides and tops of these containers with a coarse gravel layer and then placing a layer of sand or finely ground tuff on top of that. The capillary forces exerted by the sand layer would move water away from the waste containers and divert it into porous rock beneath the waste emplacement tunnel. The containers will ultimately fail from corrosion given the dampness present attributable to water vapor, but radioactivity escaping from them will diffuse so slowly along surfaces of the gravel particles (where capillary forces will be essentially absent) as to remain trapped there for hundreds of thousands of years. The effectiveness of a capillary barrier is, in principle, amenable to calculation and prediction over the period of hazard (Apted 1994; Zhou et al. 1996). This barrier could render unimportant questions about how much water will drip from tunnel ceilings, where the dripping will be greatest, and at what rate and by what manner corrosion of waste containers ultimately occurs. The DOE performed a preliminary evaluation of capillary barriers a few years ago, but a deeper, more complete review of this concept is now in order. A critical threshold for the concept is to establish that the capillary barrier would retain its integrity in an earthquake (see Carter and Pigford 2005).

The hot repository that the DOE has been planning would not be compatible with a capillary barrier (or at least not while the hot phase persists) because capillary forces weaken when temperatures approach the boiling point of water. But a cool repository is favored in any case by many technical people who follow the project, including members of the presidentially appointed Nuclear Waste Technical Review Board. The board insists that wall temperatures be kept from exceeding 80°C (well below the boiling point of water) by such means as longer aging of the spent fuel, ventilation over longer periods, and a reduced density of spent fuel emplacement (DOE 2002).

Another promising strategy has to do with the use in waste containers of *depleted uranium as a sacrificial material* that reacts with any oxygen present (Forsberg and Dole 2001; Forsberg and Haire 2002). The DOE has enormous stocks of this material—dangerously toxic both radiologically and chemically—stored in tens of thousands of aging steel cylinders outside its two inactive uranium-enrichment plants at Oak Ridge, Tennessee, and Portsmouth, Ohio, and its active plant at Paducah, Kentucky. Depleted uranium might fill voids in waste containers and be embedded in an eighteen- to twenty-five-centimeter-thick steel cermet that would fit around the container. The cermet in turn would be surrounded by an outer shell of corrosion-resistant material to block the entry of all oxygen before the outer shell begins to fail, and impede such entry for tens of thousands of years after that if (as some metallurgists believe) the shell fails gradually by cracking and pinhole corrosion. The size and number of cracks and pinholes could be determined by an accelerated experiment applying great heat and stress in a damp environment to show, possibly within a year or two, the rate at which the cracking and pinhole corrosion occurs (Pigford 2004). This should allow reliable calculation of the diffusion of air through the cracks and pinholes, and a bounding estimate of the oxidation lifetime of the depleted uranium that protects the spent fuel from corrosion. The DOE should revisit its past studies of oxygen "getters" such as depleted uranium with a new sense of priority.

The NRC might most usefully find room in a Yucca Mountain licensing proceeding for experiments that could lend confidence to the use of depleted uranium and capillary barriers for proof of safety.

## Placing the Project under Stronger, More Independent Direction

The Nuclear Waste Policy Act specified that the Office of Civilian Radioactive Waste Management was to be placed within the DOE under a director who would be appointed by the president and be directly responsible to the secretary of energy. In principle, the director was thus afforded considerable status, but in practice intermediaries have too often been interposed between the secretary and the program director, weakening the director's standing and ability to get things done.

A stronger Yucca Mountain Project should result if Congress were to reestablish the Office of Civilian Radioactive Waste Management as a special, independent federal agency or corporation to be headed by a politically resourceful administrator. This would have to be someone able to bring together an influential alliance from the array of interests who need and want the safe storage and disposal of spent fuel from the nation's power reactors, nuclear navy, and dangerous legacy of highly radioactive waste from the production of nuclear weapons.

There is precedent and considerable support among experts on public administration for placing such an independent entity under a single administrator appointed by the president, with no policymaking board to encumber the administrator's decisions. As long as the Yucca Mountain Project is struggling to survive, the administrator should be based in Washington, DC, close to the strings of political power. If the project ever reaches a phase of quiet implementation, the administrator might then move to Nevada.

Creating this new administrative entity should be coupled with a congressional initiative to remove budget constraints that are plainly contrary to the intent of the Nuclear Waste Policy Act. The act created a user fee on nuclear-generated electricity that now yields about $700 million a year for the Nuclear Waste Fund. But the Yucca Mountain Project's annual appropriation, from all sources, is less than what the user fee alone produces. More than $14 billion in user-fee revenue and investment income has been allowed to pile up unspent in the Nuclear Waste Fund as an offset against the federal budget deficit (DOE 2004). While subject to congressional review, the new corporation's financial plan should be outside the usual annual appropriations process.

The administrator could be required to consult two formally constituted boards of advisors (Dean 2002). One would be of stakeholders, with an appropriate balance struck between Nevada officials, environmental groups, the utilities, and representatives of states and communities having major accumulations of spent fuel and other highly radioactive waste. The other board would be a technical review body of eminent scientists and engineers. A Nuclear Waste Technical Review Board already exists, and is appointed by the president from a short list of scientists and engineers drawn up by the National Academy of Sciences; but if this present board is to continue, it should be refocused to raise and address proof of safety issues with greater force and clarity.

## Interim Spent Fuel Storage

Pending the availability of a geologic repository, the case for beginning storage of older spent fuel at a site away from the reactor stations is compelling. In principle, the best place for such storage may be somewhere in the vicinity of Yucca Mountain, but present law forbids choosing a site in Nevada. The earliest place to become available seems likely to be in Utah, on the reservation of the Skull Valley Band of the Goshute Indians about fifty miles southwest of Salt Lake City. Private Fuel Storage, LLC, a consortium representing roughly a third of all U.S. nuclear utilities, eight years ago filed an application for cask storage at this site. On September 9, 2005, the NRC denied the state of Utah's appeals to block this fuel storage initiative. The initial twenty-year license would be for storage of up to 40,000 metric tons of fuel, but the storage might begin and end with the first of the four planned 10,000-ton increments.

Having an assured option for interim storage would relieve the nuclear reactor stations' host states and communities of the worry that all spent fuel generated may remain with them forever despite the promises of the Nuclear Waste Policy Act. Under the act, the DOE was to have begun taking custody of spent fuel by early 1998. Since then, the department has been in default and the utilities' damage claims against it are now running into the billions of dollars. But not least among the advantages of having a surface storage site available would be to relieve managers of the Yucca Mountain Project from political pressures to push schedules beyond what proof of safety allows.

The only spent fuel that *must* be stored at the nuclear reactors is the very hot fuel recently discharged from reactor cores. Such fuel requires several years of cooling in pools or ponds near the reactors. Storage of all older spent fuel at reactor stations is now done either in the pools, or in the case of thirty of the sixty-five nuclear stations, in arrays of massive, passively cooled dry casks or monoliths. All on-site storage of older fuel could ultimately be phased out in favor of central storage in Utah or perhaps ultimately in Nevada.

According to a new study, the common practice of storing variously aged fuel in reactor cooling pools at densities several times greater than the pools' design capacity poses the risk of a radiological catastrophe of immense proportions (Alvarez et al. 2003). This study holds that if the cooling water in a pool were to be lost, as in a terrorist incident, the newest, hottest fuel could catch fire. With the conflagration then spreading to the older fuel, the release of volatile fission products (especially cesium-137) might be far greater than occurred at Chernobyl. A new report by the National Academy of Sciences' Board on Radioactive Waste Management concludes that this concern is warranted despite the fact that the NRC believes that it is not (National Research Council 2005). When opinions among responsible technical people are divided about so alarming a scenario, prudence surely dictates removing all but the newest, hottest fuel from the pools.

For older fuel, a generic safety advantage of surface storage in Utah or Nevada would be that special security precautions against terrorists can be provided far better at a remote desert location than it can be now at scores of nuclear stations sited adjacent to rivers, lakes, or coastal waters. Eighteen commercial reactors have been retired from active service over the past thirty years, and in the coming decades the number will grow. Central spent fuel storage can only make the decommissioning of such reactors and the disposition of their fuel easier and cheaper, and given the growing concern about terrorism, much safer too.

Finally, a word in a global perspective is in order about a U.S. solution to the storage and disposal of spent fuel and high-level waste (Carter and Pigford 1999, 2000). Such a solution could strengthen the increasingly troubled international nuclear nonproliferation regime. International spent fuel storage and disposal centers remain a scarcely dreamed of possibility even though they might afford a new means of control over the weapons-usable fissile material that this fuel contains. Willing host countries for such centers are less likely to be found when neither the United States nor any other country with a major nuclear program has yet to confidently demonstrate the management of even its own spent fuel from beginning to end. The United States could meet that test and even begin pondering whether limited amounts of spent fuel might be accepted from abroad when this would serve its nonproliferation policy.

## References

Alvarez, R., Beyea, J., Janberg, K., Kang, J., Lyman, E., Macfarlane, A., Thompson G., and von Hippel, F.N. (2003) Reducing the Hazards from Stored Spent Power-Reactor Fuel in the United States. *Science and Global Security* 11, pp. 1–60.

Apted, M.J. (1994) Robust EBS Design and Source Term Analysis for the Partially Saturated Yucca Mountain Site. In *Proceedings of the Fifth Annual Conference on High-Level Waste Management*. Las Vegas: DOE, p. 485.

Bredehoeft, J.D., England, A.W., Stewart, D.B., Trask, N.N., and Winograd, J.J. (1978) Geologic Disposal of High-Level Radioactive Waste: Earth Science Perspectives. *U.S. Geological Survey Circular* 779, p. 3.

Carter, L.J. (1987a) *Nuclear Imperatives and Public Trust: Dealing with Radioactive Waste*. Washington, DC: Resources for the Future, 473 p.

Carter, L.J. (1987b) Making the Nevada Test Site the Nation's Nuclear Waste Center. *Environment* 29, October, p. 28.

Carter, L.J., and Pigford, T.H. (1998) Getting Yucca Mountain Right. *Bulletin of Atomic Scientists* 54, March/April, pp. 56–61.

Carter, L.J., and Pigford, T.H. (1999) The World's Growing Inventory of Civil Spent Fuel. *Arms Control Today* 29, January/February, pp. 8–14.

Carter, L.J., and Pigford, T.H. (1999–2000) Confronting the Paradox in Plutonium Policies. *Issues in Science and Technology*, Winter, pp. 29–36.

Carter, L.J., and Pigford, T.H. (2005) Proof of Safety at Yucca Mountain. *Science* 310, 21 October, pp. 447–448.

CRWMS M&O (Civilian Radioactive Waste Management System Management and Operating Contractor) (1998) *Total System Performance Assessment–Viability Assessment (TSPA–VA) Analyses Technical Basis Document*. Las Vegas.

Dean, A.L. (2002) Interview with the author. September 30 and October 1.

Department of Energy (DOE) (2002) *Yucca Mountain Science and Engineering Report*, *Rev. 1*. Office of Civilian Radioactive Waste Management. DOE/RW-0539-1. February.

Department of Energy (DOE) (2004) Office of Civilian Nuclear Waste Management Budget and Funding Monthly Summaries. http://www.ocrwm.doe.gov/pm/budget/index.shtml.

Forsberg, C.W., and Dole, L.R. (2001) Depleted Uranium Dioxide Waste Packages for Spent Fuel Materials. Paper presented at the Research Society meeting, November 26–30, Boston.

Forsberg, C.W., and Haire, M.J. (2002) Depleted Uranium Dioxide–Steel Cermets for Spent Nuclear Fuel Multipurpose Casks. Paper presented at the American Nuclear Society meeting on DOE Spent Fuel Management, September 17–20, Charleston, South Carolina.

National Research Council (1983) *A Study of the Isolation System for Geologic Disposal of Radioactive Wastes*. Washington, DC: National Academy Press, p. 174.

National Research Council (1995) *Technical Bases for Yucca Mountain Standards*. Washington, DC: National Academy Press.

National Research Council (2005) *Safety and Security of Commercial Spent Nuclear Fuel Storage*. Washington, DC: National Academy Press, 114 p.

Pigford, T.H. (1999) Geologic Disposal of Radioactive Waste: Ethical and Technical Issues. In *Proceedings: VALDOR—Values in Decision and Risk*, ed. Anderson, K. Stockholm: pp. 113–128.

Pigford, T.H. (2004) Private communication with the author. August 27.

Roadblock at Yucca Mountain (2004) Lead Editorial. *New York Times*, August 23, p. A22.

Tetreault, S. (2004) U.S. Government to Sit out Challenge of Radiation Ruling. *Las Vegas Review Journal*, August 24, p. 1B.

U.S. Geological Survey (1983) *Review Comments on the [U.S. Department of Energy] Site Characterization Report for the Basalt Waste Isolation Project*, pp. 2–3. May 6. Cited in Carter 1987a, p. 169.

Zhou, W., Conca, J., Arthur, R., and Apted, M.J. (1996) *Analysis and Confirmation of Robust Performance for the Flow-Diversion Barrier System within the Yucca Mountain Site*. Office of Civilian Radioactive Waste Management. TR–107189.

# 24

# Uncertainty, Models, and the Way Forward in Nuclear Waste Disposal

Allison M. Macfarlane

I end this book with a discussion of uncertainty, the theme that has been woven through the chapters. In examining uncertainty, I also evoke the theme of process, in particular the process involved in the technical aspects of nuclear waste disposal policy decisions. Before going any further, I should make clear that the views in this chapter are my own and do not necessarily reflect those of the other chapter authors.

The focus of this chapter, and the book as a whole, is on two questions: Is the Earth system understood well enough to make predictions about the future behavior of radioactive waste emplaced into rock? And can the models that provide these predictions be verified or validated? Furthermore, if the answer to these two questions is no, then how can a nuclear waste repository site be evaluated?

The authors of the first chapters (1–6) of this volume have described the history and process that have brought the proposed geologic repository at Yucca Mountain nearly to the point of a license application submission to the Nuclear Regulatory Commission (NRC). The remaining chapters (7–23) have described a number of the technical issues that have been raised in the evaluation of Yucca Mountain as a suitable and safe geologic repository for high-level nuclear waste. With each of these technical issues there are uncertainties (chapter 5), some of which can be reduced by additional work and research, and some that are inherent to the extrapolation of the results of models over time and space. Some of the uncertainties may be critically important to understanding the performance of the repository, but others may be of little consequence.

The main issue confronted by advocates and detractors of Yucca Mountain is one of evaluating the safety of a repository, even in the face of some level of uncertainty that cannot be reduced. In the extreme, there are two possibilities, both undesirable. One is a positive determination of compliance by the NRC based on models that are so uncertain that the results and decision remain questionable. The other extreme possibility is that an adequate and safe repository site may be abandoned due to the large uncertainties involved. The authors of this book's chapters have raised many issues, but how does one place them into a context and process that leads to the best policy decision? In this last chapter, I will summarize the important sources of uncertainty in the analysis of the performance of Yucca Mountain and offer my own suggestions for the path forward.

## Uncertainty

The Department of Energy (DOE) contends that we now have an adequate understanding of geologic and hydrologic processes to make predictions about radionuclide transport over geologic periods of time (Abraham 2002).[1] I would rather pose the question, Has the DOE identified and does it understand all the "features, events, and processes," as the DOE calls them, that will occur over the next hundreds of thousands of years at Yucca Mountain, especially once thermally hot radioactive waste has perturbed the natural system?

The chapters in this book have noted a variety of factors that make it difficult to predict repository behavior over geologic time, including climate, saturated zone behavior, volcanism, unsaturated zone behavior, and the environment of the repository. Many chapters (12, 13, 15–22) have identified a similar uncertainty: the environmental and chemical conditions of the repository environment as it evolves over time, especially the chemistry of the water that will exist in the repository. This uncertainty arises from the difficulty of predicting interactions over tens to hundreds of thousands of years brought about by introducing a thermally and radioactively hot waste package into a complex geologic environment. The difficulty stems from the sheer number of parameters and processes to identify, many of which are coupled interactions (e.g., thermohydrologic-chemical, thermohydrologic-mechanical) that are hard to understand or even recognize. Furthermore, scientists have not yet identified some of the fundamental thermodynamic parameters and kinetic processes by which these interactions would be controlled.

The method by which the DOE and the NRC will evaluate Yucca Mountain's suitability as a repository is a probabilistic performance assessment. The DOE's version of probabilistic performance assessment, called Total System Performance Assessment (TSPA), is a huge computer model, the results of which will be used to compare to the Environmental Protection Agency (EPA) standard to determine compliance. The DOE's TSPA model "is a probabilistic analysis that identifies the features, events, and processes that might affect the performance of the repository; examines their effects on performance; and estimates the expected annual dose to the potential receptor. This method synthesizes data and information into a set of models that simulate the behavior of individual system components and then abstracts and refines this information into linked models that represent important aspects of system performance" (OCRWM 2002b, sec. 4.2).

This book discusses uncertainties in both the geologic and engineered barriers in the repository system at Yucca Mountain. But will these factors actually be detrimental to the performance of the repository? According to the DOE's weighting scheme in the TSPA model, some of these uncertainties are not significant enough to generate large radiation doses far into the future. For instance, the DOE claims that it has determined the percentage of water that travels in fast pathways compared with slow pathways, and that it is very limited: "It is estimated, however, that the fast flow component of the overall groundwater percolation flux at this horizon [the repository level] is less than a few percent of the total flow (table 4.12), which would contribute inconsequentially to releases of radionuclides from the repository" (OCRWM 2002a, sec. 4.3.4).

Knowledge about fast flow paths in the repository comes from studies that showed high levels of chlorine-36 in repository rocks. These high values most likely result from precipitation of bomb-pulse chlorine-36 from the testing of nuclear weapons over the Pacific Ocean in the 1950s. This estimate is quantified by transport models based on a set of assumptions that conclude that little water is carried in the fast flow paths (CRWMS M&O 2000). These assumptions include: bomb-pulse samples are found from only a few locations in the Exploratory Studies Facility; no significant correlation between high matrix saturation and elevated $^{36}Cl$ has been reported; these discrete fast paths are not associated with large catchment areas involving large volumes of infiltrating water; bomb-pulse signatures of $^{36}Cl$ were not found in the perched water bodies (CRWMS M&O 2000, U0085, sec. 6.6.3); and postbomb pulse tritium was detected only in one sample from the perched water (in borehole NRG-7a), but not in any of the other samples (CRWMS M&O 2000, U0085, sec. 6.6.2).

The authors of chapter 11 point out that one might not expect to find much bomb-pulse tritium because of its short half-life (12.3 years) and that bomb-pulse chlorine-36 would be highly diluted in perched water bodies. Moreover, the few locations in which bomb-pulse radionuclides were found depends entirely on the sampling methodology—a larger sample size may have indicated more instances of the presence of bomb-pulse radionuclides.

As the authors of chapter 11 suggest, the DOE's models do not include such processes as the effects from thousand-year storms. Scientists know little about which fractures may flow in the rocks, the processes that control fracture flow, and how water is partitioned between the rock matrix and the fractures. This example illustrates the degree to which the DOE's modeling results depend on assumptions, and the current level of knowledge about features, events, and processes.

Furthermore, knowledge about features, events, and processes at Yucca Mountain is continually in flux. For instance, the DOE is still collecting data on the process of water transport through the rocks at Yucca Mountain. One of the first field experiments done on repository-level rocks to understand how water flows over short distances in a fault zone was reported in 2005 (Salve et al. 2005). In addition, the DOE is still working to collect information on the diffusion process in the rocks at Yucca Mountain and to understand repository-induced processes, such as the effects of the repository's ventilation on the moisture content of the ambient atmosphere (Dyer and Peters 2004). Such data will result in further refinements in the performance assessment model.

Over the long term, such adjustments may cause substantial divergence from the original model, calling into question the use of these models in policy making (Oreskes et al. 1994). The DOE, however, disagrees: "Because uncertainty is fully integrated into the assessment of total system performance, DOE does not expect that additional information will significantly change the TSPA results or the conclusions reached in the site suitability evaluation and has confidence in the overall safety of the repository" (OCRWM 2002b, sec. 3.2.2). But the DOE is basing its opinion on the assumption that it has characterized the uncertainty correctly, and further, that it has characterized all the features, events, and processes that will occur in the repository as it evolves.

## Use of Models in Predictions

The discussion above illustrates how the DOE relies heavily on models that predict the performance of the geologic and engineered barriers at Yucca Mountain to determine site suitability. Many assumptions go into these models. To its credit, the DOE is conservative in the assumptions it makes. The problem, though, is that one cannot make assumptions about processes or features that one is not aware of.

Because the DOE's case for Yucca Mountain is based entirely on complex models of Earth system processes interacting with engineered features and processes, it is absolutely essential to examine the use of models in making detailed predictions. Others have explored the use and misuse of models in the earth sciences and technical policy decision making (see, for example, Bredehoeft et al. 1978; Winograd 1986; Winograd 1990; Konikow and Bredehoeft 1992; Oreskes et al. 1994; Ewing 1999; Ewing et al. 1999; Oreskes and Belitz 2001; Bredehoeft 2003; chapter 5, this volume). One of the main conclusions from these works is that these models cannot be validated or verified. Winograd (1990) goes so far as to say that models such as the TSPA cannot even be calibrated. Oreskes and colleagues (1994) explain that models of Earth systems cannot be validated because they attempt to simulate open systems, which exchange matter and energy with their surroundings. In open systems, there is no way to know all the input parameters or processes, or to assess the boundary conditions that might affect the system; the modeler must anticipate all input parameters for all processes that will occur over the time period modeled. For geologic timescales, this task is unfeasible because the current-day data sets are incomplete (Oreskes et al. 1994). Perhaps Secretary of Defense Donald Rumsfeld (2002) put it best when, speaking of another issue entirely (the Iraq War), he said: "There are known knowns; there are things we know we know. We also know there are known unknowns; that is to say we know there are some things we do not know. But there are also unknown unknowns—the ones we don't know we don't know." The chapters in this book underscore the fact that we are missing input data and knowledge of processes that will operate during the life of the repository.

Models of natural systems over geologic periods of time, such as the TSPA model, ignore the realities of the complexity of open systems over large timescales. Field geology is based on observations of the natural world and investigations into processes that occurred in the past. Investigations into past reactions among minerals and fluids in rocks show that equilibrium may be rarely reached, and therefore it is almost impossible to decipher the detailed history of a rock, let alone predict reactions into the geologic future. Geology has not advanced far enough yet to expect that it can do this for the rocks at Yucca Mountain.

The TSPA model is, by definition, a simplification of the natural and engineered systems at Yucca Mountain. Rare events, such as the thousand-year storms discussed earlier, are not included either because they are not recognized or not enough is known about them to construct reasonable input parameters or processes for modeling. Oreskes and Belitz (2001) suggest that models that omit low-probability events tend to produce overly optimistic results.

Sarewitz (2004) states that uncertainty should be thought about as a manifestation of a lack of coherence among scientific (and engineering) subdisciplines. Complex Earth systems problems, such as understanding the behavior of a repository at Yucca Mountain, require the cooperation and coordination of many different groups of scientists and engineers, all of whose subdisciplines have different values and diverse perspectives. The results they produce are in a sense incomparable because of their different origins (Sarewitz et al. 2000; Sarewitz 2004). It is possible, then, that a different management agency (not the DOE), different management, or different scientists and engineers might have come up with a different version of the TSPA, which might have given a different result. Interestingly, International Atomic Energy Agency scholars found that among six models prepared to study contaminant transport in agricultural fruits, there were large differences. These differences resulted from "what we call 'modeler uncertainty,' i.e., difference in problem formation, model implementation, and parameter selection originating from subjective interpretation of the problem at hand" (Linkov and Burmistov 2003, p. 1297).

A model of an open system such as the repository at Yucca Mountain can provide only a rudimentary glimpse of the system—and even that can be misleading, if certain processes and parameters are not accounted for. It is therefore unrealistic to think that the modeling of such a complex system as the Yucca Mountain repository could provide a single number that is in any way valid and represents the "truth." But this is just what the TSPA model does. According to the DOE, the TSPA model suggests that the dose from the repository ten thousand years after the waste is put in place will be 0.1 millirems per year (Office of Civilian Radioactive Waste Management 2002b).

The DOE says that it will validate models through the use of laboratory, in situ, and field tests as well as by studying analogous systems. For instance, as the DOE states, "Simulated thermal-hydrologic-chemical processes have been validated by comparison to water and gas compositions measured during thermal tests and with laboratory data" (Office of Civilian Radioactive Waste Management 2002b, sec. 3.2.1.2.3). Oreskes and colleagues indicate that although models can be confirmed through comparison with laboratory and field data, they cannot be validated:

Finally, even if a model result is consistent with present and past observational data, there is no guarantee that the model will perform at an equal level when used to predict the future. First, there may be small errors in input data that do not impact the fit of the model under the time frame for which historical data are available, but which, when extrapolated over much larger time frames, do generate significant deviations. Second, a match between model results and present observations is no guarantee that future conditions will be similar, because natural systems are dynamic and may change in unanticipated ways. (1994, p. 643)

The DOE continues to use the language of validation and verification extensively when discussing its models. As Oreskes and colleagues (1994) note, using the terms "validate" or "verify" are powerful signifiers of the truth of model results. Laypeople and policymakers tend to believe that when technical experts talk about models being validated, this means that the model provides an accurate representation of what will happen in the future (Oreskes and Belitz 2001). As pointed out above, in the case of Earth systems models, though researchers may claim that their models have been validated, this is not to say that the model can accurately predict the future behavior of the system. More disturbing is the use of model results by DOE experts in papers, discussions, and talks. In these situations, they often mistakenly refer to model results as if they represented actual data. For example, in Secretary of Energy Spencer Abraham's (2002) recommendation of the Yucca Mountain site, he remarks, "The amount of water that eventually reaches the repository level at any point in time is very small, so small that capillary forces tend to retain it in small pores and fractures in the rock. It is noteworthy that all our *observations* so far indicate that no water actually drips into the tunnels at this level and all of the water is retained within the rock" (emphasis added). Knowledge about how much water will reach the repository in the future is based on a number of model results; in this statement, what appears to be actual data are not—they are simply the predictions of a computer model. Such behavior is misleading to laypersons, policymakers, and other experts, and results in confusion about what is really known about the repository system and what is not.[2] Talking about modeling results as if they were established data leads to overconfidence in the state of technical evaluation of the situation.

Language has become an important tool in legitimizing both the Yucca Mountain site and the TSPA. Just consider the term "TSPA," which suggests that the model really does describe the "total" system. Consider further the following statement: "The code is described in the S&ER [Science and Engineering Report] Rev. 1, Section 4.2.4. Both conservative and realistic versions have been developed" (OCRWM 2002a, sec. 4.7.01). The DOE is informing us that it knows what is realistic and what is conservative when it comes to modeling. The problem is that the agency does not know, as has already been emphasized, all the features, events, and processes that will affect a repository over geologic timescales, and therefore cannot know what is realistic and what is conservative.

## Can Predictions Guide Policy Decisions?

If the TSPA is neither accurate nor a reliable method of judging the Yucca Mountain repository site, then how should the site be judged? Aren't predictions required to make sound technical policy decisions? Predictions are viewed by laypeople, policymakers, and the technical experts themselves as a way to legitimize policy decisions in an authoritative way (Sarewitz and Pielke 1999). Sarewitz and Pielke (1999) point out that scientists benefit from predictions by receiving funding to make them, while politicians benefit from deferring decision making until the prediction is completed. Herrick and Sarewitz (2000) assert that debate centered on the technical aspects of a complex policy decision often deflects attention from the "real" sources of debate—the issues of economics, values, ethics, equity, aesthetics, ideology, and regional politics that are at the base of issues such as high-level nuclear waste disposal.

Perhaps it's best to first examine how past predictions have fared. Have they been useful? Were they proven accurate? The record here is sobering: assessments of model-based predictions show that they have a poor track record (Konikow and Bredehoeft 1992; Oreskes and Belitz 2001; Bredehoeft 2003). For example, groundwater flow modeling of the Coachella Aquifer in California for the period 1974–1980 did not accurately predict water levels because of incorrect assumptions about the amount of stream recharge from tributaries in the system (Konikow and Bredehoeft 1992). Some scholars suggest that predictions are not necessary to make technical policy decisions (Sarewitz and Pielke 1999; Sarewitz et al. 2000; Oreskes and Belitz 2001; Bredehoeft 2003). If it is so difficult to get model predictions right, is there any reason to rely on models to make policy decisions? If not, then how should these decisions be made?

## Wisdom?

This book has focused predominantly on technical issues, but what is important for policy is how technical issues enter the decision-making process. Thus I pose this question: What would the U.S. nuclear waste program look like if it were done "properly"? Would it be as large as it is now? Would it cost as much as it does now? How would the technical information be used and what role would it play in policy decision making?

In an address to the Uranium Institute, Ian Duncan identified a number of ways *not* to go about siting a nuclear waste repository. They bear repeating here:

• To maximize the attraction of NIMBY, have the site decided behind closed doors and then announced and defended (DAD) . . .

• Trust only yourselves. Make sure that the oversight body is male dominated, industry loaded, with perhaps some token female and government presence.

• Rush the procedure, minimizing community involvement as much as possible.

• Pretend that the site selected will only be used for scientific evaluation, certainly not for final disposal.

• Distribute glossy brochures that depict the decay of radiotoxicity in obscure units over time, both on a log/log scale, and hope that the lay population will each have degrees in physics, chemistry, mathematics and geology . . . (Duncan 1999, p. 3–4)

In identifying a successful process of nuclear waste disposal policy, I will rely on the available information, including an examination of what other countries have done, a look back at the U.S. Nuclear Waste Policy Act, and suggestions from the literature. I will then lay out my own plan for success.

## Other Countries' Experience

Although the U.S. high-level nuclear waste program is relatively advanced, we can learn from looking at what others are doing around the world. Sweden and Finland, for example, have programs just as advanced as the one in the United States, but they are much more socially accepted. Sweden's program is run by SKB, a company formed by the four Swedish nuclear power companies. After initial attempts to characterize the geology in different areas of the country were met with public resistance in the 1980s, SKB decided to ask municipalities to volunteer for investigation (Lidskog and Sundqvist 2004). This second attempt to find a site also failed after the volunteer localities eventually rejected further investigation. Success came when SKB began its own study of the five municipalities that already hosted some type of nuclear facility. Four of the communities were asked to participate in a feasibility study, and two accepted. These two are now the focus of SKB's final investigation before an ultimate siting decision is made (Lidskog and Sundqvist 2004).

The Swedish example is interesting because after SKB's initial encounter with public resistance, the company switched its focus from finding a suitable geology to finding a site where the public would accept a facility (Lidskog and Sundqvist 2004). At this point (the late 1980s), SKB began to argue that suitable geology existed throughout the country. Thus, its technical focus shifted to developing an engineered barrier (the waste package, in this case a canister with a shell of copper) that would work well in the geologic environment typically found in the country (a reducing environment in saturated conditions). Another factor in Sweden's success so far has been that the public has a great deal of trust in SKB and the nuclear industry generally. Sixty percent of Swedish citizens trust information provided by their national organization in charge of nuclear waste disposal, as compared to 27 percent in other European Union countries (Lidskog and Sundqvist 2004).

Finland's nuclear waste program is also run by a company, Posiva Oy, established by its nuclear power companies. Posiva Oy has already settled on a single site, Eurajoki near Olkiluoto, where it is building an underground laboratory. Like Sweden, Finland intends to use a copper canister to store its waste in crystalline rock below the water table. After the country passed a law in 1987 that allowed local communities an absolute veto over a nuclear waste site, Posiva Oy investigated five sites representing Finland's different rock types (Lidskog and Andersson 2002). Though the final site at Eurajoki, which already houses a nuclear power plant, might have seemed likely from the beginning, the Finns found it useful to compare a number of sites, concluding that "without a program that included several investigation sites, it would not have been possible to say anything about the comparative advantages and disadvantages of the Olkiluoto and other candidate sites" (Vira 1996, p. 31).

Germany provides another thought-provoking example of how to dispose of nuclear waste even though it has not yet identified a site. With the election of the Socialist-Green government in 1998, a commission was established to revisit the issue of nuclear waste disposal and develop a plan to select a site. The commission issued its report in 2002. The plan is to follow a five-step process (AkEnd 2002). The first step will be to consider all possible sites in Germany and exclude some from use on the basis of five geologic criteria. To be considered suitable for nuclear waste disposal, the site:

• Must not exhibit large vertical movements above the rate of one millimeter per year

• Must contain no active fault zones

• Must have only very low seismic activity

• Must exhibit no quaternary volcanism (or the potential for future volcanism)

• Must not have young groundwater (the water must not contain carbon-14 or tritium) (AkEnd 2002)

Once sites are excluded based on these restrictions, the next step will be to identify sites with favorable geology based on a set of weighting criteria. Of the candidates that pass geologic muster, three to five will be selected for surface exploration in step three. At this time, the communities at the identified sites will be given the opportunity to reject the surface exploration studies. In the fourth step, at least two of the above sites will be selected for underground exploration; again, the local communities will be allowed a say in whether this happens. The fifth step is the site selection decision.

In the German case, strong public participation is required in the process to ensure success. Unlike the Swedish situation, the Germans plan to base their selection decision first on geology to ensure the safety of the site. They feel that geologic barriers will provide the longest-term resistance to radionuclide transport and thus they intend to emphasize the geology over the engineered barriers, depending on the site selected (AkEnd 2002). By selecting a tectonically stable site, the German time of compliance will be about one million years, give or take an order of magnitude (AkEnd 2002).

In all of these European examples, public participation in the siting effort is highly valued. In these countries' political systems, a nuclear waste repository cannot be sited without the support of the local public. These waste programs also took pains (and paid the financial costs) to characterize more than one site before the final site selection is made. Hence, their site selection does not depend on the single result of a complex model. They are using a variety of methods, including social acceptability, to make a final site selection. How does this compare with the U.S. situation?

### The U.S. Experience

The early U.S. nuclear waste program, as governed by the 1982 Nuclear Waste Policy Act, does not look terribly different from some of the European programs.

In light of this, I would like to ask what "advice" the act has for us now. The authors of the Nuclear Waste Policy Act took pains to instill a sense of fairness into the legislation because they knew how enormous the burden would be on a locale once it was chosen as the site to store such undesirable material. Thus, the act required the DOE to investigate a second site, which was tacitly understood to ensure geographic equity—one in the eastern portion of the country and one in the west (Davis 1987; Colglazier and Langum 1988). Similar to the German plan, the Nuclear Waste Policy Act directed that the DOE should first choose five sites for which environmental assessment reports would be written. From those five, the DOE was to choose at least three for in-depth characterization, including exploratory shafts to the proposed repository level. These three sites were to include at least two different rock types. As reported by Colglazier and Langum, "This requirement was to ensure sufficient diversity of sites to make possible a reasonable comparative assessment, in keeping with the intent of the National Environmental Policy Act" (1988, pp. 320–321). The act also included a provision that the selected state could veto the siting decision—a veto that would require majorities of both houses of Congress to override. (No veto power was given to the local municipality where the site was to be located.) Moreover, the act required that the federal government consult with the selected states and provide financial aid to them, which would support independent technical analysis and ease impacts resulting from the repository (Colglazier and Langum 1988).

This all changed, however, with the passage of the Nuclear Waste Policy Amendments Act in 1987. The three sites to be compared were reduced to one. The need for a second site was left in limbo, awaiting a report from the secretary of energy between 2007 and 2010 (Congress 1982). Congress even went as far in the amendments to legislate the type of rock that the DOE could investigate. The amendments included language that forbade investigation into crystalline rock (such as granite)—a constraint that in effect exempted a large part of the eastern portion of the country from even being considered as a repository site (Gerrard 1994).

With the enactment of the amendments, Congress in effect broke the covenant it had established with the country, and especially with states such as Nevada, which had potential sites. The amendments were viewed by Nevada as singularly unfair—a clear case of the "decide, announce, defend" process that Duncan (1999) decried earlier. In turn, because of the painful political process involved in the nuclear waste legislation, the DOE found itself under great pressure to deem Yucca Mountain a suitable site.

There is wisdom, though, to be found in the original Nuclear Waste Policy Act. The characterization of multiple sites, and better yet, sites with different geologies, allows one to choose not the best site but the best of the group. This gives some comfort to the public in the selected region. A Utah newspaper editorial written in 1981—before any nuclear waste legislation had yet been enacted—stated it well: "Neither Utah nor any other state can properly refuse to bear the nuclear waste burden once it (the repository site) has been established to the best of human conditions. However, the honor of making such sacrifice for time without end must confer on the luckless lamb the satisfaction of knowing first-hand that the duty couldn't have been just as well assigned elsewhere" (quoted in Colglazier and Langum 1988, p. 352).

During debate over the original Nuclear Waste Policy Act, Rep. Morris Udall (D–AZ) recommended that the states with the sites slated for characterization should be allowed to draw up their own compensation packages to negotiate with the DOE (Carter 1987). The need for two sites remains an open question, though Congress will likely prefer to stick with just one. But geographic equity argues for the existence of a second site.

**What to Do Now?**

If the U.S. nuclear waste disposal program had been developed "correctly," it would probably seem closer to what was outlined in the original Nuclear Waste Policy Act and the German plan. Sites would have been initially selected based on their geologic merit. More detailed research would have been conducted at a subset of sites (two or three). These sites would need approval by the local community and state before any research could go on. With such a plan, the method of final site selection would be measured in a relative sense, by comparing the sites, instead of an absolute sense, with no relative context as is the case now. One site might require more engineering, and one might require less. The final decision would be based on societal preferences at the time. If the United States preferred the Swedish path, it would choose the heavily engineered site; if it preferred the German path, it might select the site with the better geology. Furthermore, there would be a provision for two sites in the country—one in the East, and one in the West—as originally planned to ensure a sense of geographic equity. Then, one state would not feel solely put upon. The money spent on characterizing multiple sites simultaneously would be balanced by that saved from fewer lawsuits due to less citizen opposition to the plan.

This is not the U.S. situation, of course, and I am not trying to suggest abandoning Yucca Mountain and going back to the drawing board. Instead, I would like to put forth some ideas for improving the current situation based on this analysis. First, I suggest that policymakers, including the DOE and the NRC, de-emphasize the importance of performance assessment. The long discussion above highlights the many reasons that complex models such as the DOE's TSPA will not be able to make adequate predictions, and as a result these predictions cannot be validated or verified. The government's decision to base a policy of such significance as the disposal of high-level nuclear waste on a single result from a complex predictive model is misleading to the public and policymakers. No country should depend so heavily on the results of an uncertain computer model to make policy decisions such as these.

How, then, should policy decisions on nuclear waste be made? Let's revisit the advice of the person who first suggested that the United States consider the unsaturated zone in the Nevada desert: Ike Winograd. Winograd (1990) proposed that the DOE use "technical judgment" to evaluate the suitability of the site. Technical judgment includes the use of multiple barriers in a repository and encourages the use of multiple techniques to analyze the site, with weighting as judged reasonable by experience. This is simply a more honest way of acknowledging our inability to specify the suite of correct conceptual models for analysis.

Technical judgment can be supplemented by comparative analysis. In the current situation that relies on performance assessment, Yucca Mountain is being evaluated in isolation—that is, without comparison to any other site. This makes it difficult for policymakers, the public, and even scientists to grasp all the crucial issues that will affect the safety of a repository over geologic time. For both scientific and social reasons, comparing the site to others makes sense (Flynn and Slovic 1995). Thus, I propose that Yucca Mountain be evaluated via comparison to other existing or planned sites about which a substantial set of information has been gathered. Potential sites for comparison include the Swedish site, the Olkiluoto site in Finland, the clay site in France, and the Waste Isolation Pilot Project site in Carlsbad, New Mexico. If Yucca Mountain comes up significantly short in making such a comparison, then Congress will have to reconsider site selection.

Fortunately, an opportunity has recently emerged to make changes in the way Yucca Mountain is evaluated. Because of a July 2004 court decision (see chapter 1, this volume), the EPA must revisit the radiation protection standard for Yucca Mountain. This will cause the NRC to promulgate new regulations and the DOE to issue new guidelines. The NRC and the DOE could de-emphasize performance assessment in determining whether the Yucca Mountain site is suitable, and add in other measures such as those suggested above.

What else can be done to improve the current situation? Jasanoff (1995) has shown that the legitimacy of scientific assessments, especially those done by governmental bodies, can be improved by negotiation and compromise instead of controversy. Conflict over scientific assessments can be ameliorated by the use of the independent scientific community. Jasanoff (1995) suggests that policy decisions should be arranged so that there is continual and repeated consultation among the scientists producing the analysis, independent scientific experts, the public, and policymakers. The Nuclear Waste Technical Review Board, which oversees the DOE's nuclear waste programs, is an oversight body, with its members appointed by the president from candidates recommended by the National Academy of Sciences. This appointment process can result in a sense of political dependence that may mute the board's message. Instead, Congress or the DOE could provide funds to citizens' groups to consult with scientists not affiliated with Yucca Mountain to develop a list of independent experts to act as a review board. This was done quite successfully for the Waste Isolation Pilot Plant in New Mexico, where a federally funded, state organization, the Environmental Evaluation Group, provided credible technical oversight of the pilot plant. Such an oversight group or board would be required to have meetings open to all who are interested during its review sessions. Such a measure would supply both oversight and legitimacy to the DOE's assessments.

Finally, I believe that a false sense of urgency surrounds nuclear waste disposal in the United States. In reality, all reactor sites whose pools have filled and have required additional storage space for their spent fuel have been able to use dry cask storage (Macfarlane 2001a, 2001b). By 2005, a majority of reactor operators will have purchased dry cask storage because not all the spent fuel can be picked up at once (Macfarlane 2001a). Concerns about the security of spent fuel pools at reactor sites, highlighted in a recent National Research Council report (2005), may increase the pressure on utility companies to move spent fuel into dry casks on-site. Dry cask storage of spent fuel over a period of decades will allow both radioactivity and thermal output to decrease, making the waste easier to handle and dispose of. Older fuel provides somewhat fewer challenges in predicting repository performance.

The transport of spent nuclear fuel to a geologic repository will necessarily require decades. Hence, there is considerable time to reconsider whether Yucca Mountain is a reasonable site for the long-term storage of nuclear waste. There is little to be gained, and much to be lost, from rushing a decision of such magnitude.

## Notes

1. In its *Yucca Mountain Site Suitability Evaluation*, the DOE states that it

is satisfied with the level of treatment and understanding of these uncertainties in the current TPSA analyses supporting the site recommendation decision process. In addition to multiple barriers and the use of analogues, additional provisions are being implemented to increase confidence that the postclosure performance objectives will be met. These provisions include validation, implementation of a quality assurance program, and development of a performance confirmation program. (Office of Civilian Radioactive Waste Management 2002b, sec. 3.2.2)

2. One of my own experiences with confusion of model results and actual data stems from a talk by a person working for the DOE headquarters (a technical expert himself) who, when asked about flow rates through the unsaturated zone, responded by saying that fast pathways were a small fraction of the flow, so such flow on fast pathways would not affect the performance assessment. In doing so, he confused me as to what was actual data (not the percentage fractionation of flow in the pathways—these are model results; see discussion in text) and what was not.

## References

Abraham, S. (2002) *Recommendation by the Secretary of Energy regarding the Suitability of the Yucca Mountain Site for a Repository under the Nuclear Waste Policy Act of 1982.* Department of Energy. February.

AkEnd, Arbeitskreis Auswahlverfahren Endlagerstandorte (2002) *Site Selection Procedure for Repository Sites.* Committee on a Site Selection Procedure for Repository Sites. December. http://www.akend.de/englisch/berichte/index_1024.htm.

Bredehoeft, J.D., England, A.W., Stewart, D.B., Trask, N.J., and Winograd, I.J. (1978) Geologic Disposal of High-Level Radioactive Wastes: Earth-Science Perspectives. *U.S. Geological Survey Circular 779*, 15 p.

Bredehoeft, J.D. (2003) From Models to Performance Assessment: The Conceptualization Problem. *Ground Water* 41, pp. 571–577.

Carter, L.J. (1987) *Nuclear Imperatives and Public Trust: Dealing with Radioactive Waste.* Washington, DC: Resources for the Future, 473 p.

Colglazier, E.W., and Langum, R.B. (1988) Policy Conflicts in the Process for Siting Nuclear Waste Repositories. *Annual Review of Energy* 13, pp. 317–357.

CRWMS M&O (Civilian Radioactive Waste Management System Management and Operating Contractor) (2000) *Unsaturated Zone Flow and Transport Model Process Model Report.* Department of Energy. TDR–NBS–HS–000002 REV 00 ICN 02. August.

Davis, J.A. (1987) Nuclear Waste: An Issue That Won't Stay Buried. *Congressional Quarterly Weekly Report* 45, pp. 451–456.

Duncan, I.J. (1999) Some Aspects of the Relationship between Society and the Disposal of Radioactive Waste. In *The Uranium Institute 24th Annual Symposium.* London: pp. 1–12.

Dyer, J.R., and Peters, M.T. (2004) Progress in Permanent Geologic Disposal of Spent Nuclear Fuel and High-Level Radioactive Waste in the United States. *Proceedings of the Institution of Mechanical Engineers, Part A: Power and Energy* 218, pp. 319–334.

Ewing, R.C. (1999) Less Geology in the Geological Disposal of Nuclear Waste. *Science* 286, pp. 415–417.

Ewing, R.C., Tierney, M.S., Konikow, L.F., and Rechard, R.P. (1999) Performance Assessments of Nuclear Waste Repositories: A Dialogue on Their Value and Limitations. *Risk Analysis* 19, pp. 933–958.

Flynn, J., and Slovic, P. (1995) Yucca Mountain: A Crisis for Policy: Prospects for America's High-Level Nuclear Waste Program. *Annual Reviews of Energy and the Environment* 20, pp. 83–118.

Gerrard, M.B. (1994) *Whose Backyard, Whose Risk: Fear and Fairness in Toxic and Nuclear Waste Siting.* Cambridge: MIT Press, 335 p.

Herrick, C., and Sarewitz, D. (2000) Ex Post Evaluation: A More Effective Role for Scientific Assessments in Environmental Policy. *Science, Technology, and Human Values* 25, pp. 309–331.

Jasanoff, S. (1995) Procedural Choices in Regulatory Science. *Technology in Society* 17, pp. 279–293.

Konikow, L., and Bredehoeft, J.D. (1992) Ground-Water Models Cannot Be Validated. *Advances in Water Resources* 15, pp. 75–83.

Lidskog, R., and Andersson, A.-C. (2002) *The Management of Radioactive Waste: A Description of Ten Countries.* SKB. http://www.skb.se/upload/pulications/pdf/The%20management. pdf.

Lidskog, R., and Sundqvist, G. (2004) On the Right Track? Technology, Geology, and Society in Swedish Nuclear Waste Management. *Journal of Risk Research* 7, pp. 251–268.

Linkov, I., and Burmistov, D. (2003) Model Uncertainty and Choices Made By Modelers: Lessons Learned from the International Atomic Energy Agency Model Intercomparisons. *Risk Analysis* 23, pp. 1297–1308.

Macfarlane, A. (2001a) Heated Decision: Prospects for the Interim Storage of Spent Fuel in the United States. *Energy Policy* 29, pp. 1379–1389.

Macfarlane, A. (2001b) Interim Storage of Spent Fuel in the United States. *Annual Review of Energy and the Environment* 26, pp. 201–235.

National Research Council (2005) *Safety and Security of Commercial Spent Nuclear Fuel Storage,* Pre-publication draft. Washington, DC: National Academy Press, 114 p.

Office of Civilian Radioactive Waste Management (OCRWM) (2002a) *Site Recommendation Comment Summary Documen.* Department of Energy. DOE/RW–0548. February.

Office of Civilian Radioactive Waste Management (OCRWM) (2002b) *Yucca Mountain Site Suitability Evaluation.* Department of Energy. DOE/RW–0549. February.

Oreskes, N., and Belitz, K. (2001) Philosophical Issues in Model Assessment. In *Model Validation: Perspectives in Hydrological Science*, ed. Anderson, M.G., and Bates, P.D. New York: John Wiley and Sons, pp. 23–41.

Oreskes, N., Shrader-Frechette, K., and Belitz, K. (1994) Verification, Validation, and Confirmation of Numerical Models in the Earth Sciences. *Science* 263, pp. 641–646.

Rumsfeld, D. (2002) DOD News Briefing-Secretary Rumsfeld and Gen. Myers. February 12. http://www.defenselinkimil/transcripts/2002/to2122002_t212sdv2.htm.

Salve, R., Hudson, D., Liu, H.-H., and Wang, J.S.Y. (2005) Development of a Wet Plume Following Liquid Release Along a Fault. *Vadose Eone Journal* 4, pp. 89–100.

Sarewitz, D. (2004) How Science Makes Environmental Controversies Worse. *Environmental Science and Policy* 7, pp. 385–403.

Sarewitz, D., and Pielke, R., Jr. (1999) Prediction in Science and Policy. *Technology in Society* 21, pp. 121–133.

Sarewitz, D., Pielke, R.A., Jr., and Byerly, R., Jr. (2000) *Prediction: Science, Decision Making, and the Future of Nature*. Washington, DC: Island Press, 405 p.

Vira, J. (2001) Taking It Step by Step: Finland's Decision-in-Principle on Final Disposal of Spent Nuclear Fuel. *Radwaste Solutions*, September/October, pp. 30–35.

Winograd, I.J. (1986) Archaeology and Public Perception of a Transscientific Problem: Disposal of Toxic Wastes in the Unsaturated Zone. *U.S. Geological Survey Circular* 990, 9 p.

Winograd, I.J. (1990) The Yucca Mountain Project: Another Perspective. *Environmental Science and Technology* 24, pp. 1291–1293.

# Index